T0329571

Engineering Research

Design, Methods, and Publication

Herman Tang
Eastern Michigan University
MI, US

Registered Office(s)
John Wiley & Sons, Inc., 111 River Street, Hoboken, NJ 07030, USA
John Wiley & Sons Ltd, The Atrium, Southern Gate, Chichester, West Sussex, PO19 8SQ, UK

Editorial Office
111 River Street, Hoboken, NJ 07030, USA

For details of our global editorial offices, customer services, and more information about Wiley products visit us at www.wiley.com.

Wiley also publishes its books in a variety of electronic formats and by print-on-demand. Some content that appears in standard print versions of this book may not be available in other formats.

Library of Congress Cataloging-in-Publication Data

Names: Tang, He (Herman), author.
Title: Engineering research : design, methods, and publication / Herman
    Tang.
Description: Hoboken : Wiley, [2021] | Includes index.
Identifiers: LCCN 2020029292 (print) | LCCN 2020029293 (ebook) | ISBN
    9781119624486 (hardback) | ISBN 9781119624523 (adobe pdf) | ISBN
    9781119624530 (epub)
Subjects: LCSH: Engineering–Research–Methodology.
Classification: LCC TA160 .T36 2021 (print) | LCC TA160 (ebook) | DDC
    620.0072–dc23
LC record available at https://lccn.loc.gov/2020029292
LC ebook record available at https://lccn.loc.gov/2020029293

Cover Design: Wiley
Cover Image: © nadla/Getty Images

Contents

About the Author

He (Herman) Tang is an associate professor at the School of Engineering, Eastern Michigan University. His research experiences are in the areas of mechanical, manufacturing, quality engineering, and so on. Dr. Tang has published three technical books and many scholarly journal papers. He moreover serves as an associate editor and a reviewer for several scholarly journals and conferences, and a panelist for NSF.

Dr. Tang earned his doctorate degree in Mechanical Engineering from the University of Michigan – Ann Arbor, master's degree and bachelor's degree in Mechanical Engineering from Tianjin University, and MBA in Industrial Management from Baker College. Dr. Tang is a member of SAE International (originally Society of Automotive Engineers), SME (originally Society of Manufacturing Engineers), American Society of Mechanical Engineers (ASME), and American Society for Quality (ASQ).

Preface

Research Book?

Research Matters

Research is both one of the most important elements in the technical world and one of the most interesting learning features in higher education. Many universities have undergraduate research programs and offer research methodology courses to graduate students. The more advanced degree a student pursues, the greater the amount of research is required for success. Without research, an organization cannot develop or improve its products, processes, or services.

There are many types of engineering work, such as analysis, design, and construct. As engineers, we often ask "how" questions. On the other hand, research exploration is different as we more address "what" and "why" questions. In addition to engineering and technical knowledge, the following key traits can help researchers to be more successful. A skilled researcher is

- Innovative
- Collaborative
- Interdisciplinary
- Ethical

For applied research and engineering R&D, cost-effectiveness is critical as well. The contents of this book focus on research methodology with these imperatives.

Book Intent

This book serves a concise overview of critical aspects of research development, which includes thinking, planning, executing, administration, and disseminating of a research project. Its aim is to provide an essential body of work that is focused on the comprehension of general guidelines for new researchers. In addition, the basic doctrines, principles, procedure, methods, and considerations of research are also a valuable guide to undergraduate students who are just delving into the research field.

New researchers entering the field need systematic training on research methodology to learn the best and most effective practices. This book is a systematic guide for learning how to plan and execute a research project, report research findings, and publish research results. This text can also be useful for improving research practice and achievements for both new and experienced researchers alike.

The methodology in this book has been taught in my master's level course. Through the course learning, all students established better confidence, increased interests, and demonstrated more concrete knowledge. They were able to move forward in conducting research projects.

Book Characteristics

General Principles and Methods

While the research methods described in this book are applicable for all areas of studies, I have prepared this text primarily for the students who are in engineering and technology fields and with an assumption that readers have a basic mathematical and technical background. For every method described in this book, I have selected many examples and citations across multiple engineering disciplines that readers can reference for a better understanding of each concept.

Various practices and standards of R&D are sometimes specialized within each academic and professional field. Research in social sciences, for example, utilize the standard *ISO 20252:2019 Market, opinion and social research, including insights and data analytics – vocabulary and service requirements*. Many research methods can be applicable to almost all areas. The methods used medical sciences may have good application potentials in engineering and technology. The methods presented in this work are not exclusively limited to research in a specific engineering area. Hence, this book is comprehensive in its contents, including some uncommon approaches in engineering and technology, so that all readers have a reference of most available methods for their particular research projects.

Various Examples

One effective way to begin research work is to learn from other professionals and their current practices. The book cites over 460 scholarly papers and other types of sources in various engineering and technical areas. While not all the examples may be relevant to other specific research topics, the information, approaches, processes, and suggestions in the examples are the proven techniques that can be useful as a guide for readers perusing their own research projects.

Furthermore, as an engineer myself, I intentionally designed this book with many illustrations: 147 diagrams and 70 tables. Hopefully, you the readers can find the illustrations effective for your comprehension of research principles and methods.

As to be discussed more in Chapter 1, an emphasis on critical thinking is central when it comes to research. Readers are highly encouraged to bear critical thinking in mind when reading the text and to consider implementing these suggestions to improve the current techniques and progress with improved research methods.

Practice Guidelines

Regardless of technical discipline or research subject, one basic principle never changes: the best way to learn research is to *practice*. One cannot learn how to do research just by reading books but rather, one can use their books as guidance to make research more effective.

For learning, an old adage says that I hear and I forget. I see and I remember. I do and I understand. Referring to the learning pyramid (Figure 1) below, which is adapted from the original developed by Edgar Dale in his book *Audio-Visual Methods in Teaching* published in 1946, one can learn research better if we practice through exercise, discussion, developing research proposals, etc. When one is able to apply the knowledge learned toward conducting a real research project, one effectively learned the knowledge and improved our research capability and skills.

Along with learning research methodology, my students develop their own research plans, section by section, corresponding to the relevant chapters of this book. At the end of my class, every student will have a good proposal, ready to move forward with conducting their own research. The guidelines and sample proposals developed from the course are included in the Appendices for instruction and learning references.

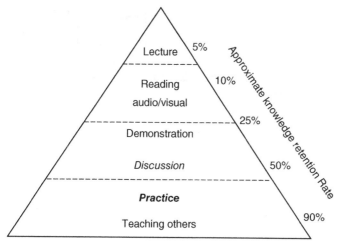

Figure 1 A learning pyramid.

Organization and Contents

Overall Information Flow

I arrange the contents of this book in typical sequence for the execution of a research project. There are three parts and nine chapters, as described in the Figure 2.

If readers want to learn research methodology systematically, it is a good idea to follow the chapters in order. If using this book as a reference manual, one can jump directly to the most relevant chapters.

Parts and Chapters

Part 1: Research introduction and development, including proposal writing and literature review.

Chapter 1 Research Overview. This first chapter introduces the concepts and building blocks of research. The chapter reviews three types (basic research, applied research, and R&D) and discusses the validity of research results. Providing a basic understanding of research, this chapter serves a starting point for preparing and executing research work.

Chapter 2 Research Proposal Development. This chapter presents the entire process of research development, focusing on a proposal development. Introducing an overall develop process, Chapter 2 explains how to draft, review, and revise a proposal, and addresses the key issues of the development process.

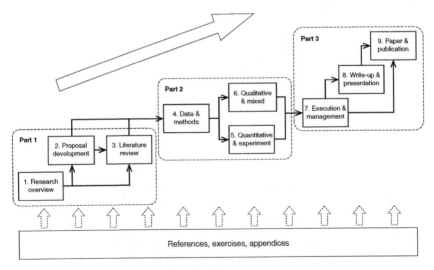

Figure 2 Parts, chapters, and subjects of this book.

Chapter 3 Literature Search and Review. This chapter talks about the techniques and considerations of a literature search and many engineering literature sources. The chapter discusses the tasks and focal points of a literature review. The chapter also offers guidance on how to prepare a review, as well as some suggestions for review writing up.

Part 2: Various types of research methods.

Chapter 4 Data Collection and Method Selection. Starting with the discussion on the roles of data in research and the types of data, later on, this chapter addresses the main aspects of data collection and sampling. Based on data and other considerations, the chapter explains the selection considerations of different research methods.

Chapter 5 Quantitative and Experimental Research. This chapter presents the methods for research as four topics. The first one is about quantitative data analysis, including statistical analysis and interpretation. The second section is on engineering quantitative research methods. The third section discusses common experimental studies, and the fourth section introduces on Design of Experiment (DOE).

Chapter 6 Qualitative and Mix Methods. The chapter reviews characteristics of qualitative research and mixed-methods. The chapter discusses main qualitative approaches, including survey, interview, and observational studies. Based on these characteristics of methods, the chapter further discusses mixed-method approaches.

Part 3: Management, scholarly writing, and publication of research.

Chapter 7 Research Execution and Management. Performing research, particularly an external-funded project, can be complex and may require more teamwork with institutional administration. This chapter introduces the functions and requirements of project management for research execution. Then, the chapter presents the main functions of institutional research administration, including the proposal, pre-award, post-award, and closeout phases. For student research, the information presented in the chapter is a useful reference.

Chapter 8 Writing Up and Presentation. Research reports and presentations are integral parts of research work. This chapter discusses them and student theses. Then, the chapter reviews report design, suggests guidance, and illustrate some considerations in technical report writing. Last, the chapter recommends some practical guidelines for preparing presentations.

Chapter 9 Scholarly Paper and Publication. This final chapter provides real-world guidance for scholarly publication. The chapter discusses an overall publication process and targeting scholarly journals. Furthermore, the chapter provides recommendations on how to prepare a manuscript and work with journal editors for publication.

Exercises and Supplemental Materials

Research is not only about theories but practice as well. As previously emphasized, exercises are vital to learning this subject.

Two types of exercises are designed at the end of each chapter: Review Questions and Mini-project Topics. There are about 20 and 10 for each chapter, respectively, total 188 and 91. Both types of exercises are complementarily designed for individual student thinking and class team-learning activities.

The Review Questions can be used for immediate classroom/online discussion, while the Mini-project Topics, requiring more time and student's effort to search materials and critically prepare a short essay, to be used to lead further in-depth discussion on a particular topic. Many Review Questions and Mini-project Topics may be exchangeable; meaning a topic in Review Questions can be expended to a Mini-project or a Mini-project topic be simplified for an immediate discussion.

There are also three supplemental materials for in class lecture, including a sample course syllabus, project development guidelines, class exercise instructions, and discussions. These can be valuable for instructors as well as students. The best exercise while studying this subject is to plan and conduct an entire research prospectus. Following these guidelines, students can find an area of interest and develop a research prospectus in six steps. Instructors and students may refer to the appendices in a classroom setting and via online learning.

In addition, three research prospectuses developed by some of my former students are included for the reader's reference. These samples differ in subject, method, research process, style, etc., which reflects the openness and diversity of research.

Acknowledgments

I am deeply grateful to the mentors and colleagues of my research work at Tianjin University, University of Michigan—Ann Arbor, Fiat Chrysler Automobiles, and Eastern Michigan University (EMU). Although not involved directly in this book preparation, these mentors deserve a valid credit of the book as they influenced my research work and scientific thinking. The EMU 2019 Faculty Research Fellowship Award supported this book manuscript preparation.

Teaching this subject is among my favorite activities here at EMU. I really enjoy working with the students in classes, as we tackle over various research topics and challenges. Many of my students are experienced professionals; their insights and experiences have enhanced the manuscript in terms of practical and broad scope.

Special thanks go to the professionals in various areas for their reviews and suggestions on individual chapters. They include Dr. Grigoris Argeros, Dr. Kathy Chu, Kelly Getz, Dr. Dorothy McAllen, and Dr. Wade Tornquist. Thanks also go to Dr. Dan Fields, Dr. Sohail Ahmed, Dr. Bryan Booker, and Dr. Joe Bauer for sharing their teaching materials. EMU students Jacob Benn, Erin Butler, and Brendan Ostrom helped manuscript proofreading. I appreciate the authors and organizations for their works cited in the book.

I am grateful to three anonymous reviewers who supported and provided constructive recommendations on the book proposal and drafts of the chapters. Wiley's acquisition, editing, and publication teams, particularly Brett Kurzman, Steven Fassioms, and Sarah Lemore, played the integral roles to the quality and effective publication of the book. I greatly appreciate their work and contribution.

Last but not least, I would like to thank my wife's understanding and full support helping to bring this volume to fruition and to our sons Boyang and Haoyang, both engineers, for their great help in reviewing and editing on the manuscript.

Research is a broad, complex, and diversified practice that no single volume can cover every aspect of research. Regardless of just beginning research or acting on years of real-world experience, readers of this book can help improve this text. Send me your comments, criticism, and suggestions to htang369@yahoo.com or htant2@emich.edu – I greatly appreciate your remarks and will be carefully reviewing them for future editions of this book.

Finally, thank you, readers, for your interest in research. I hope this book will help you gain a new understanding of research and in implementing an informed research project. I wish you the best of success in your research.

He (Herman) Tang
Ann Arbor, MI, USA
January 2020

Part I

Overview, Proposal, and Literature Review

1

Research Overview

1.1 Introduction to Research

1.1.1 What Is Research?

Research is a universal word. Professionals in almost all disciplines are prosecuting doing research, such as in science, engineering, medicine, languages, literature, history, and business. There are various definitions of research. However, the process and requirements of research in some areas, such as medical science, might be better defined than that of other fields. While difficult to define research comprehensively and precisely in one sentence, we can understand research from its various aspects.

1.1.1.1 Seeking New Knowledge

According to Merriam-Webster Dictionary, the word research is derived from the French "recherché," which means "to go about seeking." Research concerns the seeking and creation of new knowledge and understanding the principles and characteristics of a phenomenon. For example, another definition of research is "the process of finding out something that we don't already know" (Hazelrigg n.d.). The new thing should be interesting or of concern to a profession or humanity. Research can be any kind of investigation that intends to uncover new facts.

The words "what" and "why" may be used to show what research is about. Knowledge takes two forms: "know that" and "know why." The "know that" may be called declarative knowledge, which represents ideas and understanding. As such, declarative knowledge is relatively easy to teach and learn. The "know why" is about a type of functional knowledge, which varies with individual capability.

In many cases, scientific work includes applied research (R) and engineering development (D) called research and development (R&D) in short. Such R&D efforts can be either applied research, development, or a combination of both and in context of "know how." The yields of most engineering and technical R&D

Engineering Research: Design, Methods, and Publication, First Edition. Herman Tang.
© 2021 John Wiley & Sons, Inc. Published 2021 by John Wiley & Sons, Inc.

are new or improved physical artifacts, such as software, products, and processes. Sometimes, the research characteristics of R&D may be debatable as far as its contributions to new knowledge.

A key to research is an innovation to professional community at large. Think about some efforts that may look like but are not research. Here are a few common types (Leedy and Ormrod 2016):

- Simply gathering information
- Merely rummaging around for hard-to-locate information
- Transfer of facts from on location to another

For example, the term "research" is often used for describing the act of information discovery in our daily life. For example, if one is looking for a new car, he/she may do "research" on various features, models, safety records, price, etc. When looking for a job, one would "research" the websites of companies with openings. These types of everyday exploratory activities are good for an individual's purposes, but no contribution to the general knowledge of a professional community. Therefore, such acts and efforts of information search are not scientific research because the information is not new to professional community.

1.1.1.2 A Systems Viewpoint

We may view a research project as a system (Figure 1.1), which can have different objectives and tasks, such as analysis, experiment, and computer simulation. From a system viewpoint, doing research is to invest inputs, consider influencing

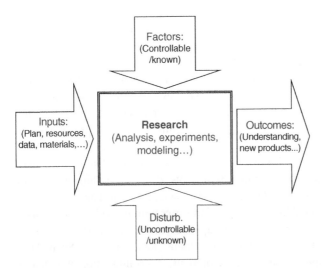

Figure 1.1 A system view of research project.

factors and distributions, and do original and diligent work for the expected outcomes. Accordingly, one of the important characteristics of research from a systems viewpoint is to deal with the complex relationship among inputs and controllable factors.

For a defined purpose or curiosity, knowledge exploration starts from observations. Figure 1.2 shows the overall process flow of systematic knowledge exploration: (i) basic research and (ii) applied research and R&D (problem-solving).

After identifying a research question or problem, we need to establish our guess, or hypotheses, for basic research. Based on the question and hypotheses defined, we then develop a detailed study plan. For a relatively simple problem, we may directly address it.

Following the plan, we conduct the study, which includes data collection, analysis, and interpretation. In most cases, we submit our findings for publications to share with others. It is often the case that research results promote new questions. Following this circular process, the understanding and knowledge on a particular subject can continuously deepen and widen.

Exploring new knowledge never ends. "Research is iterative and depends upon asking increasingly complex or new questions whose answers in turn develop additional questions or lines of inquiry in any field" (ACRL, Association of College and Research Libraries 2016). Research continuously advances to new levels of knowledge and innovation.

1.1.1.3 General Characteristics

Research in all disciplines shares common characteristics in addition to the systematic exploration:

- Scientific research is a structured study with a plan to execute and document the process and results.
- Research work always has various assumptions.
- Much research, particularly basic research, is normally hypothesis guided.
- The entire research process, or methodology, is just as important as the specific methods used for research success.
- Research methods and outcome always have limitations.
- The outcomes from research should be independently verified or recognized by other professionals.

It may be a good exercise for reader to think about known research for these general characteristics. For example, what are the assumptions for a completed research project? Are they explicitly stated? Other questions may include the nature, method, and procedure used in the research.

Engineering R&D share common principles and methods from the research in other scientific fields, such as medical and social sciences, but may have some

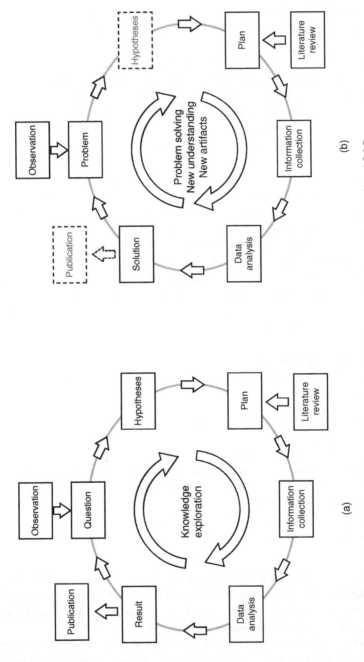

Figure 1.2 Overall processes of basic and applied research. (a) Basic research, (b) applied research and R&D.

different emphases, characteristics, and practices. Engineering researchers can and should learn from the professionals in other disciplines to improve their research methodology. More discussion on engineering R&D is in a later section of this chapter.

1.1.2 Impacts of Research

1.1.2.1 Impacts on Societies

Research work needs resources, including funding, personnel, and facilities. A question often asked is "what is the return on that investment?" Therefore, researchers must provide the justification of a research project with positive impacts and effects.

The impacts of research can be the overarching benefits for human society. Research can contribute in many ways and areas, such as technological developments, environmental impact, economic benefits, health and wellbeing, national interests, and policy change.

The impact of research also depends on the type of research. Figure 1.3 illustrates how the three main areas of research (i.e. academic, economic, and societal) can have significant impacts. Situations are also various in terms of the designed impacts of research. For example, a research project may be purely scientific without an immediate impact on society. If a R&D project is for commercialization, the product may make a significant technological advance.

The impacts from a research project depend on various factors, such as the type, objective, and size of a research project. For example, basic research, to be discussed in Section 1.3.1, is to expand humanity's knowledge. Two criteria related

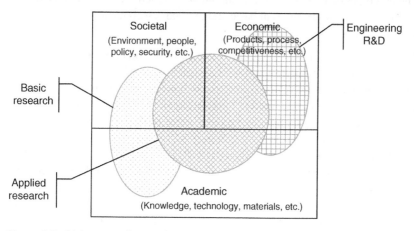

Figure 1.3 Main areas of research impacts.

to the overall evaluation of basic research are intellectual merit of a new discovery and the broader impact on society.

While applied research and R&D have different applications, the impacts of such application-oriented research may be different from those of basic research. Industrial corporations often consider commercial advantages to fund R&D projects. In addition, corporations may support research institutes for basic and applied research as well.

1.1.2.2 For Specific Objectives

As discussed previously, the general objectives and benefits of different types of research are different and with substantial overlap:

- Basic research to explore knowledge
- Applied research to solve problems
- R&D to generate new artifacts

In most cases, they may build on each other in succession through their similar goals. For example, new knowledge derived from basic research inspires theories leading to applied research and R&D. R&D in turn can raise demands to conducting basic and applied research.

Some funding sponsors focus on the solutions of particular problems or a specific area. The research projects funded by the sponsors may have very specific objectives and criteria. A granted research project should meet the sponsor's requirements, such as a better understanding of the phenomenon in question, more accurate predictions about future events, additional interventions for a better quality of environment or life. It is important to note that you as a researcher can have your idea first and then try to find and matching, funded opportunities.

The outcomes of a research project should be conducive to its predefined objectives. Here are a few examples of basic and applied research supported by the US government agencies:

- The Basic Energy Sciences program of the US Department of Energy is to "discover new materials and design new chemical processes" (DOE n.d.-a).
- The Advanced Scientific Computing Research program is to "discover, develop, and deploy computational and networking capability to analyze, model, simulate and predict complex phenomena important to the Department of Energy and the advancement of science" (DOE n.d.-b).
- The Secure and Trustworthy Cyberspace (SaTC) program of US NSF states, "The goals of the SaTC program are aligned with the Federal Cybersecurity Research and Development Strategic Plan (RDSP) and the National Privacy Research Strategy (NPRS) to protect and preserve the growing social and economic benefits of cyber systems while ensuring security and privacy" (SaTC n.d.).

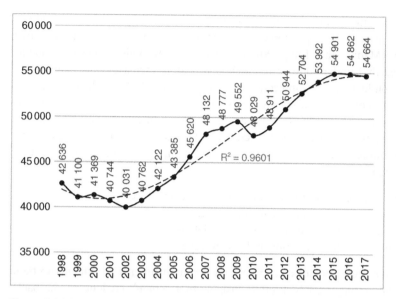

Figure 1.4 Doctorate recipients in engineering and science from US universities. Source: Data from NSF (2018).

- The Science Mission Directorate of NASA "targeted technology investments fill technology gaps, enabling NASA to build the challenging and complex missions that accomplish groundbreaking science" (NASA n.d.).

1.1.2.3 Benefits to Student Researchers

Research is a part of higher and graduate educations. Figure 1.4 shows the significant increase of doctorate degree recipients in engineering and science fields from US universities in recent 20 years (NSF 2018).

Research training and projects can start at an undergraduate or master's graduate level. Many universities offer a dedicated course on research methods for master's graduate students and some universities offer introductory courses and programs of research for undergraduate seniors (Depaola et al. 2015). A research course opens students a new opportunity to engage in creative and critical thinking that leads to hands-on engineering applications.

Some students consider doing research simply out of interest without any in-depth concern for anything else. Without much experience, they may be hesitant to choose research opportunities in curriculum. Before doing research, students should discuss with professors and experienced researchers and ask what prerequisites are needed. With the guidance and encouragement by experienced researchers and professors, students can quickly grow in motivation and understanding, start to develop their research skills, prepare to pursue a higher degree, and to become good researchers. A study showed that many undergraduate students discover their passion for research through exposure to simple research projects (Madan and Teitge 2013). A large survey with 15 000 respondents indicated that undergraduate research significantly increases understanding of how to conduct research and confidence in research skills (Russell et al. 2007).

Through completing a research thesis or capstone project, students can also deeply explore something they have a passion for and enrich their understanding of the topics. They can apply their learned skills do better future research and conduct industry projects they work on. In addition, research can help students in the following ways:

- Improve critical thinking and intellectual independence
- Develop creativity and problem-solving skills
- Have opportunities to communicate special ideas
- Enhance your determination and perseverance

Research work provides examples and accomplishments of students to their peers and employers, which offers insights to a person's credentials and background as well. In today's technical professions, if someone has a well-rounded mix of skills, he/she may stay relevant in a competitive position.

1.2 Building Blocks of Research

We refer to building blocks here as the essential elements of doing research. They include several key factors, including knowledge, competence, information, teamwork, resources, etc. These building blocks are essential to, as Dr. Richard Miller, President of Olin College of Engineering, said: "Learning things that matter; learning in context; learning in teams. Envisioning what has never been and doing whatever it takes to make it happen" (Ark 2019). In addition to proper knowledge, motivation and critical mindset play critical roles to research work and success. We address these building blocks in this section and will have more in-depth discussions in the following chapters.

1.2.1 Innovative Mind

1.2.1.1 Motivations to Research

In most cases, researchers are generally enthusiastic about what they do. For example, a researcher stated, "my research is motivated by interesting challenges arising from the growing size and complexity of modern pattern recognition problems in the sciences, engineering and social media" (Kpotufe 2014). In

contrast, if burdened with a "have to do" mindset, one will not be very successful in their research.

Intrinsic motivating factors include curiosity, determination, and/or enjoyment of solving a challenging problem. For example, researchers may have strong personal preferences for a particular subject or direction of research.

While some motivation sounds extrinsic (such as educational requirements, studying for a master's or doctoral degree and professional career requirements, such as for employment, promotion, and recognition), these are all actually intrinsic motivation because researchers determine to pursue their career interests or educational paths. It still comes down to someone being personally motivated.

As general interests change and advance along with new technologies and community demands, it is important that an individual researcher's interests be in line with overall trends. For instance, in Computer Science and Engineering fields, students would most likely to do research in one of the significant advancing subjects: networking technologies and distributed systems, embedded systems, ubiquitous computing, interoperability and data integration, object-oriented programming, human–computer interaction, software safety, security and cryptography, and so on. In other words, successful researchers focus on the future and stay in the present.

1.2.1.2 Thinking and Research

As with any learning knowledge or skill, human cognition has several levels. Refer to Figure 1.5 for the general growth of capability and contribution of professionals.

In colleges, undergraduate education focuses on the comprehension of knowledge and it is application on Levels 1 and 2. Once proficient at the first two levels, students can analyze real and complex problems with their knowledge (Level 3), which is a starting point of a professional career, research, and creativity. With

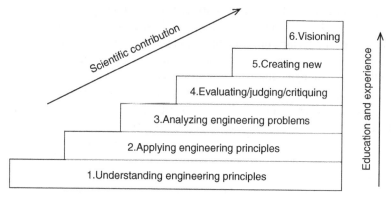

Figure 1.5 Thinking levels and research capability.

analysis and problem-solving, they become more capable to check, assess, and critique the work and achievements of themselves and others. Such capability of evaluation (Level 4) includes critical thinking and analysis, which is a foundation for creating and producing new understanding and/or artifact (Level 5). At the top level (Level 6), researchers are grown: becoming visionary leaders to predict and guide research directions. The higher level of knowledge and thinking, the more contributions we may have on the scientific and professional community.

Appropriate knowledge is often a prerequisite for critical thinking, creating, or improving solutions. Technical knowledge, along with other factors, influences cognitive capability, which plays an essential role in the success of research. We can learn particular thinking skills, such as deductive logic and inductive reasoning, from training and practice.

1.2.1.3 Critical Thinking

Critical thinking is a skill and process, which includes analyzing, assessing, and challenging an observation on a subject. Some authors defined critical thinking as "self-guided, self-disciplined thinking which attempts to reason at the highest level of quality in a fair-minded way" (Elder 2007). Critical thinking plays a more important role than that of observation, which does not necessarily generate a research question. Critical thinking questions may include

- Questions beyond norms or traditions
- Logical evaluation of evidence, process, and conclusions
- Connections between similar or different ideas
- Systematical review and consideration of all aspects and elements
- Open minded to be challenged and to different perspectives

It is true that "many ideas that were previously thought of as 'facts' or 'theories' were debunked by others who did not give up in their quest to prove that inaccuracy of those ideas. Those who continued researching against what was 'known' were thought of as crazy" (Laird 2018). Thus, critical thinking is a key for us to challenge the existing understanding and generate new ideas and/or new methodology. We can find new opportunities and initiate new studies from thinking critically and challenging the existing status or conclusions.

Practicing critical thinking in a literature review often generates new ideas and questions, which is also an effective way to initiate a new research proposal. In the literature review section of published papers, the words like "however" and "although" are often used to challenge the existing status or scenarios. For example, the authors stated, "Although there have been considerable developments in manufacturing technologies and processes, the actual scope and elements of manufacturing systems are complex and not adequately defined" (Esmaeilian et al. 2016). In critical thinking mode, we may also question ourselves on different aspects of research, such as on its assumptions, data reliability, method adopted, interpretation, conclusions, and potential applications.

In workplace, there is an accountability vs. responsibility relationship, as with superiors and subordinates. In a university, there are student researchers and faculty advisors. For research, we should try to think outside the box on research as much as we can but not constrain our critical thinking based on work relationship. In conducting research, our technical discussion should be based on the convincible ideas and facts, rather than authority. Senior researchers must encourage and guide novice and young researchers on critical thinking and innovative initiatives to challenge the existing principles and status.

1.2.2 Assumptions and Hypotheses

1.2.2.1 Assumptions

Assumptions are the foundation and conditions that affect the outcome of research. Assumptions help narrow the scope of research work, effectively drive the execution process, and guide the focus of research work. In addition, any research task, such as data collection and analysis, is under certain assumptions based on physical constraints and situations. Assumptions affect the ways the data is gathered, analyzed, and concluded. Assumptions also indicate how far we have gone to prove findings.

Assumptions may or may not precisely reflect the real world. We should avoid assumptions that are extremely restrictive. Similarly, too many assumptions may result in the research becoming over simplified. If it turns out that the assumption is not reasonably accurate, then the findings and results may not be meaningful or externally valid. In other words, successful research outcomes are conditional to appropriate assumptions.

Assumptions may be broken down into three types: epistemological, ontological, and methodological. They are about the ways to acquire the knowledge, the nature of the world and human being in social contexts, and analysis of the methods used, respectively. Most assumptions in engineering applied research and R&D are methodological, related to the methods, data, and process.

Assumptions should be an integral part of research publication to make sure that the audience is aware of them. The assumptions, in either quantitative or qualitative form, should be stated in a proposal and in a result report later on and sometimes even in the title of a proposal and paper. For instance, this study clearly stated assumption in its title: "New two-phase and three-phase Rachford-Rice algorithms based on free-water assumption" (Li and Li 2018).

The abstract of a paper may include the assumptions as well. For example, a paper stated its assumptions thus: "This paper shows the usual inconsistency made in the linear elastic fracture mechanic, which is to estimate plastic zones

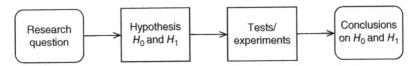

Figure 1.6 A process flow of hypothesis-driven research.

(PZ) from a linear elastic (LE) analysis with correction of the PZ size based on assumptions of equilibrium" (Sousa and Figueiredo 2017). Another paper indicated, "... satisfies certain mild assumptions which we outline below" (Razaviyayn et al. 2013).

Sometimes, researchers consider some assumptions so obvious not to even mention them in a proposal or paper. However, even though they are not explicitly stated, the assumptions do exist.

1.2.2.2 Hypothesis

Before doing a research task, we often have a specific aim or expectation of the outcomes. Such an expectation may be formulated as a hypothesis, which is a predictive statement based on our knowledge, experience, and research targets. Therefore, most research, particularly basic research, is hypothesis-driven. The corresponding tasks are to test the hypotheses and draw conclusions, refer to Figure 1.6.

For example, a paper had its hypothesis statement (Davoodi-Nasab et al. 2018):

$$H_0 : \beta_1 = \beta_2 = \beta_3 = \cdots = \beta_k = 0$$
$$H_1 : \beta_j \neq 0, \text{for at least one } j$$

Where, the null hypothesis (H_0), as a starting point or a default position: all β_j are zero. The alternative hypothesis (H_1) is stated that at least one β is not zero.

Although stating a hypothesis mathematically is typical, not all researchers do so explicitly, particularly for applied research and R&D projects. There are various ways to state the hypothesis. For example, a paper stated, "The hypothesis of rigid adherents is here assumed" (Santarsiero et al. 2017). Another example expressed, "To evaluate the hypothesis that a reduction in tensile strength could be associated with some sort of reaction and/or interaction between HALS" (Staffa et al. 2017).

For the research using statistical analysis, we should explicitly state the hypotheses in format of H_0/H_1. We should also discuss the corresponding research tasks and results based on the stated hypotheses, which will be discussed more in Chapter 2.

1.2.3 Methodology and Methods

Both methodology and methods are the keys in research and play combined roles in research success. We often use the two terms of methodology and methods interchangeably in our professional communication. However, when we prepare and conduct research, it would be better to understand their differences and relationship. In a general sense, methodology is a broad scope and overall view of research, while methods are the specific approaches to conduct the research.

1.2.3.1 Methodology

Methodology is a general research strategy and procedure. It refers to all methods used to meet research objectives and all perspectives of a research process as a whole. Research methodology may include data collection, analysis approaches, equipment and facilities, process, validation, and so on.

Overall, research methodology addresses the objectives and procedure to a research project: what, why, and how to collect and analyze data. We develop a methodology by selecting and justifying a particular method to be used. For instance, one paper summarized the methodology for systems engineering research (Caillaud et al. 2016) while another is on ergonomic product design (Dianat et al. 2018).

1.2.3.2 Methods

A method is a specific approach or tool used to do a research task. For example, we may use probability theory as a tool for data analysis. In most cases, researchers in the same field prefer the similar methods.

As a tool, research methods may themselves be research topics as well. For example,

> *"A new method for the automatic sketching of planar kinematic chains"* (Yang et al. 2018).
>
> *"A new approach to compute temperature in a liquid-gas mixture. Application to study the effect of wall nozzle temperature on a Diesel injector"* (Payri et al. 2017).
>
> *"The New EDrives Library: A Modular Tool for Engineering of Electric Drives"* (Haumer and Kral 2014).

Interestingly, there is no clear cut difference between the methodology and methods, refer to Figure 1.7.

1.2.3.3 Process

Conducting research consists of multiple tasks, such as problem identification, goal setting, hypothesis establishment, data collection and analysis, interpretation of the results, and so on. Researchers organize and execute the tasks in a systematic, scientific process. Some researchers consider research process is a main part of methodology. However, we may view a research process as a collection of main efforts or steps in a certain sequence.

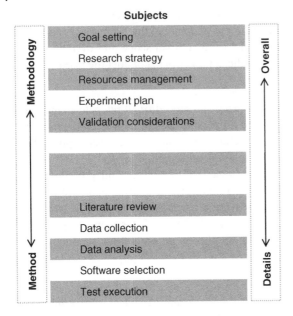

Figure 1.7   Research methods and methodology.

The process of a research project may be viewed as two segments: preparation and execution. An overall process flow is shown in Figure 1.8. We will discuss these items in depth in later chapters.

1.2.4   Research Community

1.2.4.1   Environment
Most of research is teamwork-based. Teamwork-oriented spirit and environment is fundamental for a research project initiation and execution. For example, observe the teamwork at a college, where many graduate students have advisors and work in well-established laboratories supported by technical personnel.

Experienced researchers normally have good connections with research organizations and in the professional community worldwide. The researchers may have better access to the latest research information, such as innovative methods, new progress, and development trends, from peers and facilities. Novice researchers should team up with and get help from experienced researchers.

1.2.4.2   Ethics
Ethics in research, including honesty, objectivity, respect for intellectual property, confidentiality, and so on, plays an integral part in research. Many professional associations and sponsors have developed codes and policies that outline ethical behaviors of researchers.

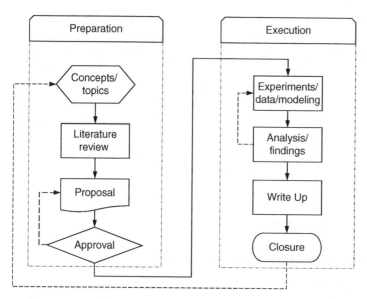

Figure 1.8 An overall process of research projects.

We must share and follow these ethical principles. For example, everyone should acknowledge the previous works of others in proposals and published papers. A common way to acknowledge others works is to have clear in-text citations and references. Inappropriate usages, purposely or unintentionally, of other's works can be grounds for accusations of plagiarism.

Another ethical consideration is related to human subjects, or obtaining data through the intervention or interaction with the individuals or from individually identifiable information. Work safety and ergonomics may be included as well. To consider human subjects, most institutions and organizations have an Institutional Review Board (IRB). An IRB is a panel of people to ensure the safety of human subjects in research and to assist in protecting human rights. A research project related to human subjects needs an IRB approval, which helps protect the institution and the researchers against potential legal implications from any behavior that may be deemed unethical. We will have more discussion on the ethical considerations in Chapters 2, 8, and 9.

1.2.4.3 Funding Sources

Research projects may be funded by various sources, including those internal to an organization, government agencies, philanthropy foundations, and industry partners. Surveys by the NSF showed that the US federal agencies provided about 44% of the $86 billion spent on basic research in 2015 (Mervis 2017). Private and philanthropy foundations also play an important role in basic research.

If a researcher is affiliated with a university, he/she should first contact the department or college research offices. Most research universities have internal funding and fellowship programs to support faculty and student research. In addition, most universities have an office for sponsored programs, which coordinates grant requests and helps researchers on external funding grant opportunities.

Depending on funding sources and sponsors, the nature and requirements of research projects vary significantly. For example, government-funded research is often on basic research. Quality publications are requested if funded by the US NSF. We will have more discussion on research funding in Chapter 2 for proposal development and on research administration in Chapter 7.

1.3 Types of Research

1.3.1 Basic Research, Applied Research, and R&D

1.3.1.1 Basic Research

Research is a creative and systematic work. We normally categorize it in three types: basic research, applied research, and R&D. A NSF document summarized the common definitions and understandings on the research categorization (Moris 2018):

> *"Basic research is experimental or theoretical work undertaken primarily to acquire new knowledge of the underlying foundations of phenomena and observable facts, without any particular application or use in view."*

Another definition from the US Department of Defense for non-DOD respondents:

> *"Basic research: Basic research is defined as experimental or theoretical work undertaken primarily to acquire new knowledge of the underlying foundations of phenomena and observable facts."*

Therefore, basic research is to expand humanity's knowledge and investigate a natural event, process, or phenomenon. Such research helps build new frontiers of knowledge of how things work but no immediate commercial value or application are expected. Sometimes basic research is called pure or theoretical research.

For example, "While there has been significant progress in understanding how such decisions should ideally be made, there is a significant gap in knowledge about how humans actually make such decisions. This gap is a barrier to improving systems engineering and design practice. In this project, basic research towards addressing this gap will be carried out" (Panchal et al. 2017). Here are a few other examples:

- High-power alternative energy conversion
- Computational thinking in biological engineering
- Modeling for cardiac tissue manufacturing
- Theoretical framework in systems engineering
- Engineering artificial cells

Research institutions are a core team for basic research. For research universities and institutes, considerable amount of research is on basic research. The research divisions of large corporations also do basic research.

1.3.1.2 Applied Research

Applied research aims to explore new knowledge as well. A NSF document (Moris 2018) stated the definitions of applied research:

> *"Applied research is original investigation undertaken in order to acquire new knowledge. It is, however, directed primarily towards a specific, practical aim or objective."*

The definition from the US Department of Defense for non-DOD respondents is:

> *"Applied research: Applied research is defined as original investigation undertaken in order to acquire new knowledge. It is, however, directed primarily towards a specific, practical aim or objective."*

Therefore, the main difference between basic research and applied research is whether a research effort has a practical objective to real-world situations and applications. Much engineering research, such as running experiments, advancing new technology, and conducting case studies, is applied research.

Here is an example of the research objective of a NASA research proposal: "the technical objective of this proposed project is to develop and experimentally validate key technologies needed for autonomous rover traversing on Mars-analog terrains. The goal is to reduce the number of sols needed to complete the MSR mission-required total traverse distance (i.e. 'fast traverse'). This will be achieved through increased rover operation 'duty cycle' and 'mean time between human interventions,' while utilizing limited onboard power and computational resources" (Gu 2017).

To address and resolve a real-world situation, applied research may be either technology-driven or problem-driven when seeking new understanding and knowledge. Researchers conducting technology-driven studies are primarily interested in a particular technique (e.g. neural networks) and looking for its development and new applications. The goal of such research is to advance techniques, create a new technology, and improve their capabilities. Most technology-driven research contributes incrementally to the technology advance. For example, a paper is titled "A facile ion imprinted synthesis of selective biosorbent for Cu^{2+} via microfluidic technology" (Zhu et al. 2017). To reach a problem's solution, we may use existing appropriate methods or construct new ones.

Some may challenge the idea that problem-solving is real research. We can determine this by examining whether the problem-solving research creates or extends our knowledge and benefits the professional community (in the form of publication), such as the paper addressed a practical problem as titled "An energy-efficient multi-objective optimization for flexible job-shop scheduling problem" (Mokhtari and Hasani 2017).

1.3.1.3 Engineering R&D
Engineering R&D is a technical invention, focusing on the form, function, and capabilities of implementation into products to the marketplace. Engineering R&D may be defined as (Moris 2018):

> "Experimental development is systematic work, drawing on knowledge gained from research and practical experience and producing additional knowledge, which is directed to producing new products or processes or to improving existing products or processes."
>
> "Development: systematic use of the knowledge and understanding gained from research for the production of useful materials, devices, systems, or methods, including the design and development of prototypes and processes."

While it is true that many R&D projects are based on experimental investigations, it takes other forms, too, such as product development and manufacturing technology. The new findings and results of R&D can advance human understanding as well. However, due to commercial confidentiality, the achievements from engineering R&D are infrequently published and sometimes patented. The discussion of engineering R&D continues into the next section.

Applied research and R&D are often overlapped and used exchangeably. Apple and General Motors spent $14.24B and $7.8B on R&D in 2018, respectively (GM 2019; Apple 2019), which included basic research, applied research, and development – directly associated with their future products. Industry research is primarily to corner specific areas of a market, in order to be the first to market or protect intellectual property through patents and trade secrets.

The NSF document (Moris 2018) also lists some R&D activities based and otherwise based on the Financial Accounting Standards Board (FASB) Accounting Standards Codification (ASC). The R&D activities include

"a. Laboratory research aimed at discovery of new knowledge
b. Searching for applications of new research findings or other knowledge
c. Conceptual formulation and design of possible product or process alternatives
d. Testing in search for or evaluation of product or process alternatives
e. Modification of the formulation or design of a product or process
f. Design, construction, and testing of preproduction prototypes and models
g. Design of tools, jigs, molds, and dies involving new technology
h. Design, construction, and operation of a pilot plant that is not of a scale economically feasible to the entity for commercial production
i. Engineering activity required to advance the design of a product to the point that it meets specific functional and economic requirements and is ready for manufacture
j. Design and development of tools used to facilitate research and development or components of a product or process that are undergoing research and development activities."

While the following activities would typically not be considered R&D:

"a. Engineering follow-through in an early phase of commercial production
b. Quality control during commercial production including routine testing of products
c. Trouble-shooting in connection with break-downs during commercial production
d. Routine, ongoing efforts to refine, enrich, or otherwise improve upon the qualities of an existing product
e. Adaptation of an existing capability to a particular requirement or customer's need as part of a continuing commercial activity
f. Seasonal or other periodic design changes to existing products
g. Routine design of tools, jigs, molds, and dies
h. Activity, including design and construction engineering, related to the construction, relocation, rearrangement, or start-up of facilities or equipment other than the following:
i. Legal work in connection with patent applications or litigation, and the sale or licensing of patents."

Table 1.1 Characteristics of basic, applied research, and R&D.

Characteristics	Basic research	Applied research	Engineering R&D
Overall goal	Exploration to new knowledge or theories	Expand or create technology/knowledge	Develop or improve products and processes
Focus	Fundamental and new understanding	New process, materials, parameters, etc. (in lab)	Specific realization (in production)
Applicability	General/universal principles	Predictable, general scope	Specific situations or problems
Contribution	Theory (maybe revolutionary)	Technology and inventions	New products, technological reference
Typical funding	Government and foundations	Government, foundations, industry	Industry internal, sometimes joint ventures
Timeframe	Long term	Midterm, fixed with certain flexibility	Short term, fixed
Practitioners	Scientists in academia and institutes	In variety of settings and combinations	Industry engineering professionals
Selection	By researchers, guided by sponsors	Selected by researchers based on necessity	Based on demand and directives
Commercial value	No or unclear	Good for a long term, influential	Short term, expected, and direct
Success rate	Low risk, high uncertainty	Medium risk and uncertainty	Low risk, high certainty
Outcome	Publication	Publication, internal reports, and/or patents	Mostly internal reports and/or patents

Comparison

The differences between three types of research can be seen in each of their goals and objectives. Table 1.1 summarizes the characteristics of the three types of research. Please note these are generalities, there are some exceptions.

It is worth noting that basic research forms the foundation for most applied research and R&D. We should understand the relevance of basic research and use its results to execute applied research and R&D. In fact, there is an overlap

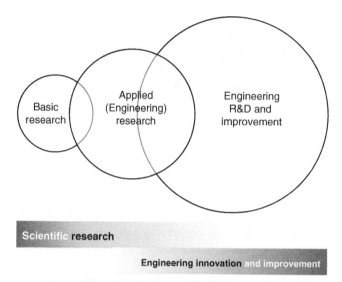

Figure 1.9 Types of research and development.

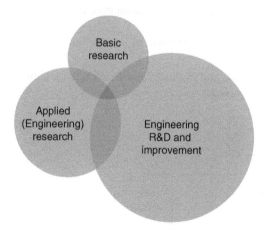

Figure 1.10 Relationship among three types of research.

between two neighboring types. Sometimes, a research project may be viewed a combination of both types, refer to Figure 1.9. In addition, applied research emphasis bridges basic research and R&D.

Figure 1.10 shows an interesting relationship among the three types of research activities. Each of the three types of supports, promotes, and drives the other two.

For R&D, companies who perform internal (funded) research projects for their own development purposes are not bound to any obligation to share learnings from research with society. Depending on the policies, some companies publish their general ideas after implementation. For example, Toyota developed a method to monitor production performance (Roser et al. 2001). They published the method only after having successfully implemented it for years. Another way of sharing the results of industrial development is via patents, which will be more discussed in Chapter 7.

1.3.2 More Discussion on R&D

1.3.2.1 Objectives of Engineering R&D
The general goal of engineering R&D is to improve the condition or functionality of particular goods. For instance, a research project may aim to increase the calculation speed of a new computer system, create a new algorithm for autonomous vehicles, or design a new software language.

A R&D objective can come in various formats and often is one of the three interconnected types or a combination of the three (Figure 1.11).

1. To explore, explain, or verify new *knowledge* or phenomenon; or synthesize existing knowledge,
2. To investigate and find *solutions* on a major problem by understanding causational relationship, or
3. To develop (conceptualization, create, and evaluate) a new *artifact* or technology, such as a tool, product, and process.

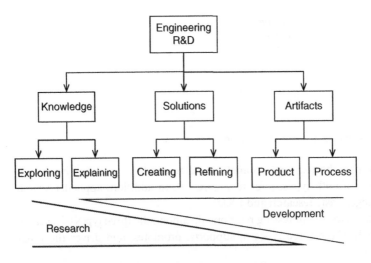

Figure 1.11 Aims and outcomes of engineering R&D.

The contributions to common knowledge from R&D projects may or may not be significant or "revolutionary." For example, a new technological development may only slightly advance contemporary knowledge and understanding questions may be raised about how originality and innovation of R&D is compared with the existing work in order to be qualified as quality research. When doing a comparison, we should include external information if possible because internal results may or may not be sufficient to validate the innovation.

1.3.2.2 Experimental and Empirical Research

Experimental and empirical methods are often used in R&D by directly collecting and analyzing data from experimentations or observations. Other times, indirect data is used as well. Figure 1.12 shows the overall process of experimental and empirical studies.

Note the differences between experimental and empirical studies. In experimental studies, we design and conduct experiments (i.e. manipulating and controlling variables, or designing procedures under certain circumstances). In empirical studies, we observe and collect data from the real world without human manipulation.

We may analyze data quantitatively or qualitatively both in experimental and empirical studies. In addition, there are different ways and methods of collecting and analyzing data for experimental and empirical studies. Based on facts or evidence, the outcomes of such a study are often considered reliable, at least for the particular case defined in a study.

Theory and principles play a guiding role in both experimental and empirical studies. Furthermore, such studies can be used to verify the results based on computer calculation, modeling, and simulation. For example, a thermo-mechanical coupling analysis of transient temperature and rolling resistance for a solid rubber tire is studied by computer simulation and verified by experiments (Li et al. 2018).

1.3.2.3 Descriptive, Exploratory, Analytical, and Predictive Research

Studies can also be categorized into four types to represent their different objectives. Table 1.2 lists these four types of research based on their goals and methods.

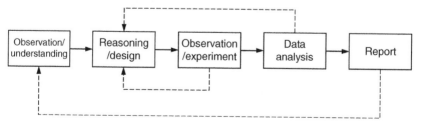

Figure 1.12 A process flow of experimental and empirical research.

Table 1.2 Characteristics of four types of research goals.

	Exploratory	Descriptive	Analytical	Predictive
Goal	To look for a new idea, pattern, or hypotheses	To identify and classify the elements or characteristics of a subject	To explain why or how, to evaluate a subject, and to compare multiple situations	To speculate intelligently on future possibilities, maybe to control
Methods	Observations, literature review, etc.	Often quantitative by observations, case studies, survey, etc.	Often based on the data collected, maybe descriptive	Based on theory and analytical studies
Example	Effects of a new technology	How a new technology affect	Data mining	Artificial intelligence

Exploratory research is to investigate a problem for better understanding, for example, a framework using deep learning constructs to generate mechanical designs (Raina et al. 2019) and an exploratory survey research in engineering ergonomics (Marcos et al. 2018). An exploratory study can be an early phase of a large research project.

Sometimes, research objectives and tasks may be viewed in different orders and levels. For example, analysis is a foundation of prediction. If a future thing can be correctively predicted, then it would mean we have the necessary expertise and tools for mastering of the subject. Figure 1.13 shows the four levels of studies of the overall advance of knowledge, which continuously becomes deeper, closer to the truth, and more difficult to investigate.

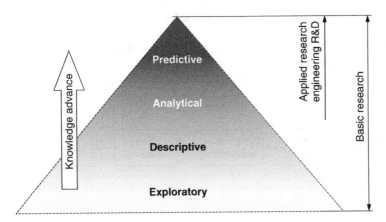

Figure 1.13 Knowledge and skill levels of research.

There is a significant overlap between basic research and applied research in terms of knowledge levels. Applied research projects are often analytical and predictive in nature. Some of them may be associated with descriptive research as well. Therefore, some parts of the four types may be jointly applied in one research project.

1.3.2.4 Case Study

A case study is used to investigate a specific phenomenon or study problem in-depth. Engineering and technical professionals in industries often use a case study as an effective approach to learn about need-to-know situations, such as the causation between input and output, process changing over time, and support for a hypothesis.

In other words, a case study can be practical and result-driven, deemed successful or meaningful only if a particular problem is resolved. Conducting a case study can be an excellent training exercise for student researchers in colleges and new researchers in industry. From a data analysis point of view, case studies can be either qualitative or quantitative, though more frequently the latter than the former. The results from a case study are often descriptive and explanatory.

Below are a few examples of case studies.

> *"Squeezing flow between rigid tilted surfaces: A general solution and case study for MEMS"* (Moy et al. 2017).
> *"A case study on failure of superheater tubes in an industrial power plant"* (Dehnavi et al. 2017).
> *"Effect of misalignments on the tribological performance of elastomeric rod lip seals: Study methodology and case study"* (Pinedo et al. 2017).

Due to their characteristics, case studies have a limitation: the findings from a case may or may not be generalizable or applicable to other situations. However, the procedure and methods used in a case study may have a good reference value for similar applications.

For the same reason, a case study may or may not be considered as research. We may identify a case study as research if it contributes to professional community at large, not only to the organization of doing the case study. Recognizing whether a case study is research can be debatable. A literature review of 31 papers concluded that in case studies can be used for research, but they cannot cover all the hallmarks of science (Josefsson 2016).

Many scholarly journals accept papers of case studies if they have practical significance and/or sufficient technical depth. There are also a few refereed journals dedicated to cases studies, such as

- Case Studies in Thermal Engineering (ISSN: 2214-157X)

- Case Studies in Construction Materials (ISSN: 2214-5095)
- Mathematics-in-Industry Case Studies (ISSN: 1913-4967)

1.4 Validity of Research Results

1.4.1 Research Validity

1.4.1.1 Concept of Validity

In general, research and its results should be based on objective facts rather than an opinion. Rigorous, precise, thorough methods and processes should be used to keep the results and conclusions objective and correct. Logical arguments and meaningful conclusions must be supported with the evidence.

One of the important properties of research outcomes, validity is about the soundness and quality of research design, methods, and conclusions and their applicability to similar situations. When addressing the quality of research, we often use the term "validity" and sometimes other similar terms, such as "reliability," "accuracy," "objectivity," "generalizability," and "credibility" for the same purpose and comparable meanings.

The purpose of validation is to convince readers that the research conclusions are legitimate and justifiable. The evaluation of research validity is based on its contents of measurement and data, analysis and criteria of the data, the methods, and processes used, the reasoning for conclusions, and so on.

There are two types of research validity: internal validity and external validity. Table 1.3 briefly shows a comparison between two types, which will be discussed more in the following subsections.

There are several factors to consider, including the researcher's personal interests, abilities, assumptions, resources, and ambitions. All these things influence

Table 1.3 A comparative view between internal and external validity.

Aspect	Internal validity	External validity
Definition	Soundness of work and outcomes	Generality to other situations and real world
Meaning	True and accurate for the study	True and accurate for similar other studies
Relationship	Essential, itself	Additional, on top of internal validity
Significance	Important to applied research and R&D	Important to basic research

the outcomes of research. Remember, controlling all possible factors that may affect the validity of research is a primary responsibility of the researcher.

1.4.1.2 Internal Validity

The question of internal validity refers to the study itself. In other words, how we logically conduct the research and interpret its results. Another way to think about internal validity is that it helps rule out any alternative explanations in relation to research findings.

A research project and result should "stand on its own," meaning that the methods and process used within the study, such as for obtaining the data, hypothesis testing, and data analysis, are implemented appropriately. The problem statement and supporting tasks should be framed correctly in order to answer the question in a logical sequence.

We may consider research internally valid if its findings accurately represent the data or phenomenon measured and claimed findings. Otherwise, the results may be subjective, incorrect, or even misleading.

The threats to internal validity include the detailed items in study procedures, experimental settings, data collection, analysis methods used, and result interpretation. To evaluate the internal validity of a research project, we may ask a few questions (refer to Table 1.4). Inviting colleagues and friends to ask such questions can be effective and provoke additional thinking. For example, a "maybe" answer to a question indicates a possible issue.

A common issue with internal validity is that researchers do not appropriately monitor or control some variables during a study. These issues due frequently to study design, execution, measurement, or a combination thereof. The resultant findings then, such as a causational relationship, could be questionable due to a problematic process or flawed data. Sometimes, the results from invalid data are not reasonably interpretable.

Table 1.4 A checklist for internal validity.

Question	Yes	Maybe	No
1. Is problem statement/hypothesis established correctly?			
2. Are methods used appropriate?			
3. Have measurement instruments been calibrated?			
4. Are the samples selected randomly?			
5. Are the findings aligned with the problem statement?			
6. Can the logic be explained?			
7. Is any interpretation lack of data support?			

One aspect of verifying internal validity is checking whether a particular research is biased in some fashion. For example, we should collect data without the influences from any preconceived outcomes in our mind if any. More importantly, a study should be independent and avoid conflicting factors. The less chance for conflict in a study, the higher its internal validity will be.

1.4.1.3 External Validity

External validity is about the generality of research outcomes or how well results and theories from one setting applicable to another. If the result from a research project is also true for other situations, either by external reviews or by actual tests, it has a good external validity. We may use the terms "generalizability" and "transferability" for the external validity as well.

If a research outcome lacks external validity, we are unable to apply the outcome to a different or larger situations and contexts. In such cases, the usefulness of the research is limited. In research reports and publications, we should acknowledge the limitation in terms of external validity.

Peer-review in a publication process is a common way to check external validity. It is ideal if the same experiments can be replicated by other professionals and yield the same outcomes. To ensure reproducibility, it is important to state clearly the analysis procedure and parameters in the paper and provide the original data in many cases. An example is a study on ignition behavior of a gas turbine (Mansfield and Wooldridge 2014). In the paper, authors provided the data source, "Supplementary data associated with this article can be found, in the online version, at https://dx.doi.org/10.1016/j.combustflame.2014.03.001."

For a laboratory research, we may assess the external validity by asking addition questions listed in Table 1.5.

One example: A new autonomous vehicle had been developed and tested well in a controlled setting. However, it is unknown if the car was operated in various real city and highway settings. In other words, without external validity, the research outcomes and conclusions can be challenged as to whether they "hold up" under a larger scope or different circumstances.

Table 1.5 A checklist for external validity.

Question	Yes	Maybe	No
1. Are the assumptions acceptable to other setting?			
2. Will the same results can be obtained from another lab?			
3. Do the conclusions apply in the real-world situations outside the lab?			
4. Can this plan carry over to other place or site and get the same results?			

Achieving external validity can be more challenging than internal validity due to various known and unknown variables in the technical world. Even approaching the same problem, different researchers may have different assumptions and/or take a different path for study and validation. Another factor is the secrecy around the innovation of applied research and R&D to allow testing external validity.

Specific to the research in the form of case studies, their value may be on the innovative ideas rather than the results of specific applications. For example, a paper in civil engineering is for building construction projects in South Korea (Cha and Kim 2018). The result may or may not be applicable to non-Korean locations. However, the systematic framework for predicting the performance proposed in the paper may have a good reference value.

1.4.2 Assessment and Advance

To assess and improve validity, it is important to identify and address the possible bias factors influencing the internal and external validities. For example, a researcher should seek, consider, and analyze possible exceptions and contradictory data, which often imply unknown factors in a study. The two tables above can be a good reference. As mentioned, seeking review and feedback from third parties can be very helpful.

1.4.2.1 To Get Validated

For all types of research, we need to evaluate the results before concluding their correctness. One way is to repeat the tasks undertaken, with the same settings, to check for similar results. If the results are reproducible, then it is highly likely the argument is true for the situation (thus proving good internal validity). Another way is to validate an outcome under similar situations outside the scope of the research originally undertaken. If getting consistent results, the research process and results can be applicable to other situations (proving good external validity).

A validation process may include four steps:

1. Set up an environment and design realistic scenarios
2. Build a prototype with the new artifact
3. Simulate the prototype against in the scenarios in the environment
4. Assess the results

The validation process is valuable to engineering research as it often yields and verifies new inventions. For example, the results of some research projects of computer science and engineering include an algorithm, method, process, technique, or device, which may be all called artifacts. In such cases, any type of artifact generated from a research project should be validated before being put into practice

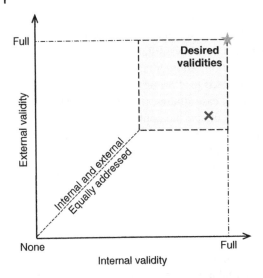

Figure 1.14 Overall validity target of research results.

or manufacturing. A new algorithm must be tested and validated before it is used in commercial software.

An ideal situation is to have research results both fully internally and externally validated, refer to Figure 1.14. It may be difficult to measure validity, but one practical way to evaluate the validity is by dividing criteria to three or five levels, such as poor, good, and full. For example, a computer simulation study, "×" in the figure, may be with good internal validity by a validation using different software and partially external validity by the reviews of domain experts.

1.4.2.2 Considerations of Validities

When we consider the validity of the research, we need to address its internal validity first. The internal validity of research is a foundation to the external validity as it extend a specific claim in a research study to be applicable in other similar situations (refer to Figure 1.15).

In many cases, internal validity is a main concern. Due to physical constraints, such as sample size or time limit, there are often bias factors, which can be any influence and condition that distort the data. For such factors, we should try to avoid them during the design, planning, and execution of a study as much as possible. In addition, we should acknowledge them based on our best knowledge.

Data collection biases include data sampling and measurement errors. Sampling bias can occur due to many known and unknown factors, which leads to nonrepresentative data samples of the entire target population. These potential issues will be discussed more in later chapters.

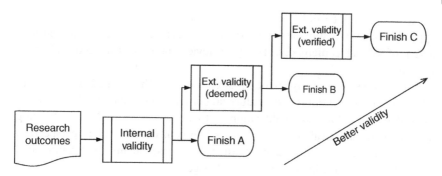

Figure 1.15 Validity levels of research results.

Another factor influencing internal validity of research may be the researchers themselves. The design and execution of a research project can be based on researcher's viewpoints or the methods and process selected by the researcher. The research conclusions can be based on the researcher's subjective interpretations as well.

Due to various reasons, external validation may be constrained or difficult for some cases. For example, research projects performed at for-profit companies and classified operations, such as military or offense projects, attempt to gain a technical or operational advantage. For such research, the validation of the study results may be achieved by putting the idea to practice in various settings "internally." In other words, research results are externally validated inside an organization.

As discussed, research is always done under certain assumptions. The assumptions play a fundamental role on the validities, particularly external validity. If an assumption is only true for a special case, for example, then the research outcome is unlikely applicable to other cases or situations.

When it is needed and feasible, a multiple-case design is a good way to improve external validity. If multiple cases can yield the same conclusions, then they have good generalizability of their findings. In addition, if we design two contrasting cases and the research findings support the contrasted hypothesis, then the results also strengthen their external validity.

Using multiple sources and forms of data is an effective way to reduce potential validity issues related to data. Regarding research methods, two or more methods may be used in a study to check the study results. If different data and/or methods yield the same conclusions, they are very likely valid. This approach is sometimes called triangulation.

1.4.2.3 Publication and Further Development

One of the requirements for research is the professional communication and dissemination of new knowledge. Publication is an important way to advance knowledge and get validation. In general, all basic research and most of the applied research results are published via peer review academic journals and professional conferences. The peer review process is a good validation process.

The findings of some types of research, such as the applied research and R&D conducted in military and industrial corporations, are of good quality but unavailable for publication. Publication is also affected by the policy of a funding agency and the approval process of a research organization. From this perspective, there are different opinions regarding the values of confidential and commercial research to the professional community at large.

After successful validation, a next step for applied research is to scale up from specific laboratory conditions to practical, real-world conditions. The scaling-up tests may be adding realistic conditions of practice, extending to larger sets of subjects, or a combination of both, refer to Figure 1.16.

Research validity and validation, in terms of the process and criteria, is a research topic itself that is not yet fully explored. Not all research papers are validated before publishing. A literature review showed that 37% of the articles in the field of research in engineering design did not have any validation (Barth et al. 2011). Studying the correlations between types of research and types of validation may lead to more research and the development of a common body of knowledge.

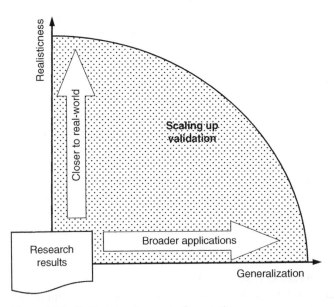

Figure 1.16 Further development of research.

Summary

Introduction to Research

1. Research is about innovation and seeking new knowledge in a systematic way.
2. Research normally follows a certain process of planning, literature review, information collection, data analysis, and results discussion, etc.
3. Research is objective driven; the outcomes of research should be conducive to its predefined objectives.
4. Research should have various types of impacts on academic, economic, and/or societal aspects.
5. Research is an integral part of higher and graduate education.

Building Blocks of Research

6. Motivation is a foundation of research success. While the motivations may include curiosity, career, and education needs.
7. Analytical, critical, and creative thinking plays an integral role in research.
8. Assumptions are the foundation and conditions for research work. Assumptions may or may not be completely and explicitly stated.
9. Research is often for hypothesis proving or problem-solving. Hypothesis, normally in format of H_0/H_1, may be predictive statements based on research aims.
10. Methodology is a general research strategy and procedure, while a method is a specific approach or tool used in research. Both terminologies sometimes are interchangeable.
11. Research is teamwork in an environment, meeting ethics requirements, and with funding sources.

Types of Research

12. There are three types of research: basic research, applied research, and R&D. They closely connect each other.
13. Basic research is fundamental or theoretical work for new knowledge. Applied research is for both new knowledge and practical aims or problems. R&D is a technical invention, focusing on new products and technologies.
14. The outcomes of engineering R&D can be new knowledge, solutions, and/or artifacts.
15. The knowledge exploration in research can be descriptive, exploratory, analytical, and predictive. Engineering research is more in the latter two types.

16. Case study is a common format of engineering research – focusing on specific phenomenon or problems.

Validity of Research Results

17. The validity is the base of research work and its results. Sometimes, validity is called in different terms, such as "reliability," "accuracy," "objectivity," "generalizability," and "credibility."
18. Internal validity is about the soundness of work and outcomes, referring to the study itself.
19. External validity is about the generality of work and outcomes to other situations and real world, or how well the work and outcomes from one setting applicable to another.
20. Internal validity is a foundation of external validity.
21. Sampling methods used and sample sizes are the two important factors to internal validity.
22. The specific assumptions largely affect the external validity of research and its outcomes.

Exercises

Review Questions

1 What is your definition on research? You may talk about it based on your experience, observation, or just current understanding. (You may save your answer and revisit it when complete this learning.)
2 Find an example that claims doing research but is actually different from scientific research.
3 How to view research from the "know that" and "know how" stand of points?
4 Select an engineering project as an example (e.g. an autonomous vehicle technology) and justify whether it may be considered as scientific research.
5 Think about the basic elements of research and select one to discuss with an example.
6 Explain why some technical efforts (e.g. gathering information and implement a practice from one place to another) may not be considered as research.
7 Based on a published research that you are familiar with, identify three characteristics and discuss why they are important to the research.
8 Explain one personal benefit and one societal benefit from completing a quality research, such as a thesis or dissertation.
9 Ask a critical thinking question to the content of this chapter.

10 Explain the role of critical thinking in research with an example.

11 Discuss the roles and significance of hypothesis in a research project.

12 Based on the discussion, find an example to distinguish methodology and method in research.

13 Explain the differences between basic research, applied research, and R&D with an example for each type of research.

14 Find an example of a type of exploratory, descriptive, analytical, or predictive research.

15 Review the meaning of empirical and experimental research with an example.

16 Find a case study and comment it whether it is considered as research.

17 Explain the significance of research validity with an example.

18 Discuss the differences between internal validity and external validity.

19 Concerning possible high costs and much time required, one might ask what level of internal validity to strive for. What is your perspective?

20 Most master's graduate programs have two options: research and nonresearch (coursework only). Which one you would prefer if the two options were available? Why? (This question could be revisited after completing this learning.)

Mini-Project Topics

1 Search different sources and compare the definitions and meanings of the word "research." In your particular discipline, please provide and justify your opinion which definitions are more appropriate.

2 There is an argument that nonpublishable research due to security reasons is not described as basic research (Thiel 2014). Would you agree? Please explain and justify your claim.

3 In your undergraduate education, master's degree study, or work, select one study and explain its
(a) Study objectives and procedure
(b) Principle and method used
(c) Conclusions and their validity
(d) What you might like to do if you could do the same study again

4 Locate a research paper related to your current or future interest from your library databases or https://scholar.google.com/, and summarize the paper on its:
(a) Overall research design
(b) Data and method used
(c) Contributions to community or knowledge

5 Interview an experienced researcher and ask the following questions:
   (a) How to find research opportunities
   (b) What methods/tools often used, why select them
   (c) Keys to his/her research success
   (d) Advice to novice researchers
6 Review a research paper on a subject you are knowledgeable from library databases or https://scholar.google.com/, and apply your critical thinking mindset to find the possible future research opportunities.
7 Research work is based on certain assumptions. Find a research work and identify whether the assumptions explicitly stated. Is there any assumption that is not explicitly stated but has some effects in the research results? Use examples to support your answer.
8 Visit the IRB website of your organization or talk with the person who is responsible to the human subjects in research and understand:
   (a) Overall requirements
   (b) Application process and approval
   (c) Required specific training
9 Assume you plan to do a research project at work place or for your degree study, please think about the project and discuss (if you choose a finished project to talk, please also think about how to improve if you could do it again):
   (a) Research objective and problem
   (b) Methods used
   (c) Internal and external validities
   (d) Contribution and impacts
10 Search the different words used in literature but with the similar meaning of research validity. Discuss their same and different focuses compared with word "validity."

References

ACRL, Association of College and Research Libraries. (2016). Framework for information literacy for higher education, adopted by the ACRL board, 11 January 2016. www.ala.org/acrl/standards/ilframework (accessed January 2019).

Apple (2019). Annual financials for apple Inc. https://www.marketwatch.com/investing/stock/aapl/financials (accessed 8 April 2019).

Ark, T.V. (2019). How to be employable forever. Forbes, 20 February 2019. https://www.forbes.com/sites/tomvanderark/2019/02/20/how-to-be-employable-(a)forever/#67fb68751ca3 (accessed November 2019).

Barth, A., Caillaud, E. and Rose, B. (2011). How to validate research in engineering design? International Conference on Engineering Design (15–18 August 2011). Technical University of Denmark.

Caillaud, E., Rose, B., and Goepp, V. (2016). Research methodology for systems engineering: some recommendations. *IFAC-PapersOnLine* 49–12: 1567–1572. https://doi.org/10.1016/j.ifacol.2016.07.803.

Cha, H.S. and Kim, K.H. (2018). Measuring project performance in consideration of optimal best management practices for building construction in South Korea. *KSCE Journal of Civil Engineering* 22 (5): 1614–1625. https://doi.org/10.1007/s12205-017-0156-2.

Davoodi-Nasab, P., Rahbar-Kelishamia, A., Safdari, J. et al. (2018). Selective separation and enrichment of neodymium and gadolinium by emulsion liquid membrane using a novel extractant CYANEX® 572. *Minerals Engineering* 117: 63–73. https://doi.org/10.1016/j.mineng.2017.11.008.

Dehnavi, F., Eslami, A., and Ashrafizadeh, F. (2017). A case study on failure of superheater tubes in an industrial power plant. *Engineering Failure Analysis* 80: 368–377. https://doi.org/10.1016/j.engfailanal.2017.07.007.

Depaola, N., Cammino, R., Haferkamp B., et al. (2015). ENGR497–An introduction to research methods course. 122nd ASEE Annual Conference and Exposition, Seattle, WA (14–17 June 2015), Paper ID #13900.

Dianat, I., Molenbroek, J., and Castellucci, H.I. (2018). A review of the methodology and applications of anthropometry in ergonomics and product design. *Ergonomics* 61 (12): 1696–1720. https://doi.org/10.1080/00140139.2018.1502817.

DOE (n.d.-a). Basic energy sciences (BES), department of energy. https://science.doe.gov/bes/research/ (accessed June 2018).

DOE (n.d.-b). Advanced scientific computing research (ASCR), department of energy. https://science.doe.gov/ascr/ (accessed June 2018).

Elder, L. (2007). Another brief conceptualization of critical thinking. www.criticalthinking.org/pages/defining-critical-thinking/766 (accessed July 2018).

Esmaeilian, B., Behdad, S., and Wang, B. (2016). The evolution and future of manufacturing: a review. *Journal of Manufacturing Systems* 39: 79–100. https://doi.org/10.1016/j.jmsy.2016.03.001.

GM (2019). Annual financials for general motors Co. https://www.marketwatch.com/investing/stock/gm/financials (accessed 8 April 2019).

Gu, Y. (2017). Fast traversing autonomous rover for mars sample collection. WV-17-EPSCoRProp-0005, 2017 EPSCOR RESEARCH PROPOSAL ABSTRACTS, NASA. https://www.nasa.gov/sites/default/files/atoms/files/2017_epscor_abstracts.pdf (accessed June 2018).

Haumer, A. and Kral, C. (2014). The new EDrives library: a modular tool for engineering of electric drives. Proceedings of the 10th International Modelica

Conference, Lund, Sweden (10–12 March 2014), 155–163. doi: 10.3384/ecp14096155.

Hazelrigg, G.A. (n.d.). Research 101 for engineers, national science foundation. 129.130.42.171/NSF2018/subfolder/RESEARCH%20101%20FOR%20ENGINEERS. (a)PDF (accessed July 2018).

Josefsson, T. (2016). *How Good are Case Studies as Scientific Products?* Sweden: Halmstad University.

Kpotufe, S. (2014). Research statement. www.princeton.edu/~samory/Papers/ResearchStatementGeneral.pdf (accessed May 2018).

Laird, N. (2018). Internal discussion in course QUAL 647 research methods, CRN 14245. Eastern Michigan University. (accessed 30 September 2018).

Leedy, P.D. and Ormrod, J.E. (2016). *Practical Research: Planning and Design*, 11ee. Pearson. ISBN: 978-0134775654.

Li, R. and Li, H.A. (2018). New two-phase and three-phase Rachford-Rice algorithms based on free-water assumption. *The Canadian Journal of Chemical Engineering* 96: 390–403. https://doi.org/10.1002/cjce.23018.

Li, F., Liu, F., Liu, J. et al. (2018). Thermo-mechanical coupling analysis of transient temperature and rolling resistance for solid rubber tire: numerical simulation and experimental verification. *Composites Science and Technology* 167: 404–410. https://doi.org/10.1016/j.compscitech.2018.08.034.

Madan, C.R. and Teitge, B.D. (2013). The benefits of undergraduate research: the student's perspective. https://dus.psu.edu/mentor/2013/05/undergraduate-research-students-perspective/ (accessed May 2018).

Mansfield, A. and Wooldridge, M. (2014). High-pressure low-temperature ignition behavior of syngas mixtures. *Combustion and Flame* 161 (9): 2242–2251. https://doi.org/10.1016/j.combustflame.2014.03.001.

Marcos, E.D.L., Silva, M.B., and Souza, J. (2018). The integrated management system (IMS) and ergonomics: an exploratory research of qualitative perception in the application of NR-17. *Journal of Ergonomics* 8 (3): 1000231. https://doi.org/10.4172/2165-7556.1000231.

Mervis, J. (2017). Data check: U.S. government share of basic research funding falls below 50%. *Science* https://doi.org/10.1126/science.aal0890.

Mokhtari, H. and Hasani, A. (2017). An energy-efficient multi-objective optimization for flexible job-shop scheduling problem. *Computers and Chemical Engineering* 104: 339–352. https://doi.org/10.1016/j.compchemeng.2017.05.004.

Moris, F. (2018). Definitions of research and development: an annotated compilation of official sources. https://www.nsf.gov/statistics/randdef/#chp3 (accessed July 2019).

Moy, A., Borca-Tasciuc, D., and Tichy, J. (2017). Squeezing flow between rigid tilted surfaces: a general solution and case study for MEMS. *Lubrication Science* 29: 531–539. https://doi.org/10.1002/ls.1386.

NASA (n.d.) Science and technology. https://science.nasa.gov/technology (accessed June 2019).

NSF (2018). Science and engineering doctorates. National Center for Science and Engineering Statistics Directorate for Social, Behavioral and Economic Sciences, NSF 19-301, December 2018. https://ncses.nsf.gov/pubs/nsf19301/report/about-this-report (accessed June 2019).

Panchal, J., Kannan, K., Helie S. et al. (2017). Understanding information acquisition decisions in systems design through behavioral experiments and bayesian analysis. Award Abstract #1662230, NSF. https://www.nsf.gov/awardsearch/showAward? AWD_ID=1662230&HistoricalAwards=false, (accessed June 2018).

Payri, R., Gimeno, J., Martí-Aldaraví, P. et al. (2017). A new approach to compute temperature in a liquid-gas mixture. Application to study the effect of wall nozzle temperature on a Diesel injector. *International Journal of Heat and Fluid Flow* 68: 79–86. https://doi.org/10.1016/j.ijheatfluidflow.2016.12.008.

Pinedo, B., Conte, M., Aguirrebeitia, J. et al. (2017). Effect of misalignments on the tribological performance of elastomeric rod lip seals: study methodology and case study. *Tribology International* 116: 9–18. https://doi.org/10.1016/j.triboint.2017.06.022.

Raina, A., McComb, C., Cagan, J. et al. (2019). Learning to design from humans: imitating human designers through deep learning. *Journal of Mechanical Design* 141 (11) https://doi.org/10.1115/1.4044256.

Razaviyayn, M., Hong, M., and Luo, Z. (2013). A unified convergence analysis of block successive minimization methods for nonsmooth optimization. *SIAM Journal on Optimization* 23 (2): 1126–1153. https://doi.org/10.1137/120891009.

Roser, C., Nakano, M. and Tanaka, M. (2001). A practical bottleneck detection method. Proceedings of the 2001 33rd Winter Simulation Conference, Phoenix, Arlington, VA, 949–953.

Russell, S.H., Hancock, M.P., and McCullough, J. (2007). Benefits of undergraduate research experiences. *Science* 316 (5824): 548–549. https://doi.org/10.1126/science.1140384.

Santarsiero, M., Louter, C., and Nussbaumer, A. (2017). Laminated connections under tensile load at different temperatures and strain rates. *International Journal of Adhesion and Adhesives* 79: 23–49. https://doi.org/10.1016/j.ijadhadh.2017.09.002.

SaTC (n.d.). Secure and trustworthy cyberspace. Program Solicitation, NSF 18-572, National Science Foundation. https://www.nsf.gov/pubs/2018/nsf18572/nsf18572.pdf (accessed September 2018).

Sousa, R.A. and Figueiredo, F.P. (2017). Correction of plastic zone estimates from linear elastic stress field: a qualitative analysis. *Fatigue and Fracture of Engineering Materials and Structures* 40: 1652–1663. https://doi.org/10.1111/ffe.12596.

Staffa, L.H., Agnelli, J.A.M., Bettini, S.H.P. et al. (2017). Evaluation of interactions between compatibilizers and photostabilizers in coir fiber reinforced polypropylene composites. *Polymer Engineering and Science* 57 (11): 1179–1185. https://doi.org/10.1002/pen.24495.

Thiel, D. (2014). Introduction to engineering research. In: *Research Methods for Engineers*. Cambridge: Cambridge University Press https://doi.org/10.1017/CBO9781139542326.003.

Yang, W., Ding, H., and Kecskeméthy, A. (2018). A new method for the automatic sketching of planar kinematic chains. *Mechanism and Machine Theory* 121: 755–768. https://doi.org/10.1016/j.mechmachtheory.2017.11.028.

Zhu, Y., Bai, Z., Luo, W. et al. (2017). A facile ion imprinted synthesis of selective biosorbent for Cu2+ via microfluidic technology. *Journal of Chemical Technology and Biotechnology* 92: 2009–2022. https://doi.org/10.1002/jctb.5193.

2

Research Proposal Development

2.1 Research Initiation

A starting point and a critical element of research success is a well-prepared proposal. Proposal development encompasses the early phases of an entire research project. We can have the best idea in the world, but if we do not propose it into a plan and start work with appropriate funding, a great idea remains at a concept level. Therefore, it is important that we address the basics of research proposal development for student and beginning researchers. We will also conduct a detailed discussion and review on proposal development for tentative external and large grants.

2.1.1 Research Proposal

2.1.1.1 Form Ideas from Problems

A research proposal, as a formal document, is a succinct summary of what we plan to do, why it should be done, how we will do it, and what the expected outcomes are for a given future research project.

Most technical research is problem-driven, meaning we identify the problems that need to be resolved as the research objectives, then develop the proposals and plans. We have numerous ways to identify problems. A typical way is to study the given situation, in order to improve or create a better solution or a new engineering artifact.

Identifying and describing a problem is conducted in two steps suggested for engineering design problems (Arciszewski 2016) and may be applicable to other types of research.

1. Briefly describe the problem. No methodology is necessary for this initial stage. In other words, researchers stay open-minded regarding possible methods and processes

Engineering Research: Design, Methods, and Publication, First Edition. Herman Tang.
© 2021 John Wiley & Sons, Inc. Published 2021 by John Wiley & Sons, Inc.

2. Prepare a detailed description of the problem, including
   - The situation in the context of a system
   - Inputs and outputs of the system
   - Functions of the system and subsystems
   - The system's structure and relationship among the subsystems
   - The system's environment

The second step is meant to provide a good understanding of the problem and corresponding solution directions a system level, which lays a foundation for research work. With a problem identified, researchers utilize with other tools, such as literature review, to generate a new idea.

A new idea may be initiated by one person or by team brainstorming. The brainstorming approach is an effective way because a research team compiles different approaches and perspectives and utilizes critical thinking to evaluate their worth. It is often true that "two heads are better than one," but a groupthink is normally disadvantageous for research brainstorming. Researchers can employ an argument analysis to evaluate brainstorming and ascribing benefits and drawbacks to methodology and to judge the value of various ideas and plans.

2.1.1.2 Idea Evaluation

To self-assess our new research ideas, we may use a simple checklist (Table 2.1) to ask ourselves a few basic questions, which are the aim and keys of a proposal. Numbers 1a, 1b, and 1c are for specific types of research. The preliminary answers for all the questions should be "yes." An unsure answer "maybe" implies an improvement opportunity. Then, we can proceed to develop the details of a proposal.

There are three cornerstones for a good research proposal. A stool model, as shown in Figure 2.1a, depicts that the three legs are generally of equal importance to a good proposal. For a specific proposal, the three cornerstones may have different levels of desired or actual importance and be measured against the goals and requirements of research funding sources (see Figure 2.1b).

1. Innovation
2. Feasibility
3. Impacts

The three elements are indispensable. A major issue on any of the three elements can postpone or even kill a proposal. Therefore, we must address all three elements in a proposal development and evaluation. There will be future discussions later in this chapter in relation to these three cornerstones.

Table 2.1 A simple checklist on research Idea in proposal development.

No.	Question	Yes	Maybe	No
1a	Will this contribute or extend human knowledge? (for basic research)			
1b	Will this generate a new method, understanding, or solve a major problem, etc.? (for applied research)			
1c	Will this invent a method, an artifact, or solve a practical problem? (for engineering R&D)			
2	Will this be a systematic study with a feasible plan?			
3	Will this be assessed for internal and external validity?			
4	Will this be meaningful and impactful to community?			

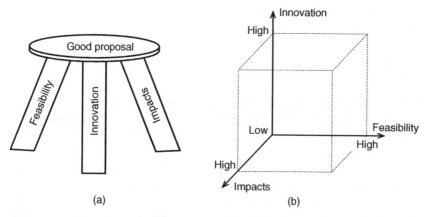

Figure 2.1 Three cornerstone of research proposal. (a) Proposal stool model. (b) Measurement of proposal quality.

Research proposal development, therefore, is a process in addition to an outcome. Preparing and conducting a research project have multiple tasks and considerations, including timing, technical difficulties, available resources, etc. Therefore, a viewpoint from a project management and systems engineering prospective helps improve the effectiveness of research development.

2.1.1.3 Student Research Development

Student research proposals may be short, but the development process is about the same and its composite parts are similar. The development of a student research proposal may be viewed as a guided independent study.

The funding sources for student research are either part of faculty research grants or supported by university programs. The various funding opportunities have different requirements, so readers should check the particular requirements of funding programs.

Figure 2.2 shows the four major phases of an entire research project and Figure 2.3 shows the main tasks in the four phases. The first phase is proposal development. We will discuss the process, phases, and tasks more in the following sections and chapters. If a research project is to be funded by an external sponsor, the overall process is more formal and needs additional reviews, administrative processes, approvals, etc., which will be discussed in Chapter 7.

In addition to seeking advice from faculty advisors and participating in workshops, students should check the reference materials posted in their university website and refer even to other university websites. For example, University of California – Berkeley has undergraduate student proposal guidelines. They suggest that a student proposal include the following sections (UCB 2015):

1. Statement of Purpose (one paragraph, 150–175 words)
2. Background and Justification (1.5–2 pages)
3. Project Plan (1.5–2 pages)
4. Qualifications (1.5 pages)
5. Budget Plan (0.5 page)

It is strongly encouraged that students search their university research opportunities and talk to faculty. One approach is "to encourage the student to research and write an interest paper as a first step to writing a proposal to seek and do organized research" (Anderson-Rowland et al. 2015).

Generally, the earlier students begin a research development, the more guidance and encouragement they will need. A study showed that engineering freshmen are "less skilled at focusing their topics in terms of breadth, use of disciplinary vocabulary, and framing their purpose in terms of an argument" (Eckel 2019). It would

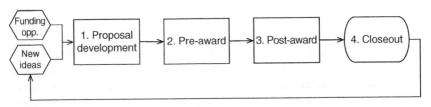

Figure 2.2 Four phases of research lifecycle.

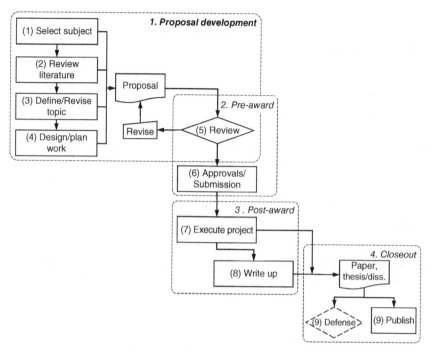

Figure 2.3 An overall process flow of research project.

be more effective if students began being involved in research when they reach their junior year of college because of the appropriate knowledge and experience they have at that point.

2.1.1.4 Proposal, Protocol, Prospectus

In some areas, such as health science, a research proposal and a research protocol are different. A research proposal is a concise summary of your research ideas, showing a study plan, justifying research, and selecting approaches. The purpose is often to seek an initial approval from a granting authority or a scientific committee. On the other hand, a research protocol is a systematic description and detailed set of activities for the research project proposed. In a research protocol, literature review, qualification to a new study, and a work timetable, among multiple other things, must be included.

In some universities and fields, students develop a research prospectus as an internal exercise, such as dissertation prospectus. A few universities require both prospectus and proposal at different stages of doctoral study, with different focuses and formats.

We may use the three terms, i.e. proposal, protocol, and prospectus, interchangeably to plan research. In this book, for simplicity, the term "proposal" is often used through the contents.

2.1.2 Hypotheses

2.1.2.1 Objective and Hypothesis

For basic and applied research, we often state a research objective to obtain new understanding or answers to predefined questions as mentioned a research hypothesis in Chapter 1. We may view a hypothesis as an educated guess and prediction for research outcomes based on research objectives. It may state, "The research objective of this project is to test the hypothesis of . . . ." for a basic research project. To establish the hypothesis is a centerpiece based on research aim and supported by method and data (refer to Figure 2.4).

In addition, an established research hypothesis should be measurable, testable, and provable (or disprovable) by the end of the study. If a hypothesis is supported and validated by research analyses, then we may use it to predict what would happen under the same situation.

2.1.2.2 Format of Hypothesis

A research hypothesis is often stated in two parts in a format of H_0/H_1:

- *Null Hypothesis* (H_0). It is the statement of a default position, assuming no statistical significance exists in a set of observations until evidence indicates otherwise.
- *Alternative Hypothesis* (H_1). The statement describes the other possible outcome, which often is set as researchers would expect and predict.

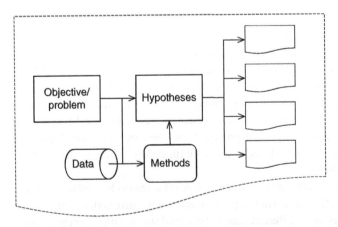

Figure 2.4 Structure of proposal description with hypothesis.

In addition, the alternative hypothesis H_1 is in one of two directions:

- *The one-direction* hypothesis predicts a specific relationship between variables. For example, the applications of collaborated robots are increasing in the next 10 years.
- *The two-direction* hypothesis predicts a specific relationship without specifying the direction of that relationship. For the same example, the applications of collaborated robots are either increasing or decreasing in the next 10 years.

In engineering research projects, the hypothesis may be stated either explicitly or implicitly. For a simple example, of testing new product performance, the hypothesis may be

- H_0 the new product has the same performance
- H_1. The new product has better performance

Researchers may have multiple hypotheses. For instance, a study on the use of heterogeneous man-machine teams set (Perelman et al. 2019):

"H_1: There should be a diversity of opinion on the generalizability and sales impact of the HF/UX tool...

H_2: There should be a high concurrence on the sales impact of the Data Fusion/ML project; ...

H_3: There should be a relatively homogenous perception of the Data Visualization investment, ...

H_4: There should be differences of perceptions on the generalizability and sales impact of the GUI Design Tool...

H_5: There should be differences in perceptions of the Mobile Training Architecture, ..."

Many engineering research proposals and scholarly papers do not state the hypothesis in the H_0/H_1 format or in an explicit way. For example, some engineering projects are solely for developing and testing models or special equipment. Research statements, not in hypothesis format, suffice for many proposals and papers.

2.1.2.3 Research Based on Hypotheses

A hypothesis may serve as a compass to provide a tentative prediction of the relationship between the variables or a statement about the value of a population parameter developed in a study. The hypothesis guides us what the research needs to do and the data and methods used to find the conclusions.

If we want to approve a hypothesis, we must examine every possible situation whether the hypothesis is true. However, if we want to disprove a hypothesis, we only need one example to discredit. Therefore, "in most cases it is easier to disprove a hypothesis" (Thiel 2014) than to prove one.

There are various possibilities that would result in invalid research hypotheses, such as imperfectly defined hypothesis; one with incorrect assumptions, data related issues, or an incorrectly applied principle. It is important to understand that a failure to prove a hypothesis is still a valuable effort because the lessons learned can be a guide for future research.

As discussed, it may be possible that we do not need a hypothesis when doing research. Such studies may have particular concentrations on subjects, for example, the connection between two concepts or two cases, without having to comply with the rigorous characteristics required of hypotheses. Another example is research on new method development. Engineering research efforts and achievements, such as inventing a new product, establishing a new principle, or making a new observation, may not be in line with the conventional statements of hypothesis.

2.2 Composition of Proposal

2.2.1 Key Elements of Proposal

2.2.1.1 Proposal Format and Structure

For student theses, universities normally have specific format requirements. In a proposal development for external funding, we need to understand the sponsor's requirements on the format of proposals. The proposal not meeting proposal requirements may be returned without review, which can be avoided. The details of format requirements vary from one sponsor to another. Below is a brief list of a few requirements based on the NASA proposal guidebook (NASA 2017).

- Paper of 8.5×11 inch with at least 1-inch (2.5 cm) margins
- Single-spaced, using one column, and 12-point font
- Text within figures and tables may use smaller fonts and should be legible without magnification
- Expository text necessary for figures or tables
- Units in the common standard for the relevant discipline
- Only nonproposal material, e.g. page numbers, section titles, in headers and footers
- All materials included

Proposal composition also varies according to funding sponsors. Figure 2.5 shows a typical example of the organization of a research project proposal. Among all the necessary elements, the proposal summary, description, and principal

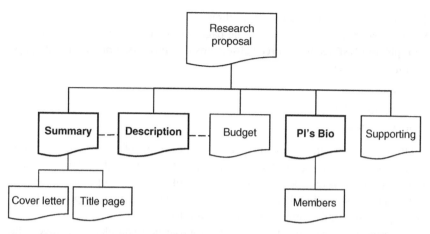

Figure 2.5 A typical organization of research proposal.

investigator (PI)'s biography sketch are the most important ones for a successful proposal evaluation from the perspective of peer and panelist review to meet the funding sponsor's mission and requirements.

2.2.1.2 Research Objectives

Research objectives are a key aspect of a proposal to the reviewers. Some funding agencies, such as the National Science Foundation (NSF), suggest the very first sentence of a research proposal should be its objectives. Thus, reviewers and panelists can have a clear picture about the research at the beginning of a review.

We may write an objective as "The objective of this research project is . . . ." The statement of a research objective should not be very long, often less than 25 words. Considering the necessity of conciseness, crafting an attractive and clear objective statement may be challenging. It is very worthwhile to try to revise an objective at least five times.

We should have a clear motivation – why this research project is needed, associated with objectives. For example, cybersecurity is currently a major concern in various programs, at the different levels and aspects. Many research projects have been conducted on this subject. For example, the NSF invested $160 million in cybersecurity basic research and education in fiscal year 2015 (NSF 2015). For engineering R&D, the objectives and motivation can be connected with and driven by a short- or long-term customer's need, profitability, market share, etc.

In addition, a research problem to be solved should be evidently linked to the existing situation, the preliminarily completed work, and the specific research tasks proposed. The objective of a research proposal should be included in both the project summary and description sections.

2.2.1.3 Proposal Summary and Description

Many research sponsors request a one-page executive summary or abstract. For example, the NSF requests three subsections of a projects summary in one page (about 450 words long):

- Overview
- Intellectual Merits
- Broader Impacts

Although an executive summary goes at the beginning of a proposal, we recommend writing or revising it last. During a proposal development, we keep thinking about the strengths of proposed research and looking at the proposal as an entity. At that point, we are in a better place to identify what are most important to the funding sponsor and summarize them in a limited space.

The project description of a research proposal is a complete narrative. The requirements for a project description vary significantly. For many government agencies, the page limit is 15 pages. For the proposals for the dissertations of doctoral students, the length could be significantly longer.

Suggested elements of a research project description are shown in Figure 2.6. The supporting materials to the Project Description include a table of contents, references, resources, etc.

1. Introduction and proposal overview
2. Background and significance of the proposal with literature review
3. Research plan (tasks and timeline of the project with justifications)

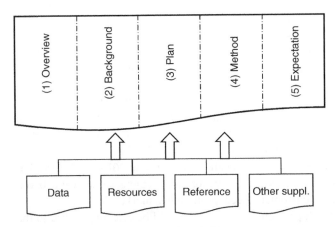

Figure 2.6 Elements of research description.

4. Methods to be used and data availability
5. Expected outcomes and summary

2.2.1.4 Student Competition and Proposals

There are national and international opportunities for innovation and research competition among undergraduate and graduate students. Such competitions can be organized by professional organizations and large corporations, sometimes sponsored by government agencies. It is strongly recommend that student researchers find and participate in such events.

For example, in airspace engineering discipline, Revolutionary Aerospace Systems Concepts – Academic Linkage (RASC-AL) organizes a new concept competition from undergraduate and graduate teams for the NASA's Artemis program (rascal.nianet.org/).

EBEC (Board of European Students of Technology) organizes an annual engineering competition for the students in European countries. The event offers the opportunity to challenge the students in solving a theoretical or a practical problem and to push their limits further by solving an interdisciplinary technological challenge (https://ebec.best.eu.org/).

Another example is Blue Sky Competition that is administrated by SME (previously the Society of Manufacturing Engineers) and funded by the NSF. The goal of Blue Sky Competition is to "influence the future of manufacturing research and education in the United States through new visionary ideas of the future." (https://www.sme.org/aboutsme/awards/blue-sky-competition/)

Each competition has not only their own aims but also specific requirements for the proposal preparation. For instance, RASC-AL requests that submitted proposals be 7–9 pages in length and include (RASC-AL 2019):

- Identification of each of the Theme Requirements and how they are addressed
- Innovative approaches/capabilities/technology
- Detailed information about the work conducted in various trades, concepts, and mission constructs.
- Indicate WHY you chose your design/configuration/system/approach in terms of VALUE in the areas of technology readiness, system performance, affordability, programmatic implementation, and risk
- Original analysis and engineering
- Mass and size estimates
- Key findings supporting the envisioned approach
- Realistic technology assumptions
- A realistic budget assessment
- Project timeline

2.2.2 Other Sections of Proposal

2.2.2.1 PI and Team

The size of research teams varies from one person to hundreds of people. The first author of a proposal is called the PI. The qualification of the PI – in terms of his/her educational background, previous research experience, publications, and funded projects, etc. – plays a critical role in the feasibility and success of assessment for external funding sources. A strong background of education and work experience may give proposal reviewers strong confidence in the success of the future research proposed.

It is required that the biographical sketches of the PI and main members are included in a proposal package. Interestingly, recommendation letters supporting the PI are often not needed. Thus, providing biographical sketches is the main way to show the researcher's credentials.

For master's theses and doctoral dissertations, graduate students are the PIs who work closely with their faculty advisors to develop proposals and conduct the research work, as student's tasks are often part of faculty advisor's research projects.

2.2.2.2 Budget Plan

A detailed budget plan is an integral part of a research proposal. Separated from technical reviews, the budget review is conducted by the financial and administrative professionals in funding agencies to ensure the eligibility, justification, and in accordance with applicable policies.

A detailed budget plan of an externally funded research project may have four subsections (Figure 2.7). In cases where it is internally funded, omit Indirect Costs (IDC). In addition, care must be exercised to ensure compliance with all sponsor and institutional policies.

1. *Research Activity Costs.* Proposed by the researchers, the costs may include experimental and laboratory work, consumable supplies and materials, equipment, computing, travel, communication, publications, etc. Most of them should be itemized. Large items need an explanation.
2. *Salary and Wages* for researchers (PI and team members) and their fringe benefits. This part of cost may also include the compensation for temporal workers, such as students and consultants. The associated costs, such as tuition, should be included, if applicable. Some funding sponsors have restrictions on this type of cost.
3. *Indirect Costs (IDC).* For external research projects, the research institute charges operating costs in addition to direct research activity costs and salary cost. IDC includes the costs of administrative and clerical services and is 20–35% of total budget. Additional discussion on IDC is in Chapter 7.

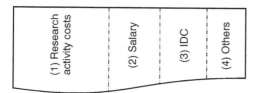

Figure 2.7 Elements of proposal budget section.

4. *Other Costs.* There may be some miscellaneous costs, such as for training and participant support, which should be specified and explained.

2.2.2.3 Supporting Materials

The supporting materials include the references cited, ongoing and pending research projects, resources (e.g. facilities, equipment, and personnel), data management plan, management support, and applicable collaboration. All of that information will be considered in the feasibility evaluation of a research proposal. It is important to ensure the main items are covered, such as specific software, laboratory work, and personnel needed. Additional items may be put into the miscellaneous category, if they are judged reasonable by peers. Some research projects need in-kind, nonmonetary resources if researchers would otherwise pay for or might not be available to buy them.

Data Management Plan is about how to handle research data during and after a research project. There are four key concepts in research data management, based on FAIR principle (Wilkinson et al. 2016):

- Findable – easy to be found
- Accessible – long term storage and accessible
- Interoperable – ready to be combined with other data
- Reusable – ready to be used for future research

Furthermore, the US federal government normally has stringent requirements to ensure the security of project data and Information Technology (IT) systems used in their funded projects. The sensitive materials or data generated, such as data involving human subjects, during research should be protected per state and/or federal laws. For example, Michigan has the Michigan Confidential Research and Investment Information Act (CRIIA) – Act 55 of 1994. Even if a research institute can act as a public body, sensitive information should be fully protected. A nondisclosure agreement (NDA) is often required to protect a sponsor's confidential information.

For some projects, data sharing is required and specified to comply with funding sponsors. For some of US federally funded government projects, the government expects the results to be publicly available. For example, the NSF has data sharing policy (NSF n.d.-a):

> *"Investigators are expected to share with other researchers, at no more than incremental cost and within a reasonable time, the primary data, samples, physical collections and other supporting materials created or gathered in the course of work under NSF grants. Grantees are expected to encourage and facilitate such sharing."*

References Cited provide a solid evidence of the innovative foundation of a proposal and its preparation efforts. Select literature carefully and cite them properly in the context of a proposal.

There is no required number of references for a proposal. For master students, their research proposals should have at least 15 references. For doctoral and external proposals, the references cited should be more than 50, often over 100. In general, an extensive reference may be advantageous to proposal review.

In addition, some general practices should be followed when using references. For example, we should use the most recent (say the previous five years) references to support a cutting-edge research proposal. We may only cite older literature (say over 15 years old) as necessary and consider the literature and research work by domain experts if appropriate.

Cost Sharing is often only applicable for external projects. That is the portion of total project costs, which is provided by the sponsor. The institute may state to provide a specific, quantified resource without asking for funding to cover the cost. Unquantified efforts may be included as well. Sometimes, internal commitment may be needed to ensure the success of a research project, even if such an arrangement is not referenced in the proposal to an external sponsor. The cost sharing may apply to an outside organization or community (e.g. a piece of equipment worth of $5000). In general, quantified cost sharing in a proposal is subject to audit. However, the Code of Federal Regulations states, "under Federal research proposals, voluntary committed cost sharing is not expected" (CFR 2018).

Appendices of a research proposal often provide important evidence and detailed data. However, reviewers or panelists may not pay much attention on the appendices. Therefore, we should be carefully evaluate and arrange what types of materials and information should be put into appendices rather than in the proposal content.

2.3 Proposal Development

2.3.1 Essential Issues

2.3.1.1 Meeting Requirements

The final version of a proposal must fully align with the mission and all requirements of the funding program. Research sponsors are normally mission-driven and have well-written proposal instructions. Different sponsors have varying terms and requirements. For example, the NSF has its specifications, Research Terms and Conditions (NSF n.d.-b). Other federal agencies, such as the US Department of Energy (DOE), have their own specific requirements. Any deviation from the applicable requirements, even in an innovative way, can result in being screened out without technical review.

Sometimes, researchers come across a grant program that appears to be interesting but has a particular emphasis on a thematic subject. In such a case, do not assume we cannot bend the research direction or adjust some details of our own goals. Flexibility can be beneficial for our career in the long term and support our work financially in the short run.

With new technologic advances, research subjects grow increasingly more diverse and specific. Even in highly academic review panels, not every panelist may know much about a specific subject and the method behind every proposal. Therefore, we should prepare a proposal to be self-explanatory, assuming reviewers are not experts in the particular research area.

2.3.1.2 Planning for Outcomes

Consistent with the research objectives, the anticipated outcomes may be stated in a proposal, which serves as solid information for the evaluation of academic value and feasibility of a project. However, it is impossible to predict all eventualities and the outcomes exactly. We may make an educated "guess" regarding the nature and scope of the outcomes and their benefits, such as the expected impact and contribution to a specific field.

The outcomes of a research project vary based on the objective and type of research. Table 2.2 provides some examples of research outcomes for your personal reference. Readers may search specific examples in your fields.

2.3.1.3 Methods Overview

If research objectives indicate the researcher's vision, then the methods to be used will confirm the researcher's capability and path to realize the vision. Therefore, research method is the means to support the success of a research project. In general, methods should be a convincing proof of the anticipated discovery. As the core of this book, research methods are discussed in multiple chapters, such as Chapters 5 and 6.

The number of methods for research is virtually unlimited. Every proposal may have different methods and approaches, even in the same field or on an identical subject. The methods used in a research project include analysis, modeling, case study, evaluation, and any possible combination of individual methods.

Table 2.2 Outcome examples of engineering research.

Outcome	Examples
Procedure	• Sensor-to-body calibration procedure for clinical motion analysis of lower limb using magnetic and inertial measurement units (Nazarahari et al. 2019) • Highly accurate and inexpensive procedures for computing chronoamperometric currents for the catalytic EC' reaction mechanism at an inlaid disk electrode (Bieniasz 2019)
Model	• The needle model: A new model for the main hydration peak of alite (Ouzia and Scrivener 2019) • A new model of shoaling and breaking waves. Part 2. Run-up and two-dimensional waves (Richard et al. 2019)
Technique or tool	• Forcespinning technique for the production of poly(D,L-lactic acid) submicrometer fibers: Process–morphology–properties relationship (Padilla-Gainza et al. 2019) • Developing a Cas9-based tool to engineer native plasmids in *Synechocystis* sp. PCC 6803 (Xiao et al. 2018)
Analysis or evaluation	• Parametric analysis of wax printing technique for fabricating microfluidic paper-based analytic devices (μPAD) for milk adulteration analysis (Younas et al. 2019) • Evaluation of Curtain Grouting Efficiency by Cloud Model – Based Fuzzy Comprehensive Evaluation Method (Zhu et al. 2019)
Problem solution	• Production of Loratadine drug nanoparticles using ultrasonic-assisted rapid expansion of supercritical solution into aqueous solution (Sodeifian et al. 2019) • Effective degradation of fenitrothion by zero-valent iron powder (Fe^0) activated persulfate in aqueous solution: Kinetic study and product identification (Liu et al. 2019)

We may try to invent a new principle, but more often, apply an existing theory for new findings in engineering research. A new principle often stems from existing theories and practice. Researchers may find the methods in other proposals through either subscription databases (refer to Chapter 7) or a review of published literature. The following are a few examples of methods used in the projects funded by the NSF.

"The next section describes the Jäger contact and provides detailed pseudo-code of its algorithm. Section 3 supplies three additional provisions (listed above) that were not part of Jäger's originalmethod but are required for an efficient DEM implementation." (Kuhn 2011)

"The method is herein revisited as a data science mechanistic approach and extended to ductile materials. To reach such an end, the mechanistic equations that SCA relies on to make predictions are reformulated for finite strain elasto-plastic materials in Sect. 2. Numerical convergence of this new method is verified in Sect. 3." (Shakoor et al. 2019)

2. Modeling methods ... a. Wave model ... b. Time-averaged currents model ... c. Variational data assimilation scheme". (Wilson 2018)

We should recognize that research normally has a certain limitations and/or under assumptions, also known as delimitations. We need to keep these in mind, when using an existing method or creating a new method, to interpret and report research outcomes. For example, a paper states, "Moreover, our results suggest that SE is a viable method to obtain porosity measurements for nanoscale systems where standard BET method might not be applicable" (Cavaliere et al. 2017).

Another interesting observation is that the limits of methods are not fixed and may be less restrictive by changing other factors or over time. The methods proposed during research planning should not be considered a fixed decision. During the execution of research, the method can either stay the same or be modified.

For most research, data plays a fundamental and critical role. Dr. Edwards Deming, a 1987 laureate of the National Medal of Technology and Innovation, said that without data you are just another person with an opinion. Data can be a decisive factor for selecting research methods, which will be discussed in depth in Chapter 4.

2.3.2 Tasks of Development

This section will discuss the main tasks or steps of proposal development. Figure 2.8 below is taken from the first phase of the entire research process, shown earlier in Figure 2.3. This development process can be applicable to student research and externally funded proposals.

1. Subject Selection
We may select a research subject from the following:

- Existing topics with advance potentials
- Continuation of previous work
- Exploring a new technology or subject
- Suggested or assigned needs

During subject selection, the interests and enthusiasms of a researcher play critical roles as part of research motivation. For example, we can be inspired by a keynote speaker or during individual discussion with a colleague in a professional conference on a new technology trend (such as propellantless electromagnetic

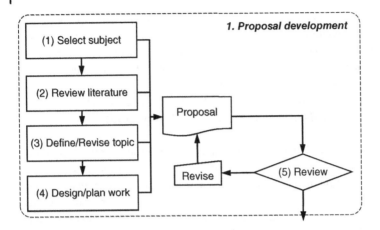

Figure 2.8 Research proposal development steps.

propulsion, virtual reality). The hopes and concerns of the future (on a subject, career, etc.) can also be an important driving force.

Within the chosen subject, a research topic may be on an unresolved controversy, a gap in understanding, or a demand from a larger project, etc. We must think about why the candidate topics are important to explore and of what benefit the findings will be. The significance of a research topic in a proposal must be convincing and promising from the viewpoint of proposal reviewers.

In the case of student research, the topics may be suggested by faculty advisors or as part of an advisors' project. For professional researchers, the subject should be consistent with the mission of the department where the researcher is working and often stems from business demands.

2. Literature Review

New research is built upon the existing knowledge and work of others. Situational awareness of a researched subject area and topics is vital. A good literature review can substantiate the statements in a proposal that explain why the new work is different, advances the body of knowledge, or fills a significant need. Thus, a literature review is the next task after we choose a research subject. This task and its outcomes serve as a solid foundation to support and justify our new research objectives.

We need to answer two key questions from a literature review:

- Status: What are the current limits of knowledge of the particular subject?
- Value: What can the proposed research contribute to extending the limits?

There may be additional tasks as part of a literature review for proposal development. One is a competition analysis, which includes researching the possible competitors and studying previous successful proposals in the field from the funding sponsor. Competition analysis can guide us to avoid too much overlap of already awarded proposals and to understand earlier proposals were successful. Competition analysis helps a proposal developer become more knowledgeable (i.e. better prepared) for a new proposal. Literature review, while not difficult from a technical sense, can be tedious and time-consuming, to be discussed in depth in Chapter 3.

3. Topic and Objective Definition

A research topic is a specific task within a selected subject. A topic often plays the decisive role in many aspects of a research project, such as objectives, method, and anticipated results. Once the topic of a research project is decided, the data, methods, and overall project design can be planned accordingly.

Of the research topic selected, we ask a few basic questions, including what, why, and how. For example, "What?" is the objective of the topic, "Why?" is the reasons and justification of value on the topic, and "How?" is about the methodology and plan, etc. When defining a topic, we need be able to convince our advisor, management, and/or the peer reviewers and panelists of a funding source on its legitimacy.

To encapsulate the essence of a research project, we suggest that a proposal topic or title be specific and self-explanatory. Often, a proposal title has more than 10 words, but the length is not important and can vary significantly. Here are a few examples of the awarded proposal topics (NSUF 2019):

"In-situ nanomechanical characterization of neutron-irradiated HT-9 steel". (7 words)

"Scanning/Transmission Electron Microscopy Characterization of Irradiated Zr-1Nb-O During Thermal Treatments". (10 words)

"The microstructure characterization of 21Cr32Ni model austenitic alloy irradiated at BOR60 reactor." (12 words)

"Role of Irradiation Damage Cascade Descriptors on ODS and Model ODS Nanocluster Evolution." (13 words)

"Atom Probe Tomography (APT) Investigation of Radiation Stability of Oxide Nanoclusters in Oxide Dispersion Strengthened (ODS) Steel Manufactured by the Cold Spray Process." (23 words)

Many experienced researchers recommend that a proposal title do not start with the phrases: "An Investigation into...," "An Application of...," or "A Study of...," etc. Such wordings provide no additional value. Later parts of this text will go into a detailed discussion about proposal and report topics.

4. Work Planning

To develop a plan to accomplish research tasks, the proposed work should be broken into meaningful and manageable subtasks. A best way to illustrate a research plan is through a Gantt chart, which identifies the tasks and displays their start and completion times in a specific order. In addition, a Gantt chart can demonstrate dependency of tasks and the critical path of a research plan. Figure 2.9 shows an example of a research proposal timetable. The corresponding Gantt chart was created using Microsoft Project. For small research projects, we may use other software, such as Microsoft Excel or web-based Innoslate (https://www.innoslate.com/), to create similar Gantt charts.

2.3.3 Development Process

2.3.3.1 Overall Proposal Development

The four steps of a proposal's development (Figure 2.8) have their own unique considerations. For example, we identify a problem to initiate a new research project. Then, we anchor the research in the existing knowledge status. In the design and planning step, we need to address project feasibility, e35qilint how to:

- Identify and secure the data sources
- Select the appropriate study type, methods, and tools
- Evaluate the potential roadblocks and issues

A research plan is similar to the plan of a construction project, e.g. building a house. We would first define what to do in what scope, prepare blueprints, identify the resources, budget, time needed, and so on.

Many research proposers are skilled writers. However, we should be aware of the differences between a research proposal and a scholarly paper. It is by coincidence that some proposals look like academic papers.

Research proposal development is a process of revision. To apply for an external funding, we should draft a proposal and conduct internal reviews with senior professionals. For a large research project, external reviews and consulting are also helpful as well. Based on review, a proposal can be revised before submission. Proposal development can take up to several months for a large research project.

Proposal development is so important, we would think of the project as "halfway done" when its proposal is well developed. A well-developed proposal provides us with a high level of confidence in success of our future research.

2.3.3.2 Three Key Aspects

As mentioned, we should address the three cornerstones: innovation, feasibility, and impacts, to convince a sponsor why they should care to fund such an endeavor. Accordingly, we may address the following questions to develop and evaluate a proposal (see Table 2.3):

These questions are equally important as they all serve integral parts of a successful research project. They like the links of a chain. Even one weak link, or an unconvincing answer to a question, may weaken or even kill the entire proposal.

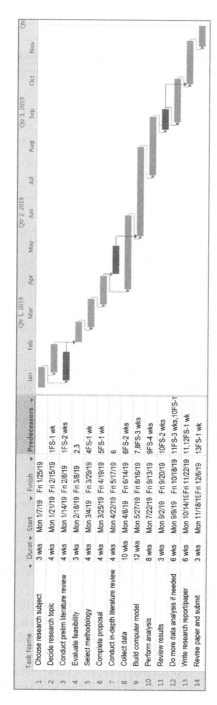

Task Name	Durat	Start	Finish	Predecessors
1	Choose research subject	3 wks	Mon 1/7/19	Fri 1/25/19	
2	Decide research topic	4 wks	Mon 1/21/19	Fri 2/15/19	1FS-1 wk
3	Conduct prelim literature review	4 wks	Mon 1/14/19	Fri 2/8/19	1FS-2 wks
4	Evaluate feasibility	3 wks	Mon 2/18/19	Fri 3/8/19	2,3
5	Select methodology	4 wks	Mon 3/4/19	Fri 3/29/19	4FS-1 wk
6	Complete proposal	4 wks	Mon 3/25/19	Fri 4/19/19	5FS-1 wk
7	Conduct in-depth literature review	4 wks	Mon 4/22/19	Fri 5/17/19	6
8	Collect data	10 wks	Mon 4/8/19	Fri 6/14/19	6FS-2 wks
9	Build computer model	12 wks	Mon 5/27/19	Fri 8/16/19	7,8FS-3 wks
10	Perform analysis	8 wks	Mon 7/22/19	Fri 9/13/19	9FS-4 wks
11	Review results	3 wks	Mon 9/2/19	Fri 9/20/19	10FS-2 wks
12	Do more data analysis if needed	6 wks	Mon 9/9/19	Fri 10/18/19	11FS-3 wks,10FS-1
13	Write research report/paper	6 wks	Mon 10/14/19	Fri 11/22/19	11,12FS-1 wk
14	Revise paper and submit	3 wks	Mon 11/18/19	Fri 12/6/19	13FS-1 wk

Figure 2.9 An example of research project timeframe.

Table 2.3 Sections and questions of proposal preparation.

Section	Type	Sample question	Innovative	Feasible	Impactful
Objective (problem statement)	What	What to do?	√	√	√
Description (background and justification)	Why	Why so important?	√		√
Background (literature review)	What	What foundation and new?	√		
Methodology (strategy, methods, procedure, and resources)	How	How to do the work?	√	√	√
Data (availability and process)	Where	Where/How to get data?		√	
Work plan (tasks and timeline)	When	When to complete?		√	
Anticipated outcomes (products, limitations, etc.)	What	What results and impact?	√	√	√

When addressing the three key aspects of a research project, we need to keep the proposal reviewers and their evaluation criteria in mind. For external funding, it is even more important that the proposal closely match our sponsor's interests. The panelists invited by the sponsors will review and judge approval or rejection of research proposals. The competition for external funding is fierce. Further discussions are in the following sections.

2.3.3.3 Two-step Development

Some funding sponsors request two steps of proposal development and submission. The first step is a conceptual and preliminary proposal or called "letter of intent." A preliminary proposal is a short and succinct project description, which gives the overall idea of where the project is headed. If a preliminary proposal is approved by the sponsor after screening, the researchers are invited to submit a full proposal as the second step.

The NSF states that preliminary proposals can help reduce the proposer's preparation effort, increase the overall quality of the full submission, and assist the NSF program staff in managing the review process when the chance of success for a full submission is very small (NSF 2018).

Please note that the two-step submission and review process may not make proposal competition less intense. For example, the NSF invitation rate of full proposals is less than one-quarter and the funded rate of full proposals is about one-third of the invitations (Mervis 2016). In other words, the overall success rate of proposals is less than 10%.

2.4 Evaluation and Revision

2.4.1 Evaluation for Success

A research proposal should not only be clear to ourselves but also convincing to reviewers and the sponsor. For external funding, research proposal development must be more thorough than for an internal funding. Therefore, the following discussion leans toward the preparation work for external funding. The principles and approaches here on the proposal evaluation may be used as a guideline for novice researchers.

2.4.1.1 Drafting and Revision

Many experienced researchers recommend that the proposal development process be broken down into multiple stages of drafting and reviews (see Figure 2.10). Accordingly, planning ahead and investing substantial time into revisions is wise.

First Draft (on main elements). The effort is on the research problem and methods in the first draft, focusing on basic ideas, logical thought, and overall structure, rather than on the details or wording.

Second Draft (a complete version). In this stage, we try to complete an entire proposal draft with added specifics, such as

- Explanation of the proposed methodology
- Execution procedure with a timeline and cost estimates
- Feasibility, data accessibility
- Pre-existing or pilot (trial) study results, if available
- PI and team credentials, etc.

Third Draft. With a complete draft, we should consult and review with internal experienced colleagues and administrative staff. They may do proofreading as well. Based on the internal review and feedback, we revise the proposal to the third version. By this time, the proposal should be exemplar, in terms of structure, format, and proofreading, and ready to submit. At the very least, we can feel certain that the proposal would not be rejected due to the format and structural requirements.

Final Version. It is suggested to consult with external experts and contact the funding sponsor if possible. The external reviews focus on the overall and technical quality. In this final stage, we should revise again according to the feedback from external professionals, double check to ensure the proposal meets the sponsor's various requirements, and submit the proposal on time.

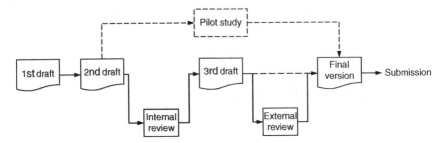

Figure 2.10   Stages of proposal development.

These stages of proposal development may be considered a minimum. Some professionals suggest drafting and revising five to six times. Individual experience will vary with circumstance.

2.4.1.2 Evaluation Overview

Research funding programs are selective and competitive. Understanding proposal evaluation criteria is very important to the success of a submission. The evaluation process has many variations and consists of several steps. Figure 2.11 shows a typical overall process. For example, the NASA basically uses the process detailed in its Guidebook for Proposers (NASA 2018) as does the NSF (Nandagopal et al. 2016).

1. The first step is an administrative review to initially screen proposals according to the minimum requirements, such as submission on time, satisfying page limit, and proposers' qualification.
2. Then the second step is a technical review, which is normally through a peer review process with a panel.
3. The third step, technical review, is the core of evaluation process as the recommendations from technical review are critical, even decisive, to the final decision.
   - Funding sponsors usually select outside reviewers or panelists. Some funding programs use in-house experts to review or a combination of in-house personnel and external experts.
   - The review conclusion can be evaluated as one of five rankings: Excellent, Very Good, Good, Fair, and Poor with comments on proposal's strength and weakness.
   - At the end of the third step, the cognizant program staff creates a recommendation list with technical comments and ranking based on the technical review. The staff evaluates the evaluation of proposals considering for funding and present their recommendations to their senior officials for decision.

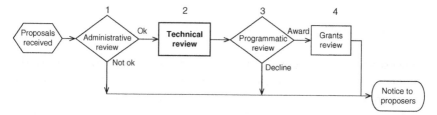

Figure 2.11   An evaluation process of research proposals.

4. After selecting the best proposals, a grant review (the fourth step) is conducted in program office for business, financial, and policy implications for the proposed work. For example, the proposal institution has an adequate grants management capacity, and there are no other outstanding issues with the institution or PI.

In many large corporations, applied research and R&D are supported internally to innovate products and processes. Such internal research proposals are reviewed through multiple levels of management and approved by a senior officer of the R&D division. Some companies may also request a type of external review. For example, a PI submits a preliminary proposal (abstract) with permission to a professional conference. If the abstract is accepted, then PI and team continue the research with internal funding and develop a paper or presentation.

2.4.1.3   Evaluation Criteria

Proposal review processes are similar, while the evaluation criteria are program-dependent. The NSF's evaluation criteria for basic research are based mainly on the intellectual merit and broader impact. The former is about the contributions to a particular engineering and technical field; the latter about a contribution to society or the community at large. Five questions for both criteria to evaluate NSF proposals (Nandagopal et al. 2016):

> *"1. What is the potential for the proposed activity to:*
> *a. advance knowledge and understanding within its own field or across different fields (Intellectual Merit); and*
> *b. benefit society or advance desired societal outcomes (Broader Impacts)?*
> *2. To what extent do the proposed activities suggest and explore creative, original, or potentially transformative concepts?*
> *3. Is the plan for carrying out the proposed activities well-reasoned, well- organized, and based on a sound rationale? Does the plan incorporate a mechanism to assess success?*
> *4. How well qualified is the individual, team, or institution to conduct the proposed activities?*

5. Are there adequate resources available to the PI (either at the home institution or through collaborations) to carry out the proposed activities?"

Some studies compare different review criteria, for example (Falk-Krzesinski and Tobin 2015). There are a few more example of the US federal agencies for research. National Institutes of Health (NIH) considers five core aspects: project significance, innovation, research approach, investigator qualification, and contribution to environment (NIH 2019).

One of DOE review criteria is scientific and/or technical merit. The review criteria of Institute of Education Sciences of the US Department of Education are in four aspects (ED 2006):

1. Significance – a compelling case for the potential contribution to solving a problem
2. Research Plan – appropriate for answering research questions or testing hypotheses
3. Personnel – PI and key personnel possess training and experience and commit sufficient time to competently implement research
4. Resources – Having facilities, equipment, supplies, and other resources required to support

It is paramount that we understand and directly address the criteria in a proposal, which makes reviewers' evaluation easier and lets them concentrate on the technical aspects of a proposal.

Some sponsors may emphasize the solutions of particular problems, while others may consider commercial impacts to fund application – oriented research projects. In a proposal, we may state, "If successful, the benefits of this research will be . . . ." The criteria factors of research may include the impacts from new knowledge, technology advance, economy, health and well-being, society, or environment. Some of the impacts may be difficult to explicitly and quantitatively state or measure.

2.4.2 Self-assessment

2.4.2.1 Two Key Factors to Address

As discussed, the three cornerstone factors are innovation, feasibility, and impacts. In general, we believe our work meaningful and impactful, which is discussed above. Then we may think more about the remaining two key factors: innovation and feasibility, for our proposals.

Innovation. The innovation or novel nature of a project is to describe the potential for advancing society in a particular area, call it "intellectual merit." In most cases, the evaluation of research innovation is based on expected outcomes. Sometimes, the novelty can be on the advancement of methodology and approaches. In other words, we will find something new to contribute, in addition to busy work.

In addition to the expected innovation, we need to address two more elements associated with innovative outcomes:

- Meeting the program objectives
- Validity of the outcomes

Feasibility. We may view the feasibility as a sum of concerns on technical competence, various resources, predicted timeline, and requested budget to achieve the objectives.

For a large proposal, we may have a dedicated section to address project feasibility, considering all influencing internal and external factors, supported by a comprehensive literature review. For novice researchers, a good way to assess the feasibility of a new proposal is to discuss with an experienced mentor or advisor.

Particularly for applied research and R&D, the consideration on feasibility is related to the potential impacts as well. For example, cost-effectiveness plays an integral role. Meaningful research may be deemed unfeasible due to a high cost at the time.

We may find out that somethings are not under our control and hold us back from reaching a research goal. If such a factor is a major issue, then the research project may be deemed unfeasible. We need to take time and perseverance as researchers to resolve the issues and/or create an environment for the improved feasibility.

Both innovation and feasibility are correlated with the technical competence of the research PI and their team. Reviewers assess the PI's educational credentials, previous related experience, and quality of their publications.

2.4.2.2 A Review Checklist

There are several aspects to address in the review and revision of a proposal. Researchers may use a checklist, like Table 2.4, containing the main items for evaluation made by the researchers themselves, as well as by internal/external reviewers. We render the assessment in three levels: very good (VG), good, and fair.

As a student, you may want to have your friends and fellow students review and critique your work, based on Table 2.4, before meeting advisors to present your proposal.

The checklist can be used as a guide to improve proposal development. The items in the checklist are worth thinking about and seeking direct feedback on from others. It may be more effective if we ask a colleague to review our proposal

Table 2.4 A checklist of proposal peer- and self-assessment.

Item	VG	Good	Fair
Overall impression			
Informative and impressive title and abstract			
Clear overall picture, concise, and logical			
Technical merits to a professional field			
Feasible with manageable ambition and risks			
Modest level of confidence (tone in writing)			
Appropriate format and readability			
Objective and problem			
Well-defined, focused research problem			
Realistic and practical objectives			
Justified significance of a research problem			
Defined hypothesis if applicable and proper assumptions			
Based on the latest advance (adequate literature review)			
Methodology and design			
Proper methods to the problem			
Data plan (accessibility; permission if applicable)			
Right timeframe and project management			
Experiment plan (equipment, prelim results, etc.)			
Appropriate analysis method and computer software			
Compliance and IP/dissemination plan			
Possible issues/limitations addressed (contingency plan)			
PI and team			
Sufficient credentials of education and skills			
Good previous accomplishments and publications			
Appropriate time to devote to the project tasks			
Task and teamwork arrangement			
Career development if applicable			
Resources and budget			
Good institutional setting, facilities, and equipment			
Sufficient support staff and other (in-kind) resources			
Reasonable and justified budget estimation			
All budget items allowable			

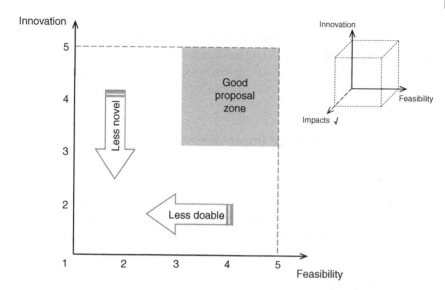

Figure 2.12 An overall evaluation of research proposal.

using the checklist. Readers can modify the checklist to fit their particular situation.

2.4.2.3 A Simple Evaluation

To assess the two key factors of innovation and feasibility, review angles can come from a wide variety of perspectives, such as from researchers, industry experts, or society in general. As researchers, we can assess the technical methods, experiments, analysis, and process. As industry practitioners, we may focus on the applicability of research from a business perspective. From a societal perspective, we can determine the potential benefits that may make a difference. Depending on the type and nature of research, the viewpoints from different angles may have distinct weight.

When self-assessing, the two key factors (innovation and feasibility) may rate in a scale of 1–5: (1) incomplete, (2) poor, (3) fair, (4) good, and (5) very good, based on our current understanding and circumstance. To be a good proposal, the ratings should be better than (3) good for both factors. Figure 2.12 visualizes the rankings of research proposals.

We may consider the product of the two ratings as an indicator of overall proposal quality, that is,

Proposal Quality Indicator = Feasibility × Innovation

For example, if the innovation and feasibility of a proposal are rated 3.5 and 4, respectively, then the quality indicator is 14. This quantitative evaluation can be useful for the comparison of alternative proposals. Obviously, the evaluation is simple, experience-related, and subjective. To have a reliable evaluation, it is better to invite several colleagues and mentors to assess a proposal. Then, we may average their review ratings into an aggregate score. In most cases, the reviewers can also identify issues and offer suggestions to improve the two key factors.

2.5 Considerations For Improvement

2.5.1 Paying Attention

It may not be difficult to know the basic requirements and composition of a research proposal. However, it can be challenging to develop a research proposal for a novice researcher. Here are some more discussion points and suggestions for beginners.

2.5.1.1 Research Aims

Research intends to discover something new and move a body of knowledge and technology forward. There are many unknowns in research. How much of the body of knowledge we attempt to carve out based on expected findings must be moderated to an appropriate level.

From time to time, the aims of the proposals prepared by novice researchers are either too broad or too aggressive, which suggests foreseeable difficulties due to various constraints, such as budget, resources, or time. For such a research project, we may have a good start-up but will have issues and difficulty reaching predefined research goals. Therefore, it is often necessary to adjust the scope of work to be more realistic.

During a proposal development, we do a lot of literature review and thinking. We must keep our proposal focused. Our aim may change along with a literature review as does a learning curve. When revising over research aim, we need to remember to keep the overall plan on track to meet the deadline.

Another common issue is that the scope of work is increased during the proposal and planning phase, which is called *scope creep* in project management. Novice researchers may feel keeping research to a manageable size is challenging for the given time and resources. Seeking guidance and encouragement from experienced researchers can be helpful.

2.5.1.2 Detail Level of Proposal

Maintaining an appropriate level of detail is another challenge to proposal writing. Often, there are length constraints for a proposal, such as 1 page for a summary and 15 pages for a project description. In a proposal, some sections must be cut and pulled back with detail to an appropriate length, while other sections may need to be stretched to provide a sufficient amount of information.

It is a common practice that the length of the final version of a proposal is close to the maximum number of pages allowed. Even a half page longer than the maximum can result in a rejection without review. Those noticeably shorter than the maximum may give a negative impression to reviewers, implying that proposers did not provide sufficient detail.

Moreover, a piece of information may fit in multiple sections. In such cases, it is important to have the information in the best place and avoid redundancy, which can also save writing space.

2.5.1.3 Other Concerns

Even then, there are possible issues with research proposals from the viewpoint of reviewers, advisors, or panelists. The issues can result in the proposal being declined. They include the following:

- Not directly addressing the research subjects required by the funding sponsor
- Deviation from the format of the proposal desired by the funding sponsor
- Insignificant need for the research topic
- Vague statement of the problem
- Incompletely described methodology
- Lack of logic in the argument of the proposal
- Major concerns with information accessibility or availability
- Items in the budget not allowed by the funding sponsor
- Failure to appropriately cite sources

The reasons for these issues are related to not thoroughly revising, typically due to a "rush mode" of proposal preparation or lack of a third-party review.

2.5.1.4 Additional Preparation Tips

Revising a proposal draft can be time-consuming. During the revising process, we may utilize a few tips to improve the outcome of revision efforts effectively. For example, writing concise sentences provides an easily understandable proposal for reviewers.

During each step of proposal revision, experienced researchers recommend setting the proposal aside for a few days without touching it. It is good practice to clear the mind and have a fresh start before reading and revising the proposal again. The total time for a large proposal development can take three months or even longer.

Reading aloud a printed copy of our proposal draft can make us think on the clarity and flow of statements; look for possible problems with wording, continuity, coherence, logic, organized thoughts, and consistency in terminology. During the reading, we can mark the manuscript to better scrutinize issues later and revise accordingly.

We should be careful when wording a proposal. For example, if a proposal is targeted to basic research, some words, such as "develop," "design," and "optimize," may imply that the research sounds more like applied research. Many reviewers are sensitive to the different types of research proposals, so we should try to avoid a potential misinterpretation.

As mentioned, the aims and scope of work may change during development based on literature review, study tryout, and feedback. We should keep each draft version for later reference.

2.5.2 More Considerations

2.5.2.1 Pilot Study

A pilot study, sometimes called a trial study, is an initial small-scale exploratory investigation before or during proposal development. Purposes of a pilot study include trying out a particular investigation procedure, using new instrument of measurement, or implementing a new analysis method. Based on the results from a pilot study, we can gain a better understanding of the project feasibility, improve the research design, and identify or verify main concerns. We can revise the research proposal in accordance with these findings.

The outcomes from a pilot study can be a solid support to the proposed tasks and anticipated achievements. From the viewpoint of proposal reviewers, the pilot study results and previous achievements related to the proposal are vital credentials to, and indicators of, a strong likelihood of research success. Therefore, a pilot study is strongly recommended if feasible. Furthermore, a pilot study may generate results that are worthy of publishing in of themselves. A published pilot study is a big plus for a proposal application. Here are a few examples of published pilot studies:

> *"Pilot study about the micro hydropower generation by use of flow induced vibration phenomenon"*. (Yokoi 2018)
> *"Potential use of blast furnace slag for filtration membranes preparation: A pilot study."* (Bílek et al. 2018)
> *"A Pilot Study: Advances in Robotic Hand Illusion and its Subjective Experience."* (Abrams et al. 2018)

2.5.2.2 Cross Disciplinary

Readers may agree that more and more research projects involve multiple academic disciplines. Many new research achievements are related to

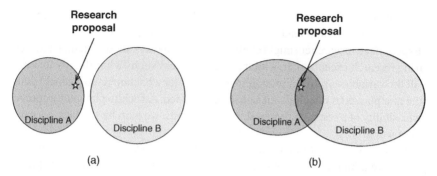

Figure 2.13 An illustration of cross-disciplinary research. (a) Proposal in a discipline. (b) Proposal in a multidiscipline.

cross-disciplinary work. They may be called multidisciplinary, interdisciplinary, etc. These names are used with various connotations.

Research funding agencies encourage researchers to move to new fields. For example, the NSF states,

> *"NSF has long recognized the value of interdisciplinary research in pushing fields forward and accelerating scientific discovery. ... Thus, NSF gives high priority to promoting interdisciplinary research and supports it through a number of specific solicitations."* (NSF n.d.-c)

Many new disciplines are already cross-disciplinary, for example, biomechanical engineering and electromechanical engineering. For the research to explore new knowledge and artifacts, the question is how to integrate two or more disciplinary concepts, processes, mechanisms, methods, etc. Figure 2.13a shows a nonmultidisciplinary research and Figure 2.13b illustrates a research involving two disciplines.

The challenges to cross-disciplinary research also depend on the principles from different disciplines. One effective way to execute cross-disciplinary research is to build your team of research across multiple disciplines. The possible limitations of cross-disciplinary research may include lack of time to collaborate on conceptualization and implementation, time to share discipline knowledge, etc. (CohenMiller and Pate 2019).

2.5.2.3 Backup Plan

We diligently and thoroughly plan research activities and reasonably predict intermediate results and final findings. However, there is no guarantee that a research project will be fully successful and completed on time. In other words, we should carefully consider the risky nature of exploring new subjects during the proposal development phase. As researchers, we should have the mindset of willingness to learn from mistakes and readiness to admit only partial success.

Experienced researchers might challenge new proposers about potential failures during research execution. It is interesting to discuss such "what-ifs." By looking at all the variables up-front, proposers may consider a backup (contingency) plan if the first plan does not work out exactly as intended. A backup plan may improve the feasibility of a research proposal. In other words, we plan for the best, prepare for the worst.

A backup plan is not often required by funding sponsors as they assume proposers have a high level of confidence in the planned research. However, it may be a good idea for proposers to evaluate the possible risk factors in their proposal and have an internal backup plan.

2.5.2.4 Unsuccessful Proposals

Remember, research proposals, including those of student research, and funded projects, both internal and external, can be rejected. Some funding sponsors have about a 10% acceptance rate. Being turned down indicates that the proposal is not competitive enough in the particular application pool, but does not imply the proposal is no good. For example, the Los Alamos National Laboratory summarizes the reasons for proposal rejection (LANL n.d.):

> *"The proposal was good but could not be funded because of insufficient CSES funds.*
> *Objectives and background are unclear or inadequately argued.*
> *Ideas are not innovative.*
> *Methods are inadequately described or do not reflect state-of-the-art.*
> *Approach is not convincing enough to satisfy the objectives.*
> *Did not align with Laboratory National Security Mission Objectives.*
> *For research proposals involving University-LANL collaborations, collaborations are weak.*
> *Important and relevant Los Alamos National Laboratory facilities are not considered or exploited.*
> *Lack of commitment by the Los Alamos National Laboratory Principal Investigator to the research or mentoring process.*
> *Subject matter is not relevant to CSES's and Institutional scientific interests or Laboratory research priorities."*

For student research, rejection means a revision is needed. For internal and some external programs, we have an opportunity to communicate with the review committee or program officers and find out where the gap is and potential improvements for next time. For some external projects, we can request review comment and appraisal. US federal agencies normally provide them, but private foundations are not obligated to do so. Regardless of the type of proposal, the next step is to rectify your plan accordingly and submit a proposal revision if allowed or consider the proposal for another opportunity. The chances for success can be much greater for the second time around.

Summary

Research Initiation

1. A research proposal is a formal document, consisting of research objective, idea, and plan.
2. When developing a new research idea, we need to know what type of research the idea is line with.
3. The three indispensable cornerstones for a good research proposal are innovation, feasibility, and impacts.
4. Research proposal development is a structured process with multiple steps and considerations.
5. The terms of research proposal, protocol, and prospectus are often used interchangeably.
6. The objective of research, sometimes in the format of a hypothesis, is fundamentally important.

Composition of Proposal

7. The key elements of a formal research proposal include summary, project description, budget, researcher credentials, etc.
8. The objective of research should be stated clearly and concisely at the beginning of a proposal.
9. The project description of research should provide information on the overview, background, plan, methods to use, and outcome expectations of the research.
10. Research costs include multiple categories: activity cost, personal compensation, IDC, and miscellaneous.
11. A formal research proposal includes supporting materials, such as data management plan, references cited, cost-sharing arrangement, and appendices if applicable.

Proposal Development

12. Research funding agencies have their own specific requirements. A proposal must fully meet these requirements.
13. The anticipated outcomes must be consistent with the objectives.
14. Methods to be used confirm the researcher's capability and technical path to realize the vision.
15. The main steps of a proposal development include subject selection, literature review, objective definition, work plan, and so on.
16. The criteria of proposal evaluations vary in funding agencies but are primarily orientated on the three cornerstones: innovation, feasibility, and impact.

Evaluation and Revision

17. After drafting, a proposal may need four to six revisions. Each revision may have its own focus.
18. Research funding agencies have rigorous review and evaluation processes.
19. In proposal development, two key factors should be central: innovation and feasibility.
20. Self-assessment and peer review (checklist) are effective tools for proposal improvement.

Considerations for Improvement

21. The aims of a proposal prepared should be specific or appropriate to available techniques and resources.
22. A pilot study or pre-proposal tryout can significantly improve the possibility of authorization.
23. Cross-disciplinary research in terms of aims and methods is often supported and encouraged by funding agencies.
24. Unsuccessful proposals may still mean additional work and other opportunities.

Exercises

Review Questions

1 The three key elements of a proposal are innovation, feasibility, and impacts. Please select one factor to discuss with an example.

2 Review the overall development process of a research proposal and discuss one or two steps with an example.

3 Discuss the differences between a research proposal and a scholarly paper.

4 Briefly explain the main elements of a formal research proposal.

5 Discuss various formats of research hypotheses.

6 Explain how you select a method and collect data for a project.

7 Discuss how to assess a proposal from a reviewer's perspective regarding the significance of research objectives.

8 Discuss how to assess a proposal from a reviewer's perspective regarding the success probability of a research project.

9 One states we should discuss research limitations to justify the technical value of research work or proposals. Do you agree? Why?

10 Draft the title of your research proposal in 10–15 words and get feedback from your classmates or coworkers.

11 Develop your research objective in one sentence (<30 words) and get feedback from your classmates or coworkers.

12 Explain the purposes of running a pilot study.

13 Use an example of a research plan, preferably your own, to discuss how to assess its feasibility.

14 When developing a budget plan, what types of costs you need to consider based on your proposal?

15 One states that a good proposal is the entire project "half done." Do you agree?

16 As a new researcher, what is/are your challenges to develop a research proposal?

17 Review the benefits and potential issues when formulating a research problem with an intended solution to the problem in mind.

18 What if you run into an issue of data availability? Should we have an alternate plan prepared in case of such a scenario?

19 Discuss the benefits of cross-disciplinary research with an example.

Miniproject Topics

1 Think about two potential research problems in your field and compare their needs, feasibility, and benefits to study them. Decide and justify which one you would consider to study. The problems could be yours or others – just a critical thinking practice.

2 Find a recent journal paper in your field and review the paper. Answer the following questions with appropriate citations:
   (a) What is the problem that is being explored?

   (b) What is the methodology and/or methods applied?

   (c) How would you conduct follow-up research?

3 In your discipline, identify a problem to resolve as your research project and discuss:

   (a) Importance justification for addressing this particular problem

   (b) Technical challenges to the problem-solving

   (c) How long to resolve the problem to within a half-year

4 Find a recent journal paper that established and proved or disproved the hypotheses and summarize:

   (a) The relationship between the research problem (or objective) and the hypotheses

   (b) The research approach and process associated with the hypotheses.

5 Find a recent journal paper in your field, analyze the paper, and answer the following questions with appropriate citations:

   (a) How should you assess the strengths of research reported in the paper?

   (b) If you were conducting this study, what you would change and how would you manage those changes?

6 Develop your research (e.g. master's thesis or a project at work) process flow (similar to Figure 2.3 in this chapter) and discuss the most technically challenging part if you were to conduct the project.

7 From your research topic selection (problem identification), explain the main reasons, such as requirements, interested parties, motivations, potential issue(s), possible limitations, or a combination of multiple factors, all in relation to why you have selected this research topic.

8 Use Table 2.4 Checklist to assess a research proposal of yours or others with a brief explanation.

9 Analyze a cross-disciplinary research paper – how it integrates two disciplines in terms of its objective, concept, methods, and achievements.

References

Abrams A.M. and Beckerle, P. et al. (2018). A pilot study: advances in robotic hand illusion and its subjective experience. Proceeding of 2018 ACM/IEEE International Conference on Human-Robot Interaction (5–8 March 2018). https://doi.org/10.1145/3173386.3176915

Anderson-Rowland, M.R., Rodriguez, A.A. and Grierson, A.E. (2015). Helping undergraduate engineering students discover their interests (and reduce their fear of research. 2015 IEEE Frontiers in Education Conference, El Paso, TX, 1–8. https://doi.org/10.1109/FIE.2015.7344275

Arciszewski, T. (2016). *Inventive Engineering – Knowledge and Skills for Creative Engineers*, ISBN: 978-1-4987-1124-1. Boca Raton, FL: CRC Press.

Bieniasz, L.K. (2019). Highly accurate and inexpensive procedures for computing chronoamperometric currents for the catalytic EC' reaction mechanism at an inlaid disk electrode. *Electrochimica Acta* 298: 924–933. https://doi.org/10.1016/j.electacta.2018.12.113.

Bílek, V., Bulejkoet, P., and Kejíkal, P. (2018). Potential use of blast furnace slag for filtration membranes preparation: a pilot study. *IOP Conference Series: Materials Science and Engineering* 379: 012012. https://doi.org/10.1088/1757-899X/379/1/012012.

Cavaliere, E., Benetti, G., Van Bael, M. et al. (2017). Exploring the optical and morphological properties of Ag and Ag/TiO_2 nanocomposites grown by supersonic cluster beam deposition. *Nanomaterials* 7 (12): 442. https://doi.org/10.3390/nano7120442.

CFR (2018). Grants and agreements §200.306 cost sharing or matching. https://www.govinfo.gov/content/pkg/CFR-2018-title2-vol1/xml/CFR-2018-title2-vol1-sec200-306.xml (accessed 27 May 2019).

CohenMiller, A.S. and Pate, P.E. (2019). A model for developing interdisciplinary research theoretical frameworks. *The Qualitative Report* 24 (6): 1211–1226. https://nsuworks.nova.edu/tqr/vol24/iss6/2.

Eckel, E.J. (2019). Topic development in the freshman engineering paper: finding a focus. Science & Technology Libraries. https://doi.org/10.1080/0194262X.2019.1566042

ED (2006). Procedures for peer review of grant applications. Institute of Education Sciences, Adopted by the National Board for Education Sciences on 24 January 2006. https://ies.ed.gov/director/pdf/SRO_grant_peerreview.pdf (accessed July 2019).

Falk-Krzesinski, H.J. and Tobin, S.C. (2015). How do I review thee? Let me count the ways: a comparison of research grant proposal review criteria across US federal funding agencies. *The Journal of Research Administration* 46 (2): 79–94.

Kuhn, M.R. (2011). Implementation of the Jäger contact model for discrete element simulations. *International Journal For Numerical Methods In Engineering* 88: 66–82. https://doi.org/10.1002/nme.3166.

LANL (n.d.). Acceptance and rejection of CSES proposals. Los Alamos National Laboratory. https://www.lanl.gov/projects/national-security-education-center/space-earth-center/proposal-call/acceptance-rejecting-proposals.php (accessed 27 May 2019).

Liu, H., Yao, J., Wang, L. et al. (2019). Effective degradation of fenitrothion by zero-valent iron powder (Fe^0) activated persulfate in aqueous solution: kinetic study and product identification. *Chemical Engineering Journal* 358: 1479–1488. https://doi.org/10.1016/j.cej.2018.10.153.

Mervis, J. (2016). NSF tries two-step review, drawing praise—and darts. *Science* 353 (6299): 528–529. https://doi.org/10.1126/science.353.6299.528.

Nandagopal, T. Santoro, K., Sharp, N. et al. (2016). Merit review process. National Science Foundation Grants Conference, Portland, Oregon (29 February–1 March). https://www.nsf.gov/bfa/dias/policy/outreach/grantsconf/meritreview_feb16.pdf. (Accessed July 2019).

NASA (2017). Guidebook for proposers responding to a NASA funding announcement. Revised as of April 2017. https://www.hq.nasa.gov/office/procurement/nraguidebook/proposer2017.pdf (accessed June 2019).

NASA (2018). Guidebook for proposers responding to a NASA funding announcement. Revised as of March 2018. https://www.hq.nasa.gov/office/procurement/nraguidebook/proposer2018.pdf (accessed July 2019).

Nazarahari, M., Noamani, A., Ahmadian, N. et al. (2019). Sensor-to-body calibration procedure for clinical motion analysis of lower limb using magnetic and inertial measurement units. *Journal of Biomechanics* 85: 224–229. https://doi.org/10.1016/j.jbiomech.2019.01.027.

NIH (2019). Review criteria at a glance. https://grants.nih.gov/grants/peer/guidelines_general/Review_Criteria_at_a_glance.pdf (accessed July 2019).

NSF (2015). NSF awards $74.5 million to support interdisciplinary cybersecurity research. News Release 15-126, National Science Foundation. https://nsf.gov/news/news_summ.jsp?cntn_id=136481&org=NSF&from=news (accessed April 2019).

NSF (2018). PART I: PROPOSAL PREPARATION AND SUBMISSION GUIDELINES – Chapter I: pre-submission information. NSF 18-1, National Science foundation. https://www.nsf.gov/pubs/policydocs/pappg18_1/pappg_1.jsp (accessed 30 May 2019).

NSF (n.d.-a). Dissemination and sharing of research results. National Science foundation. https://www.nsf.gov/bfa/dias/policy/dmp.jsp (accessed 7 May 2019).

NSF (n.d.-b). Research terms and conditions. National Science foundation. https://www.nsf.gov/awards/managing/rtc.jsp (accessed 7 May 2019).

NSF (n.d.-c). Introduction to interdisciplinary research. National Science foundation. https://www.nsf.gov/od/oia/additional_resources/interdisciplinary_research/ (accessed August 2019).

NSUF (2019). RTE 2nd call awards announced, The U.S. Department of Energy – Office of Nuclear Energy (DOE-NE) Nuclear Science User Facilities (NSUF). https://nsuf.inl.gov/Page/19rte2ndcallannouncement (accessed 30 May 2019).

Ouzia, A. and Scrivener, K. (2019). The needle model: a new model for the main hydration peak of alite. *Cement and Concrete Research* 115: 339–360. https://doi.org/10.1016/j.cemconres.2018.08.005.

Padilla-Gainza, V., Morales, G., Rodríguez-Tobías, H. et al. (2019). Forcespinning technique for the production of poly(D,L-lactic acid) submicrometer fibers: process–morphology–properties relationship. *Journal of Applied Polymer Science* 136 (22): 47643. https://doi.org/10.1002/app.47643.

Perelman, B. S., Dorton, S. L., Harper, S. et al. (2019). Identifying consensus in heterogeneous multidisciplinary professional teams. 2019 IEEE Conference on Cognitive and Computational Aspects of Situation Management (CogSIMA), Las Vegas, NV, USA, 127–133. https://doi.org/10.1109/COGSIMA.2019.8724143

RASC-AL (2019). Proposal guidelines. rascal.nianet.org/forms/ (accessed 29 December 2019).

Richard, G., Duran, A., and Fabrèges, B. (2019). A new model of shoaling and breaking waves. Part 2. *Run-Up and Two-Dimensional Waves* 867: 146–194. https://doi.org/10.1017/jfm.2019.125.

Shakoor, M., Kafka, O.L., Yu, C. et al. (2019). Data science for finite strain mechanical science of ductile materials. *Computational Mechanics* 64: 33–45. https://doi.org/10.1007/s00466-018-1655-9.

Sodeifian, G., Sajadian, S.A., Ardestani, N.S. et al. (2019). Production of Loratadine drug nanoparticles using ultrasonic-assisted rapid expansion of supercritical solution into aqueous solution (US-RESSAS). *The Journal of Supercritical Fluids* 147: 241–253. https://doi.org/10.1016/j.supflu.2018.11.007.

Thiel, D. (2014). Introduction to engineering research. In: *Research Methods for Engineers*. Cambridge: Cambridge University Press https://doi.org/10.1017/CBO9781139542326.003.

UCB (2015). Writing a research proposal Haas scholars guidelines. How to Write a Research Proposal, The Office of Undergraduate Research and Scholarships, University of California Berkeley. hsp.berkeley.edu/sites/default/files/WritingAProposal-HaasScholGuidelinesAug%2715_2_0_0.pdf

Wilkinson, M., Dumontier, M., Aalbersberg, I. et al. (2016). The FAIR guiding principles for scientific data management and stewardship. *Scientific Data* 3: 160018. https://doi.org/10.1038/sdata.2016.18.

Wilson, G. (2018). Surfzone state estimation, with applications to quadcopter-based remote sensing data. *Journal of Atmospheric and Oceanic Technology* 35 (10): 1881–1896. https://doi.org/10.1175/JTECH-D-17-0205.1.

Xiao, Y. et al. (2018). Developing a Cas9-based tool to engineer native plasmids in Synechocystis sp. PCC 6803. *Biotechnology and Bioengineering* 115 (9): 2305–2314. https://doi.org/10.1002/bit.26747.

Yokoi, Y. (2018). Pilot study about the micro hydropower generation by use of flow induced vibration phenomenon. *EPJ Web of Conferences* 180: 02122. https://doi.org/10.1051/epjconf/201818002122.

Younas, M., Maryam, A., Khan, M. et al. (2019). Parametric analysis of wax printing technique for fabricating microfluidic paper-based analytic devices (µPAD) for milk adulteration analysis. *Microfluidics and Nanofluidics* 23 (38): 10. https://doi .org/10.1007/s10404-019-2208-z.

Zhu, Y., Wang, X., Deng, S. et al. (2019). Evaluation of curtain grouting efficiency by cloud model – based fuzzy comprehensive evaluation method. *KSCE Journal of Civil Engineering* 23 (7): 2852–2866. https://doi.org/10.1007/s12205-019-0519-y.

3

Literature Search and Review

3.1 Introduction to Literature Review

3.1.1 Overview of Literature Review

3.1.1.1 What Is Literature Review

Research is about advancing knowledge and developing new technologies. Picture a research effort like a relay race where one researcher takes the torch from the previous runner. Doing the groundwork of reviewing relevant literature is a foundation of new research and become a value additive in a technical field.

"Literature" here refers to any collection of materials in a subject. "Review" can be the discussion, critique, synthesis, and summary of selected literature materials. From a review, we are able to have an understanding of the status quo and seek out new opportunities. Thus, a literature review is a search and evaluation of available literature to present the state-of-the-art with respect to a particular subject.

Literature review may not be new, as it is introduced in a high school curriculum. For our purposes though, a literature review goes more in-depth and becomes critical as a tool and process to qualify the status of knowledge and needs for moving forward.

3.1.1.2 The Formats of Literature Review

Discussed in Chapter 2, a literature review often is a specific section in research proposals. For technical papers, a literature review may either be embedded into the introduction section of a paper or stand as a separated section. When in a separated section, a literature review may be organized into three or four subsections or paragraphs depending on the subtopics and detail. In addition, a literature review can be a standalone article, which is discussed in the Section 3.3.3 of this chapter.

Engineering Research: Design, Methods, and Publication, First Edition. Herman Tang.
© 2021 John Wiley & Sons, Inc. Published 2021 by John Wiley & Sons, Inc.

As a practice, a literature review is integrated into the introduction section of a scholarly paper. Here is an example (Amershi et al. 2019):

> *"In recent years, teams have increased their abilities to analyze diagnostics-based customer application behavior, prioritize bugs, estimate failure rates, and understand performance regressions through the addition of data scientists [7], [8], ...*
>
> *One commonly used machine learning workflow at Microsoft has been depicted in various forms across industry and research [1], [9], [10], [11]. It has commonalities with prior workflows defined in the context of data science and data mining, such as TDSP [12], KDD [13], and CRISP-DM [14]."*

Some papers have a separated literature review section. In such a structure, the literature review is often arranged into a few subsections. Figure 3.1 shows an example (Diabat et al. 2019).

3.1.2 Purposes of Literature Review

The overall objective of a literature review is to know the status of relevant knowledge on the theoretical, methodological, and/or practical achievements to a particular topic. Figure 3.2 shows five objectives and benefits of conducting a literature review. Depending on the nature of research, the objectives of a specific literature review can have different emphases. For example, assessing research methods can be more important than learning from others in some situations.

3.1.2.1 To Understand Status Quo

We conduct a literature review to track down the latest knowledge, such as what has been studied and how the knowledge was obtained by other professionals.

1. Introduction
2. Literature review
   2.1. Humanitarian supply chains
   2.2. Blood supply chain
   2.3. Robust optimization
3. Problem definition
4. Computational results
5. Conclusion

Figure 3.1 A section of literature review in a paper.

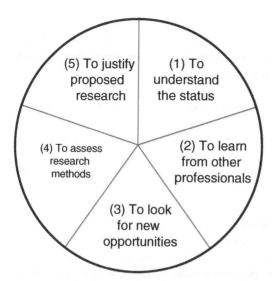

Figure 3.2 Purposes and benefits of literature review.

Then, we can do a new, enhance study, rather than "reinvent the wheel." Generally, the more we understand the existing status and perspectives in our research subject, the more effectively we can tackle our research.

Thus, reporting the latest status and labeling as such is important. Here are three examples of literature reviews:

> *"The latest preload technology of machine tool spindles: A review"* (Lee et al. 2017)
> *"A literature review on latest developments of Harmony Search and its applications to intelligent manufacturing"* (Yi et al. 2019)
> *"Bitumen and Bitumen Modification: A Review on Latest Advances"* (Porto et al. 2019)

3.1.2.2 To Learn from Other Professionals

Doing a literature review is a learning process from the works of domain experts and professionals. We can learn from our peers who have conducted the same or similar research projects. For example, some papers state,

> *"Once such systems are in place, one can better design threat mitigation techniques that are tailored to those systems by leveraging on lessons learned from other fields, such as autonomous vehicles."* (Schmale 2019)
> *"Lessons learned from evidence-based practice in other domains, especially law and medicine, are described and incorporated into evidence-based systems engineering."* (Hybertson et al. 2018)

"This paper describes research aimed at optimising the performance of metallic processes, e.g. selective laser melting (SLM) and direct metal laser sintering (DMLS); but, will first review some of the lessons learnt from other processes and materials." (Morgan et al. 2016)

Different from learning from textbooks and in classrooms, reviewing papers can effectively update and broaden our knowledge. Textbooks tell us background, basic principles, and methods. While the recent scholarly papers provide us with the latest research achievements, various viewpoints, analysis approaches, particular subject evolution, and so on. Occasionally, older (more than 10 years old) literature is valuable as well.

3.1.2.3 To Look for New Opportunities

Literature reviews play as important a role in leading our research to new directions to the current research. Through a literature review, we may gain insight into problems, recognize the imperfections within existing publications, and identify research gaps or deficiencies of a subject. For example, we can find some new research opportunities on the topics not well studied because:

- Not many people yet recognized them
- Prior advance challenges due to known reasons (technical, resources, etc.)
- Low interest (probably due to funding) etc.

Then, we may propose new research, to build our bridge to cover the knowledge need or gap on top of something that has already been worked upon. Hence, a new research proposal intends to advance by

- Applying a proven technology to a new subject or field
- Using a different method to prove an important theory
- Resolving a conflict or controversy in differing research results
- Creating a new concept or different perspective

Many papers briefly state the plan for a continuation of their research and additional studies at the end. Such statements shed a light on what the authors believe should be the next direction for the particular research. For example:

"Our future work involves extending our evaluation to other open source and commercial projects. We also plan to extend DeepJIT using attention neural network so that our model can explain its predictions to software practitioners." (Hoang et al. 2019)

"First, we shall test the proposed method with datasets other than Titanic, especially, with a much larger dataset. Second, we shall improve our method to support numerical features. Third, we require to support transformations

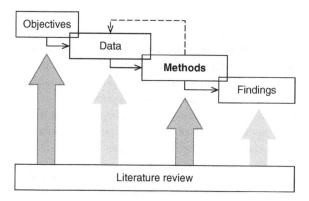

Figure 3.3 Literature review and other parts of research.

combining multiple features. We also plan to have the synthesized program in a way that users can modify it." (Narita and Igarashi 2019)

Based on a review, we may adjust or revise our research direction, emphasis, or methodology to reposition our research for an increased potential of success.

3.1.2.4 To Assess Research Methods

Literature reviews can support all parts of research; many reviews focus on research method (refer to Figure 3.3). When reviewing methods, researchers should think about the methods from different angles, such as objectives, data used, and findings, to understand better why the methods were used in a particular research project.

There are numerous research methods and approaches in a subject's domain. Different methods are used in the same subjects and for solving similar problems. By analyzing method, we can know the best one for a new research project, which is essential to the effectiveness of a research project.

In many cases, the purpose of review is to assess the methods used in published literature and to propose and justify a new method. In these cases, we need to identify the issue or gap of the existing methods. Here are examples of statements about new methods after a literature review section.

"Looking at the statistics of the translated model, the advantage of the proposed concept becomes apparent: There is no large non-linear equation system at all." (Zimmer et al. 2018)

"In this paper, we proposed a new approach to developing p-charts that does not make control limits more precise than it really is." (Ahmed et al. 2019)

"To overcome these shortcomings, a simple and systematic unifying approach is proposed in this study to extend our previous results in Refs. 3–5 to solve a complex operational amplifier." (Chen et al. 2019)

3.1.2.5 To Justify Proposed Research

We can also use literature review to justify the motivation for doing new research. In other words, a literature review shows the significance of the research project and the incomplete, limited, or unsatisfactory nature of the existing research. Accordingly, the connection between the previous research and new research can be established and then new research can be justified and supported.

We can generate new research ideas and directions with the support of literature reviews. Reading the literature review section of research papers or standalone review articles, we can find such statements. Here are examples:

> *"Given these findings in the literature on open software, open data, and open innovation, combined with the fact that the importance of data is growing for several types of applications, lead us to think of OCD as a potential means for spending less on commodity features and more on differentiating and innovative features in data-driven applications."* (Runeson 2019)

> *"In our approach, we follow the methodology proposed in TSQL2 book [8]. We specify aspects (1), (2) and (3) in a formal way, so that the necessary properties of reducibility and consistent extension can be proved ([8, 17])."* (Anselma et al. 2013)

> *"Previous studies offer the core algorithm and an implementation as a tool for industrial parts traceability in factory use [8]. However, no practical tool has been proposed for implementing our framework yet. Now, we have developed a functional prototype of such a tool."* (Bergen et al. 2019)

As a sum, a literature review is an integral value of a research project and significantly benefits new research in various ways. It gives us a base to build on and can bolster our confidence to do new and challenging work. Therefore, it is well worth our time and efforts.

3.1.3 Keys of Literature Review

3.1.3.1 Review Process

We may treat a literature review as a small project, with a few steps of planning and execution. Before conducting a literature review, we should clearly define a few items, such as

- Subject, topic, and scope
- Number of papers to review
- Timeframe of read and review

The subject and topic are determined based on research need. The scope of search can be based on experience, resources, others advice, and/or own target. The number of papers to review should be preliminarily determined, which is

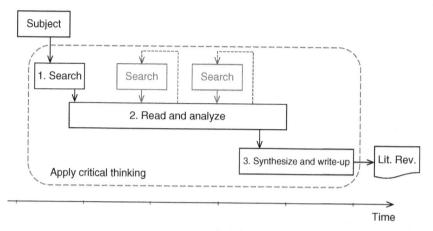

Figure 3.4 Common steps of literature review.

associated with time constraints. Depending on the width and depth of a search and review, we may arrange the time of a literature review. If you have three weeks for a literature review of 10 papers, you may plan three tasks with time overlap (see Figure 3.4):

1. Subject and search (three days)
2. Read and analysis (15 days)
3. Synthesis and write-up (five days)

During literature review, critical thinking and open mindedness are important to analyze the literatures. In addition, regular discussion meetings with an adviser and/or team members can be helpful. During the review, we may reiterate the steps of Search and Analyze a few times. We have more discussion on effective search in the next section.

3.1.3.2 Information Processing

Going through the steps of a literature review, we make the review to become more directly relevant and useful to our research subject. The narrowing down of information looks like a multiple-level filtering process (see Figure 3.5).

A literature review is not simply a collection of information and results from literature. The main effort is on "review" to

- Compare, critique, and assess the existing achievements
- Synthesize disparate or inconsistent ideas and results
- Summarize the state-of-the-arts
- Comment on possible issues and gaps
- Recommend new directions or topics for future study

Literature

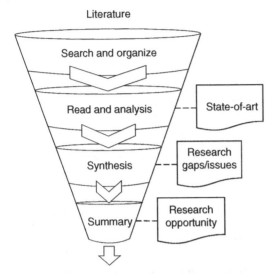

Figure 3.5 Funnel view of literature review.

The analysis process in a literature review first scans and then organizes the papers found into a few categories. We may consider this task as a prerequisite to our next step: critical analysis. With well-organized literature, the analysis, synthesis, and write-up can be streamlined.

The synthesis and write-up of a literature review is similar to writing an essay. We often organize a review section into several subsections, based on the topic and purpose of a literature review rather than around the source list. For example, a subsection may be on the research methods. We further discuss the writing of a literature review in the next section of this chapter.

3.1.3.3 Focuses and Structure of Review

In addition to knowing the overall process and purpose, understanding the characteristics of a literature review can be conductive to a quality review. The Association of College and Research Libraries published the Framework for Information Literacy for Higher Education. Readers may refer to it for the details (ACRL, Association of College and Research Libraries 2016).

A review can be on one of the following focuses. Clearly, the focus is associated with the goals of a literature review and the type of research (basic research, applied research, or R&D).

- Research finding and outcomes
- Research methods and approaches used
- Development of theories
- Applications of known principles and methods

We may organize a literature review one of three different ways, listed below. Sometimes we use a primary style, combining others with it.

- *Methodological.* To focus on the details of methods used in research papers, rather than with the content of the papers. This type of arrangement is often in research proposals and papers in engineering fields.
- *Thematic.* To organize review around a topic, issue, and opportunities. Within this type of review, progression of time may be an important factor. For example, reviews are on the issues and development of autonomous vehicles over time.
- *By Technical Trends.* To examine the sources through the lens of a developmental trend. This type of review follows the progression of time and is orientated around technical details as well. On this point, by trend and thematic reviews are similar. A review may have subsections according to the era within the period being referenced.

3.1.3.4 Items of Significance
It is important to understand the following parts of a literature search and review:

- *Scholarly Value.* Understanding the contributions of the original researcher's work to the field, which is the main purpose of a literature review. Please note that research results reflect researcher's expertise and credibility, but are not necessarily proportional to the author's status or reputation.
- *Objectivity.* On the conclusions of a particular works. Reviewers may check the persuasiveness of author's perspectives – how they are supported by data, analysis, reasoning, etc. A critical review may reveal imperfect or incomplete tasks for further work. For example, looking for certain pertinent information being ignored or inconsistent data.
- *Provenance or Authority.* About the original authors' research credentials and their affiliations. A novice researcher may rely on basic indicators of authority, such as the reputation of journals, author's institute, and/or research support sponsors.
- *Selectivity.* The information we choose to mention and comment should be related directly to our research focus. General information on a subject is not necessary.
- *Clarity.* A literature review should not only be clear to ourselves but also to our readers. Therefore, readers will see our research is backed up and justified by literature review.

For the reliability of literature sources, some librarians consider CRAAP, or five factors: currency, relevance, authority, accuracy, and purpose (Herrero-Diz et al. 2019).

Occasionally, we may experience the challenge of not having many scholarly articles available. There may be two possibilities when we cannot find enough literature on a specific topic:

- The topic is rarely studied (interesting to think about why, if not very new)
- The search range, defined by keywords and constraints, may be too specific.

3.2 Literature Sources and Search

3.2.1 Information and Process

3.2.1.1 Search Process

A literature search sounds straightforward but can be time-consuming. Two decades ago, researchers spent much time in "brick and mortar" libraries to find useful literature materials. Nowadays, the situation is different – more and more information, including scholarly papers and books, is available online. We can search remotely from anywhere to get literature, which is much more efficient and extensive than it used to be.

Even with well-designed search engines of the Internet and electronic databases, a literature search can still take significant time. A challenge is how to work effectively through the thousands (or even millions) of documents that turn up in seconds using Google Scholar or from scientific databases. The chore of finding the most useful sources to meet a research project need is still left to us.

Before beginning a literature search, we should have a basic plan and target. Do we want to do an exhaustive search or have a selected target group? A target group can be defined on its pivotal items, selective research projects by certain sponsors, conflicting objects, and so on. In general, we may consider two filters in a search process, refer to Figure 3.6.

The first task is to determine the *relevance* of the literature. Based on the search criteria selected, there may be thousands of articles available. It is necessary to review them quickly for their relevance to the research topic based solely on their titles. Then, we need to refine our search criteria, say additional keywords. We

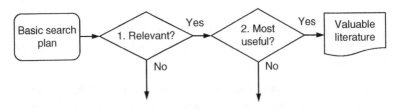

Figure 3.6 Two filters in literature search process.

may need to re-run the search again until the papers on the resultant list are directly relevant and the number of papers looks manageable, say several hundred items.

The second review filter is on the *usefulness* of the literature based on the methods and/or conclusions of the papers. Usefulness is research objective dependent. Here, we may not only add additional search constraints, such as time, but also scan the abstracts or the conclusions of papers. A systematic review reports 19 most relevant and useful papers from 1511 works with several screening steps (Veiga et al. 2019). This second step takes more time than the first one.

One practice is to check the potential research opportunities that may or may not be explicitly stated in the papers. Such a check may be on the defined research problem, imperfect executions, not fully agreeable conclusions, or study limitations mentioned in the literature. Any such items may imply opportunities for new research topics. If the proclaimed research in an article is well studied, it may be considered a good example and cited. However, such literature may or may not provide a direct clue for new research.

3.2.1.2 General Information Sources

The articles of newspapers, trade magazines, and general websites (.com) may have useful information but should not be used in research and scholarly papers. Trade magazines, for example, do not necessarily include sufficient scientific or technical details. A trade journal article may be written by a subject matter expert but may be used to promote their solution or technology over competitor's solutions.

Newspaper's articles lack detail and are written to meet reader's interest, as newspapers rely on advertising as a source of income. The undisclosed third-party influence in such sources is likely prevalent. Due to advertisement, an inherent bias could be associated with the source, whether it is intended or not. Similarly, not all websites can be trusted for a complete and accurate information without verification. For new technologies, we may find valuable information from company websites (such as organizations in ".com" domain), as they may promote their ideas, products, and/or technology. In such cases, we need to evaluate for possible commercial bias and technical validity. However, we may visit .com sites as a reference to lead us to the sources that are more academic.

Wikipedia is a good reference for general information. It is an online encyclopedia with exclusively free content and allows anyone to edit the text and publish things that may or may not be validated. The items of Wikipedia are not permanent and can be edited at any moment. Wikipedia itself states, "Wikipedia is not a reliable source for academic writing or research" (Wiki n.d.). We may check Wikipedia at the beginning to gather basic and preliminary information. However, we should not cite Wikipedia as a valid reference for scholarly works. It is extremely rare that scholarly papers cite Wikipedia. Most professors do not allow their students to use Wikipedia even for learning purposes in the classroom.

However, we may check Wikipedia for basic information and look at the references listed in Wikipedia to see if any of the scholarly ones can be useful.

3.2.1.3 Scholarly Publications

The target sources of a literature review are the papers published in refereed scholarly journals and conferences.

We can identify scholarly journals based on their names. For example, a periodical with name "Journal of..." or "... Journal" sounds scholarly. However, there are many exceptions that scholarly journals have no "journal" in their titles, such as "Computers and Operations Research" published by ELSEVIER, "Advanced Materials" published by Wiley Online, and "Research in Engineering Design" published by Springer-Verlag. If the journal name seems to not clearly show its scholarly nature, we should check the details of the journal, such as its introduction and history.

The history of a journal, for example, in terms of the volume number, is an important indicator. It is positive if a journal has been circulated in the professional field for many years. However, new technology keeps developing with new journals born every year. For example, Manufacturing Letters was created in October 2013. It is a scholarly journal of the Society of Manufacturing Engineers (SME) and considered a high-quality journal.

A key factor is that a journal must be refereed. Scholarly journals have a rigorous peer review process in place. The acceptance rates for journals of good reputation are normally less than 40%, which denotes the quality of papers published in the journals.

Considering these factors, novice researchers can search, filter, and find valuable academic research information. Refer to Figure 3.7.

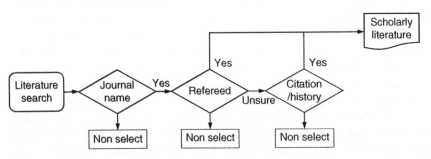

Figure 3.7 Selection considerations for scholarly journals.

3.2.2 Literature Sources

For the researchers and students of academia, the best channels are the online databases of university libraries. Access to most technical and scholarly literature databases is moderated via paid subscriptions. An institute library subscribes to many databases. The researchers and students can log in to the library databases, search the literature, and download them free of charge. If a paper is not available in full-text from a library, we may ask a librarian to obtain the paper via an inter-library loan. Consulting with a librarian can be very helpful for all researchers, particularly for student researchers.

3.2.2.1 Scholarly Databases

There are many database sources or platforms in engineering and technical fields; some are broad, others are more dedicated to specific areas, such as electrical engineering. Databases can be owned and operated by a professional society, for example IEEE, or a government agency; some are platform vendors, such as ProQuest. Table 3.1 lists the commonly used ones.

Every university has a depository for its students' theses and dissertations. In addition to ProQuest, you may visit other databases through library or via public Internet for theses and dissertation:

- Networked Digital Library of Theses and Dissertations https://www.ndltd.org/
- DART-Europe E-theses Portal https://www.dart-europe.eu
- WorldCat https://www.worldcat.org/
- Open Access Theses and Dissertations https://oatd.org/

3.2.2.2 Public Domain Internet

Using a web search engine, such as Google Scholar (https://scholar.google.com), we may find specific papers. Millions of peer-reviewed academic journals and books, conference papers, theses and dissertations, technical reports, and other scholarly literature are available, either in full-text or abstract-only to the public.

We should use the Advance Search function of Google Scholar with multiple search constraints. Figure 3.8 shows an example. We hope that one day Google Scholar will have more filtering functions, like additional constraints on title, abstract, journal name, publisher, etc., in advance search.

It is good practice for new researchers to use Google Scholar jointly with professional databases. We may use Google Scholar for a wide-ranging search to find relevant papers. Many papers found by Google Scholar are not available in full-text. Therefore, we must use professional databases for the full-length texts of the selected papers.

Table 3.1 Engineering databases and sources.

Database	Information
EI	EI Compendex is an interdisciplinary engineering bibliographic database published by Elsevier. The database has over 11 million of journal articles and around 7 million conference papers and proceedings of 190 engineering disciplines. For example, EI Compendex materials include about 29% in electrical engineering, 14% in civil engineering, 13% in chemical engineering, and 9% mechanical engineering. https://www.elsevier.com/solutions/engineering-village/content/compendex.
IEEE Xplore	It provides full-text journals and conference proceedings in electrical engineering, computer science, and electronics from the Institute of Electrical and Electronics Engineers (IEEE). IEEE Xplore provides over 4.5 million documents, comprising over 195 scholarly journals, over 1800 conference proceedings, more than 6200 technical standards, etc. https://ieeexplore.ieee.org/Xplore/home.jsp.
INSPEC	INSPEC, owned by The Institution of Engineering and Technology (IET), is extensive in the fields of physics and engineering, over 17 million of records. https://www.theiet.org/resources/inspec/index.cfm
Knovel	It includes engineering and science reference handbooks, databases, and conference proceedings and data. This library content includes more than 7000 reference works and databases from over 120 leading technical publishers and professional societies. Knovel has a sound content selection methodology to provide the more relevant information from the authoritative sources. https://app.knovel.com/web/index.v
NTIS	National Technical Information Service is US government-sponsored research and worldwide scientific, technical, engineering, and business-related information. Having over three million bibliographic records, including research reports, computer products, software, video cassettes, audio cassettes, etc. Its National Technical Reports Library (NTRL). (https://classic.ntis.gov/products/ntrl/) is freely available worldwide.
Scopus	It covers scholarly journal articles and conference papers on topics in science, technology, medicine, social sciences, and arts & humanities of about 35 000 peer-reviewed journals in top-level subject fields. The journals are listed on the SCImago Journal Rank website. https://www.scopus.com/
ASCE Library	It provides literature across the disciplines of civil engineering. The ASCE Research Library includes the full-text papers published in 36 journals from 1983, conference proceedings from 2000, and full-text ASCE standards and e-books. The Library offers free access to abstracts. https://ascelibrary.org/

Table 3.1 (Continued)

Database	Information
NTRS	NASA Technical Reports Server displays citation (not full-text) information from NASA's Scientific and Technical Information (STI) Program. It include research reports, journal articles, conference and meeting papers, technical videos, mission-related operational documents, and preliminary data. https://www.sti.nasa.gov/
ProQuest	ProQuest Dissertations & Theses Global is a comprehensive index to dissertations and theses from all fields of study, deposits from universities in 88 countries. The full-text version includes 2.4 million dissertation and theses with over 1 million full-text dissertations that are available to download. https://www.proquest.com/products-services/pqdtglobal.html
SAE Mobilus	SAE database covers thousands of SAE standards in three categories: Ground Vehicle Standards (J-Reports), Aerospace Standards, and Aerospace Material Specifications (AMS). The thousands of SAE technical papers covering the latest advances and research in all areas of mobility engineering including ground vehicle, aerospace, off-highway, and manufacturing technology. https://saemobilus.sae.org
SAFARI	As subscription-based, the digital library that includes technical reference books, videos, short-form content, and evolving manuscripts, etc. The topics include networking, Java, Linux/Unix, Perl, NET, desktop productivity, web development and more from O'Reilly and other publishers of IT books. https://learning.oreilly.com
SCI	Science Citation Index includes worldwide research literature in the sciences. It covers 8500 notable and significant journals across 150 scientific disciplines. The journals selected by SCI are often viewed as the world's leading journals of science and technology because of the rigorous selection process of SCI. http://mjl.clarivate.com/cgi-bin/jrnlst/jloptions.cgi?PC=K
EBSCO	EBSCO Applied Science & Technology Source database offers a diverse array of full-text and indexed contents, which cover the applied sciences and computing disciplines of over 4000 journals and magazines. https://www.ebsco.com/products/research-databases/applied-science-technology-source

3.2.2.3 Open Access

Open access (OA) is a relatively new business model for academic publication and is getting more popular. With OA, research papers are distributed online, free of charge to readers. Interestingly, studies showed that two thirds of the journals and three quarters of all OA publications are in scientific, technical, and medical

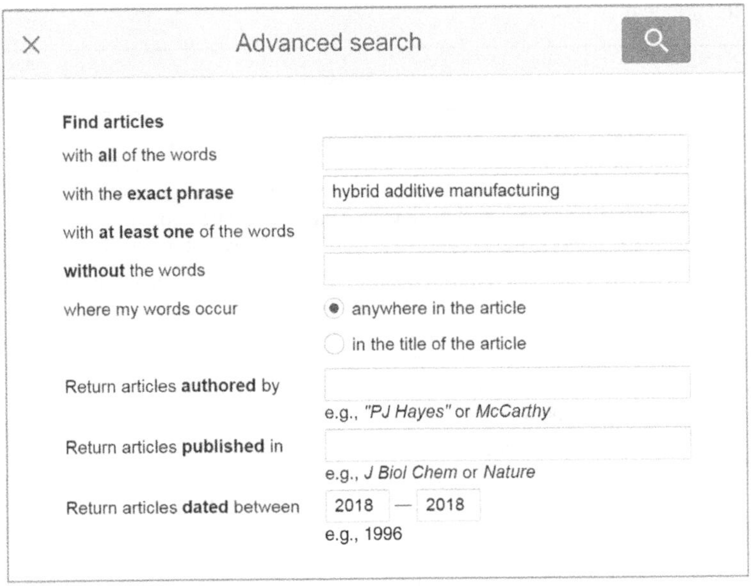

Figure 3.8 Advance search window of Google Scholar. Source: Google Scholar.

subject fields (Dallmeier-Tiessen et al. 2010). Figure 3.9 shows an example of OA in manufacturing engineering.

Table 3.2 lists some publishers for their OA publications. There is a website called Directory of Open Access Journals (DOAJ) across all disciplines (https://www.doaj.org/).

OA publication is under Creative Commons (CC) license, which has several types with different conditions for copyright material distributions. Two common versions are CC BY 4.0 and CC BY-NC-ND 4.0. The latter version does not allow using the materials for commercial purposes.

On top of its free access and download, OA publishes papers faster than conventional print journals. Many quality journals offer publication options, including regular publication, OA, and a "hybrid" format combining print and online.

3.2.2.4 Patents

A patent is a detailed report of research results for real or potential industrial applications and of commercial value. A granted patent entitles a monopoly for a fixed period, say 20 years, to the inventors, meaning other researchers cannot directly use the ideas or concepts in patents. Therefore, a patent can provide a protection for an invention. We have more discussion on patents in Chapter 7.

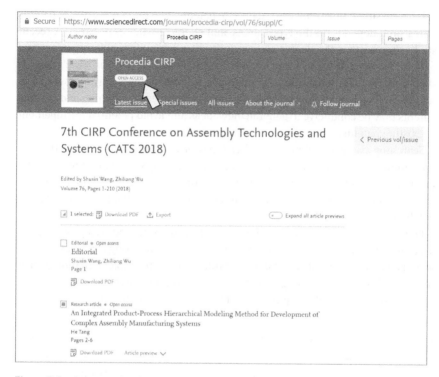

Figure 3.9 An example of open access publication. Source: Science direct.

Table 3.2 Open-access publications by some large publishers.

Name	Website
Elsevier's OA Journals	https://www.elsevier.com/about/open-science/open-access/open-access-journals
Springer Open	https://www.springeropen.com/journals
Taylor & Francis OA Journals	https://www.tandfonline.com/openaccess/openjournals
Wiley Open Access	https://authorservices.wiley.com/open-research/open-access/index.html
Royal Society Open Science	https://royalsociety.org/journals/authors/open-access/

It is occasional that scholarly papers cite patents. The reference value of patents to research depends on the types, areas, and characteristics of research. A recent study showed that the degree of popularity of a topic for scholarly papers and patents are very different from 1995 to 2015 (Qi et al. 2018). For basic research that aims for new knowledge, the reference value of patents may be low or less relevant. However, for applied research and R&D projects, searching and reading patents may be conducive to generate new ideas and avoid potential infringement of existing patents.

Searching patents is similar to other types of literature search. The first and most important way is keyword searching when keeping synonyms in mind. Some patent databases have multiple categories, such as by organization, which are helpful for searching by competitors, for example. Table 3.3 lists commonly used patent databases.

One characteristic of patents is that no verification is needed, as a patent can be a claim of a new concept only. A study estimated, "a surprising 27% of all patents would be at least partially invalidated" (Miller 2013). The implementation of peer reviews on patents is an on-going research. Several institutes and agencies have been studying it for patent applications to improve the validity of patents.

Table 3.3 Main patent databases.

Access	Database	Website
Free search (some services with fees)	Google Patents	https://patents.google.com/
	Espacenet (European Patent Organisation)	https://worldwide.espacenet.com/
	The Lens (Cambia and Queensland University of Technology)	https://www.lens.org/
	FPO (A SumoBrain Solutions Company)	http://www.freepatentsonline.com/search.html
	USPTO (US Department of Commerce)	http://patft.uspto.gov/
	WIPO (United Nations)	https://patentscope.wipo.int/search/en/search.jsf
Subscription required	Derwent Innovations Index	https://clarivate.libguides.com/webofscienceplatform/dii
	IEEE/IET Electronic Library	https://innovate.ieee.org/ieee-iet-electronic-library-iel/
	ScienceDirect	https://www.sciencedirect.com/

3.2.3 Considerations in Search

Due to the uniqueness and complexity of databases, search effectiveness is related to searching skills and experience. It is advisable to consult with experienced professionals, and/or librarians before searching. Here are practice tactics and tips for searching reference.

Bear in mind both literature search and literature review, as two phases of the process, are strongly connected and overlap (Figure 3.4). Inspection is often necessary on search results during search.

3.2.3.1 Using Keywords

A common way to search is using keywords, which are derived from our review subject and scheme. Figure 3.10 shows an overall process flow of using keywords to search interesting literature. The first step is to identify the keywords to use. Reading a couple of papers can help you identify the keywords and phrases to search. Seeking counsel from advisors and colleagues is also a good idea.

The reference section of papers is an excellent starting point. It is a good idea to examine the citations and reference lists of the papers. From these lists, you may identify additional literature of interest, possibly a bit old, and find new hints and keywords for a further search.

It can be tricky to select keywords. A search may start with two or three keywords. Commonly, thousands of items may show up in a search based on three keywords. For example, when using three keywords: hybrid, additive, manufacturing, Google Scholar yielded 132 000 results on 3 June 2019, which is unrealistic to review all of them.

One way to reduce the number of search results is to add more keywords and use two-word phrases that are more specific. There are two other tips: one is using acronyms, for instance, product lifecycle management (PLM) as it infrequently

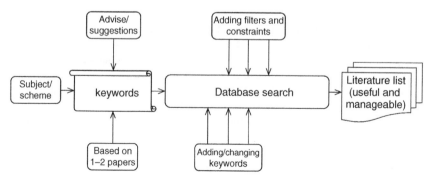

Figure 3.10 A process flow of literature search using keywords.

"spells out." The other is to use alternate spellings in American English and British English as different authors use different spellings.

For searching a very new or rare topic, we may have a hard time trying to locate literature relative to the topic. Such relevant literature may be sporadic. In such cases, try less-direct keywords. Once we find one or two papers, we have some hints from the contents and references of the papers.

3.2.3.2 Search with Constraints

In addition to using more and unique keywords in a search to narrow down search results, we may add some constraints. For example, the publication years can be limited to an appropriate range, say, the last five years, of literature. For example, the three keywords of hybrid, additive, and manufacturing within the period of 2018, Google Scholar yielded 18 800 results, about 14% of that without the time constraint.

Another way to add constraints is to group a few keywords using quotation marks, which restricts a search to the exact string inside the quotes. If using "hybrid additive manufacturing" as a whole phase for searching, Google Scholar got 149 results, which is further down from the 18 800. A total of 149 papers are not difficult to scan, so we may start reviewing their titles and details.

We may also apply direct constraints on keywords themselves and using Boolean operators, such as AND, OR, and NOT, refer to Table 3.4. For example, if excluding a word, put a minus sign "−" before a keyword, can significantly narrow from the records selected to improve the relevancy of search results.

As an exercise, we used Boolean operators for three keywords: "artificial intelligence" AND "quality control" AND "production" in recent five years (see Figure 3.11). As a results, three doctoral dissertations were found.

Another example is where we used "Systematic Literature Review" as search keywords, with the constraints of "Full Text Online," "Journal Article," "Engineering"

Table 3.4 Boolean operator applications in literature search.

Operator	Function	Examples
AND	Containing both phrases	"Artificial intelligence" AND "quality control"
OR	Containing one phrase of two	"Artificial intelligence" OR "machine learning"
NOT (−)	Excluding the phrase	−"Expert system"

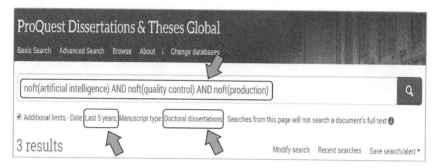

Figure 3.11 A search example with constraints in a database. Source: ProQuest dissertations and theses global.

discipline, from "6/29/2018 to 6/29/2019", and "Add results beyond your library's collection." This search came with 888 results (see Figure 3.12).

More constraints or functions are available in Google and databases. For example, we can select a particular language, such as Chinese or Japanese. Even though the papers not published in English, they normally have abstracts in English. We may also search a specific website domain, such as .edu or .gov, or to add the specific domain later to narrow down the quantity of search results. Another way to narrow down the search results is to limit the file type. For instance, if only literature in a PDF format is desired, we can have the phase "filetype:pdf" after keywords in Google search window.

We may use a sorting function, when available, to rearrange the results based on either date or number of citations. This may help bring up the most current or most prominent research first.

3.2.3.3 Currency of Literature

Many research areas evolve quickly. It is imperative that research uses the latest information, particularly for fast advancing technologies. Therefore, the publishing date of literature is important. In general, most literature to be used should be published within several years, even within 12 months for some subjects.

A publication process takes several months between the initial submission and publication date for journal papers. Thanks to the OA scenario of many journals, the OA publication time can be significantly shortened. Note: journal papers are timelier than textbooks.

There is about six months between the submission of an abstract and the presentation for the papers presented in professional conferences. Not all conferences publish the full length of papers in their proceedings. The access to these papers can be either OA, limited to the conference participants, or with

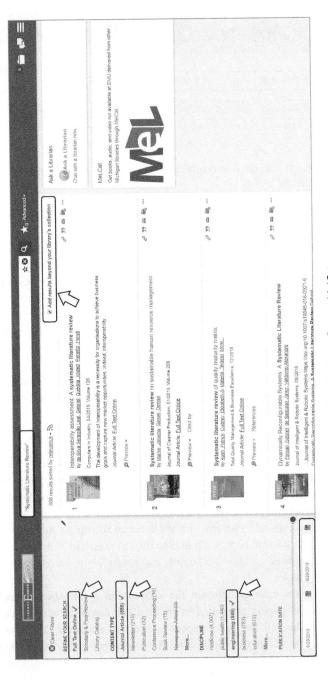

Figure 3.12 Another search example with constraints in a database. Source: MelCat.

a required subscription. Some papers presented in a conference are revised and later published by scholarly journals, which will be discussed more in Chapter 9.

After a research achievement is published, the authors and other researchers may have moved forward with their research. In other words, the information may not be contemporary. Thus, reviewers should be cautious if following the author's original suggestions.

3.2.3.4 When to Stop

We need to know when to be satisfied with our search and move forward. In theory, we should have a saturation point where we are seeing similar viewpoints repeated over and over again. Unfortunately, there is no clear, objective criteria to decide where and when to stop. For example, some master's theses have 20 references and some have over 50. You may also refer to similar published papers regarding the number of papers to be used. Students should consult with their advisors for a common practice in their field.

What if we keep finding significantly new information? That may mean the subject is very broad or we need to adjust our focus. Preferably, such focus adjustment is in an early phase of the search.

At some point, we have to stop and feel righteous in doing so. For a specific topic, we feel confident with the sought information to conduct a new research project. It is a good idea that we do a bit more search than needed. As suggested (refer to Figure 3.4), we may resume searching when you need more information during review.

When being close to the end of literature search, we may draw a tree-like map, using Figure 3.13 (to continue to Figure 3.14) as a plan, to summarize what you have and keep updating when finding new papers. With the literature map, we can know the search status and identify which topic you need to do more search.

Sometimes we would have to stop due to a time constraint, which it is not the best reason to stop, as we may not be fully ready to move on. Good planning and time management are important to literature search and review, which again can be a small project in of itself.

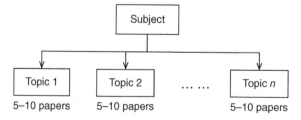

Figure 3.13 A literature tree for literature search.

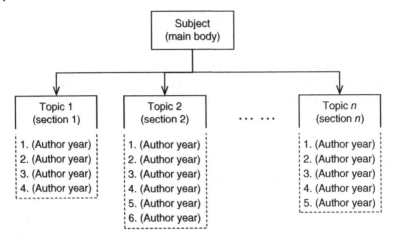

Figure 3.14 A structure of literature review.

3.3 Conducting Literature Review

3.3.1 Basic Tasks

3.3.1.1 Overall Attention

As discussed, we have two main tasks in a literature review. One is to search for current, relevant information. The other, more importantly, is to analyze and evaluate critically the ideas, methods, and conclusions contained therein.

Review itself requires careful reading and in-depth thinking that can be time-consuming and seemingly tedious work at times, especially during proposal development. Here are a few suggestions and hints for effectively preparing a literature review report.

Literature reviews should not only be a summary of the literature sought but also an exposition of the existing knowledge and reasoning to lead to and support a new research project. In other words, a literature review should involve synthesis after searching and analysis of the new research topic. In preparing a literature review, we should

- Emphasize relative context
- Use logical transitions
- Compare one work with others in the same topic
- Use appropriate paraphrases and citations
- Summarize at the end of the review

Therefore, our voice should remain front and center. For example, we may start a paragraph with our own ideas, followed by the sources supporting the point we want to make.

Another lesson learned from practice is to stay on the defined topic or problem. With so much information and details in literature, it is possible to let a review run "off-track" from the original scope of work and problem. A good practice is to pause and take stock after given literature is reviewed. The pause and in-process summary is helpful to guide further literature search and review.

As discussed in Chapter 1, critical thinking is a foundation of finding new research opportunities. Learning from literature and other professional's viewpoints can be inspiring but they should not be considered prescriptive. Understanding other research results, assumptions, and limitations can lead to the identification of the lack of perfection and deficiency of the research. Along with identifying and presenting the research status, we analyze and evaluate the literature contents for the new research. Base such evaluation on

- Reflection on the status quo
- Logical strength of research process
- Convincing strength of results
- Different viewpoints on the particular contents
- Questions on any review impression

3.3.1.2 Organizing Analysis

When reviewing 20 or more papers, we need to organize well our review activities, such as reading, analysis, synopsis, and composing a review section. We can group literature into topic groups based on the review scheme, which is a continuous effort to update the literature map (Figure 3.13). One effective way to organize literature review work is to use a table (Colon-Rivera 2019), like Table 3.5.

When reading a paper, we put a concise statement or key points of the paper into the cells of the table. Some focal points, such as methods, vary with research theme. This tabularized summary of literature information, once completed, provides a solid foundation to compose an integrated literature review section of a paper or a standalone review article.

Generally, the keywords and title of an article state what it is about. The abstract of a paper offers a short description of the study and a brief overview of the contents of the paper. Therefore, we should read the abstract once a paper is preliminarily

Table 3.5 A template of literature review summary.

Paper	Main point	Method	Research gap	Others
(1)				
(2)				
(3)				

identified relevant. In most cases, we can determine whether a paper has a good reference value from reading its abstract. If yes, we can spend more time on the content of the paper and cite it in our literature review.

Clearly, not every paper is equally useful on a specific topic. In addition, we may find a paper difficult to understand. In this case, unless article is super relevant, we may skip it for a bit while and read it again later. After analysis, a literature review frame evolves to Figure 3.14 from Figure 3.13. Now, it is about time to compose a good review.

3.3.1.3 Making Good Argument

Please keep in mind, when making a summary, comment, or recommendation in a literature review, we in fact make an argument. To have a good argument, it is suggested that we use the Toulmin method (Toulmin 1958), which involves the data, claim, and warrant.

The data in the Toulmin method are the evidence to prove or support a claim. A claim is a statement of our position on the issue. Furthermore, any claim or perspective in a review should be based on data and facts, which can be from another literature or source. Without support, an argument sounds like an assertion.

A warrant is the principle and assumptions to bridge the data and the claim (see Figure 3.15). In other words, the warrant interprets the data and supports the claim. The three elements of data, claim, and warrant form a sturdy logical structure, even when the warrant or interpretation is debatable. We may have additional statements (backing) to support or prove the warrant.

Here is an example for readers to identify the data, claim, and warrant in the concluding remarks below.

"It is worth noting that the academic production country ranking is far from KPMG's Readiness Index ranking (Table 2), due to the analyzed key

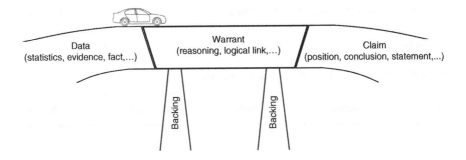

Figure 3.15 A diagram for the Toulmin method.

performance factors. This also implies that the introduction of AVs on the market is highly locally oriented, indicating that many adaptations in technology as well as in business models must be addressed." (Gandia et al. 2019)

Sometimes, we may write a nonargumentative analysis report. In such cases, we mainly synthesize what we have found out about a research topic. Our review may look like a summary of previous research with or without the comments on a future direction. Such reviews may be less interesting to readers than an argumentative literature review, but still has a value to research community.

3.3.2 Focal Points

Based on our assessment of the strengths and weaknesses of literature, we may propose new research. In some cases, we can even challenge the existing conclusions. To perform a literature review, we may keep in mind the following key points.

3.3.2.1 On Methods

Our review may focus on research methods. In addition to the methods commonly used in our field, we may search and review the methods implemented in different areas to generate new ideas for our research. Three examples of using the method of discrete event simulation (DES) in different areas are listed below. If we will use DES in our project, we want to know how the method is used in other areas.

"Simulation of auto design performance in the market to meet fuel efficiency standards." (Duff et al. 2015)
"Analysis of Oil and Gas Supply Chain Using Continuous-Time Discrete-Event Simulation." (Kbah et al. 2016)
"Modelling and Simulation of a River-Crossing Operation via Discrete Event Simulation with Engineering Details." (Jung et al. 2015)

In addition to the same methods used in different areas, we may consider different methods applied to the same subjects, individually and combined. There are trends that more untraditional methods consistently introduced to new applications and different combinations of methods are increasingly used in research. Such methodological changes may have unique advantageous potential for breakthrough in research. Here are three examples of combining different methods in research projects:

"Optimization under uncertainty in chemical engineering: Comparative evaluation of unscented transformation methods and cubature rules." (Maußner and Freund 2018)

"Throughput Analysis of Manufacturing Systems with Buffers Considering Reliability and Cycle Time Using DES and DOE." (Imseitif et al. 2019)
"Deep learning-based feature engineering methods for improved building energy prediction." (Fan et al. 2019)

In addition, there are standalone literature reviews focusing on research methods; three review of machine learning methods are listed below. This type of review can save our time when we focus on a particular method.

"A systematic literature review of machine learning methods applied to predictive maintenance." (Carvalho et al. 2019)
"Big Data Creates New Opportunities for Materials Research: A Review on Methods and Applications of Machine Learning for Materials Design." (Zhou et al. 2019)
"Machine Learning Methods for Pipeline Surveillance Systems Based on Distributed Acoustic Sensing: A Review." (Tejedor et al. 2017)

3.3.2.2 Exploring Trends

We may need to study the development and the latest advances of a give subject. Such reviews provide an evolutionary perspective for a particular research subject, which has a special value to support new research efforts. Here is an example,

"One example is the latest study carried out by Kai et al. in which tubes with pre-folded origami patterns were tested numerically." (Alkhatib et al. 2017)

In addition, a new trend may be stated in a review. For instance,

"The continuous effort in academia and industry will contribute to emerge integrated MBR in treatment and valorization of wastewater." (Neoh et al. 2016)

We should cite multiple papers on the same subject to show comprehensively their development and different aspects or perspectives.

3.3.3 Standalone Review Articles

3.3.3.1 Review Articles

As discussed, a literature review is often a section of a research proposal or paper. Sometimes, researchers write a systematic literature review as a standalone article.

Such a review article is a comprehensive synthesis or analysis of the published research and has good reference value. Interestingly, review articles are more often downloaded and cited than original research papers, which implies contribution by providing insight and comprehensive information.

In a review article, authors often provide their comments and predict further development. Sometimes, authors do not focus on a specific topic. Instead, they provide an assessment of a state of knowledge in a domain, and identify the issues for further research. Some authors engage in comparison between the two types of literature reviews (Boell and Cecez-Kecmanovic 2015).

Review articles can save time for the researchers who are new to the topic. Reading a review article, readers can learn:

- Recent advances and discoveries
- Current research focuses and gaps
- Different viewpoints and research aspects
- Researchers (reviewers) and their institutes

Because of academic value of review articles, highly reputable scholarly journals publish review articles. Often, authors often use a term "systematic" in the titles of such review articles.

3.3.3.2 To Prepare Review Article

We do not have to be a domain expert to write a review article. Most Ph.D. students need to do an intensive literature review before proposing their research. When preparing a new research grant application, we also need to do a rigorous literature review. After a review, we may spend additional time, e.g. a few weeks, to write up for publication to tell the professionals in the world what we have learned.

When considering publication of a review article, we need to think a bit broadly. For example,

- Who is the audience of our review?
- What is the central thesis of the review?
- Is the review distinct from similar reviews?
- What is the additional insight beyond the collection of literature?
- What are journals that accept review articles?

During preparing a review article, we may discover something that we did not realize or pay attention before. Keeping readers in mind, we also make an effort to coordinate the information from literature to deal with a large variety of type, style, focus, etc. For example, we may compare the definitions and create a glossary list.

Here are several examples in different technical areas, and specific focuses, such as a technique, method, modeling, applications.

- Muhammad Ilyas Azeem, et al. (2019). Machine learning techniques for code smell detection: A systematic literature review and meta-analysis, Information and Software Technology, Vol. 108, p. 115–138.
- Mickaël Begon et al. (2018). Multibody Kinematics Optimization for the Estimation of Upper and Lower Limb Human Joint Kinematics: A Systematized Methodological Review, Journal of Biomechanical Engineering, Vol. 140, Iss. 3, 11 pages.
- Hajihassani, M. et al. (2018). Applications of Particle Swarm Optimization in Geotechnical Engineering: A Comprehensive Review. Geotechnical and Geological Engineering, Vol. 36, Iss. 2, p. 705–722.
- Eva-Maria Schön, et al. (2017). Agile Requirements Engineering: A Systematic Literature Review, Computer Standards & Interfaces, Vol. 49, p. 79–91.
- Marttunen, M. et al. (2017), Structuring problems for Multi-Criteria Decision Analysis in practice: A literature review of method combinations, European Journal of Operational Research Vol. 263, Iss. 1, p. 1–17.
- Costabile, G. et al. (2017). Cost models of additive manufacturing: A literature review, International Journal of Industrial Engineering Computations, Vol. 8, Iss. 2, p. 263–282.

3.3.3.3 Structure of Literature Review

Writing a review article, we normally organize an article into three sections (refer to Figure 3.16). Each section may have two or four subsections. For the

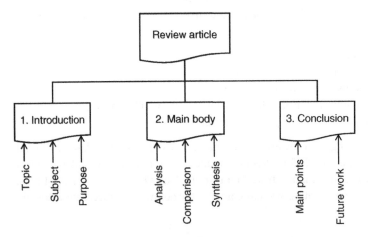

Figure 3.16 A typical structure of literature review article.

literature review section of a research paper, we may organize the review based on subtopics.

1. *Introduction.* In this section, we identify the topic and concerns, explain the purpose of the review, and provide our intention. This section may be only about 15% of the total length.
2. *Main Body.* We analyze literatures and comment on their main points. We prescribe the state-of-the-art knowledge and point out the gaps in current research. This section is about 80% of the article. We may consider:
   - Grouping topics into subsections
   - Arranging one idea or aspect per paragraph
   - Citing five or more references for each topic
   - Organizing in a priority or chronological order
3. *Conclusion.* We summarize the main viewpoints from the analysis of the literature and indicate future directions and opportunities.

As an example, Figure 3.17 shows the structure of the main body of a review article (Begon et al. 2018).

In a literature review article, we should select a large number of primary research articles. Many review articles cite over 100 papers. The length of a review article often varies from 5 to 10 printed journal pages.

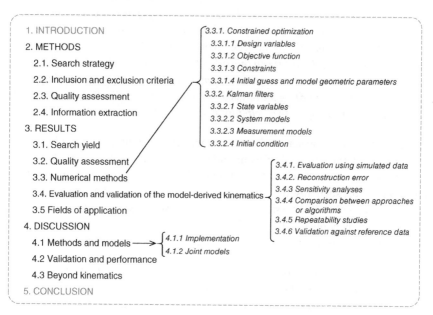

Figure 3.17 An example of main body of literature review article.

Sometimes, a review article is invited by journal editors based on the academic work of a reviewer. Without invitation, we may send an enquiry letter to journal editors about our review idea and plan before we finalize a review manuscript.

3.3.4 Writing Considerations

3.3.4.1 Professional Tone

When presenting a critical or different viewpoint, we should be professional and polite to represent the original information accurately first. Then, we analyze the research information of the original authors and point out possible areas for improvement.

We should take care when commenting on other works to avoid "putting down" a professional colleague who has worked on a similar project. We should make a criticism constructive and objective for a healthy atmosphere among cooperative professionals. Following are a few examples of critical comments.

> *"Haoli et al. (2009) showed that the previous method of Choudhury et al. (2007) may give incorrect or inconclusive confirmations in the case of interacting multi-input multi-output systems."* (Capaci and Scali 2018)
>
> *"Except when it is explicitly stated, it is unclear whether all the researchers have considered the open world assumption."* (Batres 2017)
>
> *"Such approach is very useful in design of spintronics as it eliminates large energy loses caused by Ohmic heating. However, epitaxial growth of $BiFeO_3$ films over Si substrate poses a formidable challenge for integration within conventional semiconductor devices."* (Petrovic et al. 2015)
>
> *"A review of the literature on this subject has shown that there have been occasionally incorrect or ambiguous statements [7], [8], [18], [38] – including by this author – regarding the physical origins of the flux of hydrogen to the crack tip resulting in hydride precipitation and growth."* (Puls 2009)

If the criticism of earlier work is necessary, try to be direct and mild, instead of negative manner. Instead of using judgmental words such as "no," "not," or "never," we may use "few" or "very limited." Our literature search may not be thorough. Thus, such absolute statements may be true to our knowledge but not necessarily to reality. It is possible that we do not discover some sources for various reasons, such as published in a small conference, in a different language, or in a publication process. Here are another three examples for readers review.

> *"To this end, comparisons between the Rytov and Born approximations with QMH and SPM at these thicknesses have not been done."* (Tajik et al. 2018).
>
> *"Author found that comparative study on twisting strength of different notches i.e. square, V and U shapes have not been done yet."* (Dhakar et al. 2018).

"This is generally expected for the range measurements (although often disregarded) but never considered for the angular measurements." (Medic 2019)

3.3.4.2 Citation and Format

In the context of literature review, the commonly used citation formats include

- "X's study [reference year] showed that…"
- "An approach in this subject was studied by X is that…" [reference year]
- "X concluded [reference year] that…"

The citation from one particular paper in a review should be considerably small, for example less than 10%, of the original paper. An excessive citation can be an issue of copyright breach.

We may use direct quotation if we want to emphasize a point or to have a comment after the original statement. However, many direct quotes can be distracting from the reviewer's standpoint, becoming front and center. In many cases, it may be better not using direct quotes, particularly if the original statement or discussion is long or detailed. We would need to rephrase other's ideas in our own words with the reference mentioned. For such situations, we also use caution when paraphrasing to represent the original author's statements accurately.

3.3.4.3 Common Concerns

The tenses can vary in the context of a review. In the introduction and conclusion sections of a review, we can use the present tense as general statements. Depending on the situations, in the main body of a review we

- Use the present tense to state other's thoughts or knowledge
- Use the past tense to refer to a specific work or result
- Use present tense for the implications of results
- May use the present perfect tense when referring to multiple researcher's work

There are common issues in a literature review. Some of them are already discussed above. Some issues are listed below for reader's awareness and avoidance if applicable.

- Descriptive review, but lacking in criticism to identify a research gap
- Organized around individual papers, instead of on a research topic
- No smooth transition flow between topics or subsections
- Focused on results, but not going over the methodology sections of reviewed articles
- Narrow literature sources, say only a few journals, insufficient diversity
- Using nonscholarly, nonrefereed sources
- Citing many relatively old literatures, for example over 10 years old

Summary

Introduction to Literature Review

1. Literature reviews may serve multiple purposes, such as situational awareness, learning from others, looking for new opportunities, assessing methods, and justifying a new proposal.
2. A literature review can be either a section of a scholarly document or a standalone systematic review article.
3. Literature review itself is a study process, consisting of searching, reading, analysis, and writing up.
4. Review may be primarily on comparison, critique, assessment, synthesis, summary, recommendation, and so on.
5. A literature review may be organized in three ways: methodological, thematic, or by technical trend.

Literature Sources and Search

6. Use keywords when searching literature normally.
7. Additional filter constraints, e.g. time limit and Boolean operators, can help search efficiency.
8. Literature sources should be scholarly journals and professional databases. Information from media and ".com" websites, including Wikipedia, are normally not allowed as a research reference.
9. Open access scholarly papers are normally of good value, but it depends on the journals.
10. In addition to relevance, the currency of literature is important as well.
11. For a systematic review, there may be a saturation point to stop a literature search.

Conducting Literature Review

12. The literature sought should be well organized for an effective review.
13. A literature review may have different focal points: methods, a latest trend, etc.
14. A literature review article normally consists of three parts: introduction, main review body, and conclusion.
15. The Toulmin method can be used to make an argument.
16. When judging others work, wording should be professional and carefully selected.
17. There are some common practices when writing up a literature review, such as tone, tense, and citation formats.

Exercises

Review Questions

1 Use an example to discuss one of the objectives of literature reviews.

2 Please explain the significance and one of the benefits of conducting a thorough review of existing literature before writing a research proposal, using an example.

3 Think and talk about your plan of literature search and review.

4 Discuss the importance of searching and referring to recent literature.

5 Explain how to select a focal point for a literature review based on your practice.

6 How do you evaluate literature value in your research project?

7 Identify 5–10 keywords for your research problem and consider using some of them for a literature search.

8 How do you decide whether your literature review is finished? Is that based on available time or the number of papers found?

9 Conducting a search, a researcher has a lengthy list of research articles. She must shorten the list to make a review manageable. She wanted to eliminate the articles that are not published in prestigious research journals. What are your suggestions to help her select articles?

10 A student asked, when conducting a literature review, how important is it to not only search the specific topic at hand, but also adjacent or opposing researches? Does this add value onto the topic?

11 What you would suggest if one cannot find many scholarly papers on a subject?

12 A literature review may be organized in three ways (methodological, thematic, and by technical trend). Which one you likely use based on your research objective?

13 The articles from newspapers, trade magazines, and .com websites are normally not allowed for a research project. Can you explain why?

14 Review the reference value of patents to research with an example.

15 Use an example to show how to remove less *relevant* literature during a search process.

16 Use an example to show how to remove less *useful* literature during a search process.

17 Based on your study major or professional field, which scholarly databases should be used for your literature search?

18 Execute a search with Boolean operators and discuss the difference of search results.

19 Discuss the characteristics of OA from a viewpoint of readers.

20 How should you perform critical thinking when reviewing papers and writing a review?

21 From a literature review, you likely find research gaps or opportunities for conjecture. Can you use the Toulmin method to make a convincing argument?

22 In the text, a few common issues of literature review are listed (Section 3.1.3). Provide your comments on one of these issues.

23 Comment on the differences between direct and indirect citations in literature review.

24 Please discuss the challenges of doing literature search and review. You may talk through it as a general concern or specifically in a particular field or subject. Include a supporting example.

Mini-Project Topics

1 Search your institute library databases or Google Scholar to find five journal articles (full text, interesting and related to your future research) and report how to find the papers, such as using which database, what keywords, and constraint filters. Comment on your search practice and how to improve search efficiency.

2 Search a research topic using your institute library databases and Google Scholar in parallel. Compare the search outcomes and discuss the usefulness of the two searches.

3 Critical thinking and open mindedness are important for a literature review. Find an example to show the importance of critical thinking and open mindedness.

4 Evaluate the literature review section of two papers in your field. Compare and comment the similarities and differences of their literature review in terms of supporting the research work of the papers.

5 Search three to five papers in your field and review them. Identify a possible new research topic just based on the results of the papers. Please justify your insight.

6 Search three to five papers using the same method but on different topics. Analyze how the method is used on the different topics.

7 Review three to five papers in the same topic, using different methods. Analyze how the different methods are used on the same topic.

8 Review three to five journal papers and discuss their research thinking and writing style.

9 After completing a simple literature review, talk, about your approach of literature search and analysis (refer to Figure 3.4) and potential improvement of effectiveness.

10 Find a standalone review article and analyze its structure and key points.
11 Analyze a paper to address three of the four topics below.
   a. Research contribution
   b. Suggestion for future work
   c. Research work timeline
   d. Roles of researchers/authors

References

ACRL, Association of College and Research Libraries (2016). Framework for information literacy for higher education. Adopted by the ACRL Board, 11 January 2016. www.ala.org/acrl/standards/ilframework (accessed January 2019).

Ahmed, S., Kibria, G. and Zaman, K. (2019). A new approach to constructing control chart for inspecting attribute type quality parameters under limited sample information. Proceedings of the International Conference on Industrial Engineering and Operations Management, Bangkok, Thailand (5–7 March 2019).

Alkhatib, S.E., Tarlochan, F., Eyvazian, A. et al. (2017). Collapse behavior of thin-walled corrugated tapered tubes. *Engineering Structures* 150: 674–692. https://doi.org/10.1016/j.engstruct.2017.07.081.

Amershi, S., Begel, A., Bird, C. et al. (2019). Software engineering for machine learning: a case study. ICSE-SEIP '19 Proceedings of the 41st International Conference on Software Engineering: Software Engineering in Practice, Montreal, Quebec, Canada (27 May 2019), 291–300.

Anselma, L., Bottrighi, A., Montani, S. et al. (2013). Extending BCDM to cope with proposals and evaluations of updates. *IEEE Transactions on Knowledge and Data Engineering* 25 (3): 556–570. https://doi.org/10.1109/TKDE.2011.170.

Batres, R. (2017). Ontologies in process systems engineering. *Chemie Ingenieur Technik* 89 (11): 1421–1431. https://doi.org/10.1002/cite.201700037.

Begon, M., Andersen, M.S., and Dumas, R. (2018). Multibody kinematics optimization for the estimation of upper and lower limb human joint kinematics: a systematized methodological review. *Journal of Biomechanical Engineering* 140 (3): 11. https://doi.org/10.1115/1.4038741.

Bergen T.v., Ishiyama, R., Makino, K. et al. (2019). Indexing and retrieving voice recordings by instantly tagging mentioned objects with dots. IEEE 5th World Forum on Internet of Things, Limerick, Ireland (15–18 April 2019).

Boell, S. and Cecez-Kecmanovic, D. (2015). On being 'systematic' in literature reviews in IS. *Journal of Information Technology Vo.* 30: 161–173. https://doi.org/10.1057/jit.2014.26.

Capaci, R.B. and Scali, C. (2018). Review and comparison of techniques of analysis of valve stiction: from modeling to smart diagnosis. *Chemical Engineering Research and Design* 130: 230–265. https://doi.org/10.1016/j.cherd.2017.12.038.

Carvalho, T.P., Soares, F., Vita, R. et al. (2019). A systematic literature review of machine learning methods applied to predictive maintenance. *Computers and Industrial Engineering* 137: 106024. https://doi.org/10.1016/j.cie.2019.106024.

Chen, C., Chen, Y., and Li, J. (2019). New method of solving complicated operational amplifier systems and application to online electrocardiograph. *Sensors and Materials* 31 (6): 1973–2012. https://doi.org/10.18494/SAM.2019.2336.

Colon-Rivera, W. (2019). An effective way to search literature. Internal Discussion in Course QUAL 655 Technical Six Sigma, CRN 25894, Eastern Michigan University (24 January 2019).

Dallmeier-Tiessen, S., Suenje, B., Darby, R. et al. (2010). Open access publishing – models and attributes. Max Planck Digital Library/SOAP, 62. http://edoc.mpg.de/478647 (accessed 22 June 2019).

Dhakar, P.S., Chauhan, P.S., and Sengar, V.S. (2018). Analysis of effect of shape and depth of notch on twisting strength of medium carbon steel 40C8 Bar. *Asian Journal of Innovative Research* 3 (2): 1–13.

Diabat, A., Jabbarzadeh, A., and Khosrojerdi, A. (2019). A perishable product supply chain network design problem with reliability and disruption considerations. *International Journal of Production Economics* 212: 125–138. https://doi.org/10.1016/j.ijpe.2018.09.018.

Duff, W.S., Dowling, R., Hung, B. et al. (2015). Simulation of auto design performance in the market to meet fuel efficiency standards. 2015 Winter Simulation Conference (WSC), Huntington Beach, CA, 3160–3161. doi: 10.1109/WSC.2015.7408449.

Fan, C., Sun, Y., Zhao, Y. et al. (2019). Deep learning-based feature engineering methods for improved building energy prediction. *Applied Energy* 240: 35–45. https://doi.org/10.1016/j.apenergy.2019.02.052.

Gandia, R.M., Antonialli, F., Cavazza, B.H. et al. (2019). Autonomous vehicles: scientometric and bibliometric review. *Transport Reviews*: 1–20. https://doi.org/10.1080/01441647.2018.1518937.

Herrero-Diz, P., Conde-Jiménez, J., Tapia-Frade, A. et al. (2019). The credibility of online news: an evaluation of the information by university students. *Cultura y Educación* 31 (2): 407–435. https://doi.org/10.1080/11356405.2019.1601937.

Hoang, T., Khanh Dam, H., Kamei, Y. et al. (2019). DeepJIT: an end-to-end deep learning framework for just-in-time defect prediction. MSR '19 Proceedings of the 16th International Conference on Mining Software Repositories, 34–45. doi: 10.1109/MSR.2019.00016.

Hybertson, D., Hailegiorghis, M., Griesi, K. et al. (2018). Evidence-based systems engineering. *Systems Engineering* 21 (3): 243–258. https://doi.org/10.1002/sys .21427.

Imseitif, J., Tang, H. and Smith, M. (2019). Throughput analysis of manufacturing systems with buffers considering reliability and cycle time using DES and DOE. 25th International Conference on Production Research Manufacturing Innovation: Cyber Physical Manufacturing, Chicago, Illinois (12–15 August 2019). https://doi .org/10.1016/j.promfg.2020.01.423.

Jung, C., Yun, W., Moon, I.C. et al. (2015). Modelling and simulation of a river-crossing operation via discrete event simulation with engineering details. *Defence Science Journal* 65 (2): 135–143. https://doi.org/10.14429/dsj.65.8141.

Kbah, Z., Erdil, N.O., and Aqlan, F. (2016). Analysis of oil and gas supply chain using continuous-time discrete-event simulation. Industrial and Systems Engineering Research Conference (ISERC), Anaheim, CA (21–24 May 2016).

Lee, C., Woo, W., and Kim, D. (2017). The latest preload technology of machine tool spindles: A review. *International Journal of Precision Engineering and Manufacturing* 18: 1669–1679. https://doi.org/10.1007/s12541-017-0195-0.

Maußner, J. and Freund, H. (2018). Optimization under uncertainty in chemical engineering: comparative evaluation of unscented transformation methods and cubature rules. *Chemical Engineering Science* 183: 329–345. https://doi.org/10 .1016/j.ces.2018.02.002.

Medic, T. (2019). Improving the results of terrestrial laser scanner calibration by an optimized calibration process. Photogrammetrie, Laserscanning, Optische 3D-Messtechnik, Beiträge der Oldenburger 3D-Tage 2019, Wichmann Verlag, Berlin, Oldenburg, Germany, 36–50.

Miller, S. (2013). Where's the innovation: an analysis of the quantity and qualities of anticipated and obvious patents. *Virginia Journal of Law and Technology* 18 (1): 1–58.

Morgan, H.D., Cherry, J.A., Jonnalagadda, S. et al. (2016). Part orientation optimisation for the additive layer manufacture of metal components. *The International Journal of Advanced Manufacturing Technology* 86 (5–8): 1679–1687. https://doi.org/10.1007/s00170-015-8151-6.

Narita, N. and Igarashi, T. (2019). Programming-by-example for data transformation to improve machine learning performance. IUI '19 Proceedings of the 24th International Conference on Intelligent User Interfaces: Companion, 113–114. doi: 10.1145/3308557.3308683.

Neoh, C.H., Noor, Z.Z., Ahmad, N.S. et al. (2016). Green technology in wastewater treatment technologies: integration of membrane bioreactor with various wastewater treatment systems. *Chemical Engineering Journal* 283: 582–594. https://doi.org/10.1016/j.cej.2015.07.060.

Petrovic, M., Chellappan, V., and Ramakrishna, S. (2015). Perovskites: solar cells & engineering applications–materials and device developments. *Solar Energy* 122: 678–699. https://doi.org/10.1016/j.solener.2015.09.041.

Porto, M., Caputo, P., Loise, V. et al. (2019). Bitumen and bitumen modification: a review on latest advances. *Applied Sciences* 9 (4): 742. https://doi.org/10.3390/app9040742.

Puls, M.P. (2009). Review of the thermodynamic basis for models of delayed hydride cracking rate in zirconium alloys. *Journal of Nuclear Materials* 393 (2): 350–367. https://doi.org/10.1016/j.jnucmat.2009.06.022.

Qi, Y., Zhu, N., Zhai, Y. et al. (2018). The mutually beneficial relationship of patents and scientific literature: topic evolution in nanoscience. *Scientometrics* 115 (2): 893–911. https://doi.org/10.1007/s11192-018-2693-y.

Runeson, P. (2019). Open collaborative data – using OSS principles to share data in SW engineering. Proceedings of the International Conference on Software Engineering, 41st International Conference on Software Engineering (ICSE), Montreal, Canada (25–31 May 2019).

Schmale, D.G. (2019). Perspectives on harmful algal blooms (HABs) and the cyberbiosecurity of freshwater systems. *Frontiers in Bioengineering and Biotechnology* 7: 128. https://doi.org/10.3389/fbioe.2019.00128.

Tajik, D., Foroutan, F., Shumakov, D.S. et al. (2018). Real-time microwave imaging of a compressed breast phantom with planar scanning. *IEEE Journal of Electromagnetics, RF and Microwaves in Medicine and Biology* 2 (3): 154–162. https://doi.org/10.1109/JERM.2018.2841380.

Tejedor, J., Macias-Guarasa, J., Martins, H.F. et al. (2017). Machine learning methods for pipeline surveillance systems based on distributed acoustic sensing: a review. *Applied Science* 7 (8): 841. https://doi.org/10.3390/app7080841.

Toulmin, S.E. (1958). *The Uses of Argument*. Cambridge: Cambridge University Press https://doi.org/10.1017/S0031819100037220.

Veiga, G., Lima, E., Aken, E. et al. (2019) Efficiency frontier identification on the context of operations strategy – a study on representative constructs and variables. 25th International Conference on Production Research Manufacturing Innovation, Chicago, Illinois (9–14 August 2019).

Wiki (n.d.) Wikipedia: academic use. https://en.wikipedia.org/wiki/Wikipedia:Academic_use (accessed 12 April 2019).

Yi, J., Lu, C., and Li, G. (2019). A literature review on latest developments of harmony search and its applications to intelligent manufacturing. *Mathematical Biosciences and Engineering* 16 (4): 2086–2117. https://doi.org/10.3934/mbe.2019102.

Zhou, T., Song, Z., and Sundmacher, K. (2019). Big data creates new opportunities for materials research: a review on methods and applications of machine learning for materials design. *Engineering* 5 (6): 1017–1026. https://doi.org/10.1016/j.eng.2019 .02.011.

Zimmer, D., Bender, D. and Pollok, A. (2018). Robust modeling of directed thermofluid flows in complex networks. Proceedings of the 2nd Japanese Modelica Conference, Tokyo, Japan (17–18 May 2018). doi: 10.3384/ecp1814839.

Part II

Quantitative and Qualitative Methods

4

Research Data and Method Selection

4.1 Data in Research

4.1.1 Data Overview

4.1.1.1 Data and Research

Data are fundamental for much of research because they are raw materials and can provide the connection between the real-world problems and the formalization of a study model, hypothesis, or theory. Without appropriate and reliable data, research outcomes remain unverified regardless how perfect the model, hypothesis, or process of research are.

Most research projects are data-driven. We use data to discover, explain, prove, or disprove new phenomena and principles. For engineering and technology research, almost all tasks are related to data, such as collection, analysis, interpretation, and validation. Without data, it is an opinion.

The data for a research project must be relevant and representative. In other words, the data obtained should reflect the situation, patterns, and dynamics of the measurement targets. Without valid data, the following research efforts, such as analysis, interpretation, and implementation, can be meaningless or even misleading. The quality of data has several factors, such as precision, accuracy, consistency, completeness, and timeliness. Most of them are discussed in the next two sections.

The data in research are associated with objectives, tasks, and processes. The date and the associated activities should have clear purposes to support research objectives. For example, before collecting data, we should know how to analyze and present the obtained data. Furthermore, data collection itself may lead to further data collection as a data analysis may require additional or different data.

4.1.1.2 Data Management

Research data management concerns the entire process of the generation, usages, and organization of data in research execution. For a large research project, data

Engineering Research: Design, Methods, and Publication, First Edition. Herman Tang.
© 2021 John Wiley & Sons, Inc. Published 2021 by John Wiley & Sons, Inc.

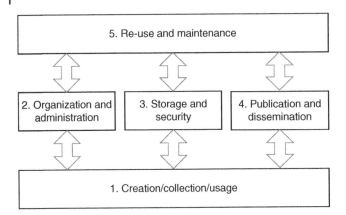

Figure 4.1 Main elements of research data management.

management is an integral part of a research proposal and execution. We must also follow the guidelines of funding sponsors and provide a data management plan. In general, there are five elements on a data management plan (refer to Figure 4.1):

1. *Data Creation and Usage.* This can be the main effort in a research project.
2. *Data Organization and Administration.* This element is to specify how data are organized and managed.
3. *Storage, Sharing, Security, and Back Up.* Data storage, associated issues with sharing, accessibility, and security controls, must be specifically designed.
4. *Publication and Dissemination.* Research data and results may be disseminated in different ways. The range and timing of dissemination and approval process are defined.
5. *Re-use and Maintenance.* Data generated from a research project can have a long-term reference value, in addition to its importance to research validity. During and after a research project, data maintenance should be planned.

Funding sponsors may have specific requirements on data management. For example, a data management plan is a required section of a proposal to the NSF. A data management plan should describe how the proposal conforms to funding agent policy on the dissemination and sharing of research results, including data sources, format, storage, sharing/dissemination, security, and intellectual property right. The NSF provides a guideline of data management plans for proposers. The five components of a data management plan in the proposals to the NSF Engineering Division include (NSF, a n.d.):

1. *Products of Research.* The types of data, samples, physical collections, models, software, materials, and other materials will be collected and/or generated.

2. *Data Formats and Standards.* The format and/or media in which the data or products are stored, accessible format, etc.
3. *Dissemination, Access, and Sharing of Data.* The plans to provide access to data, such as websites, to maintain data, and contributions to public databases/repositories.
4. *Re-use, Re-distribution and Production of Derivatives.* Policies regarding the use of data provided via general access or sharing.
5. *Archiving of Data.* The plan of data to be archived and accessed over time.

If a research project is more data focused rather than using data for a process or methodology, then data management is in the core of research. Accordingly, a more comprehensive and detailed data management plan may be required. Some universities provide comprehensive data management services. For example, University of California – Berkeley has such a service (https://researchdata.berkeley .edu/services), which includes data management tools/software, active data management, data backup, and licensing.

4.1.1.3 Data Science

Data can be a scientific field and research topic itself. There is an interdisciplinary subject called Data Science, which involves multiple fields of sciences, such as mathematics, statistics, information science, and computer science. Data Science focuses on the methods, processes, algorithms, and systems to study data. As a cross-disciplinary science, it shares the common objectives of research, i.e. to understand and analyze actual phenomena with data. For example, machine learning and big data are very active research areas in computer engineering.

Data Science may also be viewed as an advance and extension of applied statistics or statistical engineering. The statistical engineering is defined as "The study of systematic integration of statistical concepts, methods, and tools, often with other relevant disciplines, to solve important problems sustainably." By the International Statistical Engineering Association (https://isea-change.org/). The new extensions include neural networks, support vector machines, cloud computing, data processing, etc.

In a well-known paper by a statistics professor at Stanford University, the author stated (Donoho 2017),

> "The activities of Greater Data Science are classified into 6 divisions:
> *1) Data Exploration and Preparation*
> *2) Data Representation and Transformation*
> *3) Computing with Data*
> *4) Data Modeling*

5) Data Visualization and Presentation
6) Science about Data Science"

The author summarized his vision about the Data Science,

> *"The larger field cares about each and every step that the professional must take, from getting acquainted with the data all the way to delivering results based upon it, and extending even to that professional's continual review of the evidence about best practices of the whole field itself."*

In other words, Data Science is about all spectrum of data, applying various approaches on top of statistics, aims at solving complex problems. There are many books on the Data Science and its engineering applications, such as,

> *"Data Mining for Scientific and Engineering Applications"* edited by Grossman, etc.
> *"Data-Driven Science and Engineering: Machine Learning, Dynamical Systems, and Control"* by Brunton and Kutz
> *"Data Science in Practice"* by Said and Torra

Readers may refer to them for the detailed info and consider a possible application in their research.

Many engineering research projects are on data-related processes. Here are a few examples of research primarily on data:

> *"Data-driven process reengineering and optimization using a simulation and verification technique"* (Khan et al. 2018)
> *"Validation of thermophysical data for scientific and engineering applications"* (Diky et al. 2019)
> *"A Situational Approach for the Definition and Tailoring of a Data-Driven Software Evolution Method"* (Franch et al. 2018)
> *"Data map – method for the specification of data flows within production"* (Joppen et al. 2019)
> *"Towards an integrated process model for new product development with data-driven features (NPD³)"* (Li 2019)

4.1.2 Characteristics of Data

There are many characteristics of research data. Even though not all characteristics may seem significant or applicable to a particular research project, it is a good idea to understand all of them.

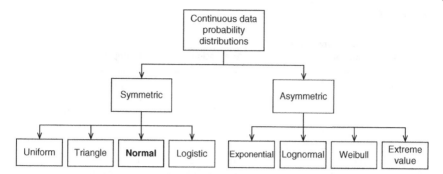

Figure 4.2 Common distribution types of continuous data.

4.1.2.1 Data Distributions

Data probability distribution is important as a research assumption. Many data analyses are valid only if the assumptions on the data distribution are true. There are also a lot of research on the characteristics and modeling of data distributions. For continuous data, the basic information of common distributions is shown in Figure 4.2 and Table 4.1.

There are interesting relationships between these various distributions. For instance, the Erlang distribution is a simple case of an exponential distribution when $k = 1$. While, exponential distribution belongs to the Gamma $\Gamma(\alpha, \beta)$ distribution family.

In most cases, we assume data following a normal distribution. In real-world cases, data may be skewed (distorted from the symmetry) or have excess kurtosis (outliers or tails) from an ideal normal distribution, which can reduce the accuracy of analysis results and predictions. Determining and validating the distribution of a set of data may be challenging.

4.1.2.2 Considerations in Research Data

Source. Data source is also a source of variation. For example, data may be collected from either a single experiment or in multiple settings. If data are collected from (even slightly) different conditions, such as multiple pieces of test equipment and production shifts, the data most likely have different characteristics, say mean values. The characteristics can be both a challenge and an opportunity to study the variation and its causes.

Elusiveness. Data are explicitly presented in such a way that the meaning is sometimes obvious. However, in many cases, data need to be thoroughly studied using different approaches to reveal the real meanings. That is often a major challenge in research.

Conditions. A key point should be kept in mind is about the conditions of collection when doing research. Data of any phenomenon are limited due to sources,

Table 4.1 Common data distributions.

Distribution	Probability density function	Application	Example
Uniform	$f(x) = \dfrac{1}{b-a}$	For the cases with a range (a, b) of equally likely values	Strength and thermal conductivity of metal matrix composites (Mazloum et al. 2019)
Triangle	$f(x) = \begin{cases} \dfrac{2(x-a)}{(c-a)(b-a)} \\ \dfrac{2(b-x)}{(b-c)(b-a)} \end{cases}$	For the analysis of risk and stochastic processes with min (a), max (b), and mode (c)	Dual phase nano-particulate AlN composite (Zhao et al. 2019)
Normal (Gaussian)	$f(x) = \dfrac{1}{\sqrt{2\pi}\sigma} e^{\frac{-(x-\mu)^2}{2\sigma^2}}$	Common statistical distribution; typical assumption for unknown random variables	Modeling for fragment-size distribution (Kim and No 2019)
Logistic	$f(x) = \dfrac{e^{-\frac{x-\mu}{\sigma}}}{\sigma\left(1 + e^{-\frac{x-\mu}{\sigma}}\right)^2}$	For the situations with longer tails and higher kurtosis than the normal distribution	Bootstrap confidence intervals of CNpk (Gadde et al. 2019)
Exponential	$f(x) = \lambda e^{-\lambda x}$	To model the time between events in a continuous Poisson process	Modified chain sampling plan (Jeyadurga et al. 2018)
Erlang	$f(x) = \dfrac{\lambda^k x^{k-1}}{(k-1)!} e^{-\lambda x}$	For the simulation of queuing systems	Operations simulation (Agostino et al. 2018)
Lognormal	$f(x) = \dfrac{1}{\sqrt{2\pi}\sigma x} e^{\frac{-(\ln(x)-\mu)^2}{2\sigma^2}}$	For the situations of the logarithm of the random variable in normally distribution	Simulation and empirical implementation (Prasojo and Prasetyoputra 2019)
Weibull	$f(x) = \left[\dfrac{k}{\lambda}\left(\dfrac{x}{\lambda}\right)^{k-1}\right] e^{-\left(\frac{x}{\lambda}\right)^k}$	Adaptable to different applications: engineering, medical research, and quality control	Wind energy prospect for power generation (Aririguzo and Ekwe 2019)
Extreme value	$f(x) = \dfrac{1}{\sigma} e^{\left[\frac{x-\mu}{\sigma} - e^{\frac{x-\mu}{\sigma}}\right]}$	To model the min or max values from a distribution of random observations	An approach to multi-choice multi-objective stochastic transportation problems (Qahtani et al. 2019)

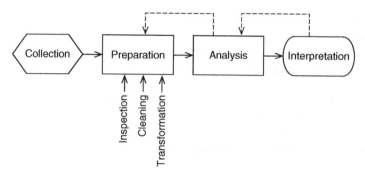

Figure 4.3 Main steps of data handling in research.

conditions, data analysis, and time. Therefore, research outcomes based on the limited data can be conditional. Again, it can be both a challenge and an opportunity based on limited data to draw a general conclusion.

Ephemerality. Data may be available for a short time in a particular place. The observations may be different at different times due to known and unknown factors. For example, observations on a machine's operation are dynamic and transient in nature. If data have such a short-lived characteristic, even if valuable, using them can be controversial because they are difficult to validate later on.

4.1.3 Data Analysis

As a process, data handling in research may include four steps, shown as in Figure 4.3, where the core part is analysis. During the process, it is possible we need to go back for additional work.

4.1.3.1 Prep to Data Analysis

First, we should have a clear and logical rationale for collecting data and using them. The process and approach we use will affect the analysis results and interpretation of the data.

Before doing a data analysis, there may be a few preparation tasks needed:

- *Data inspection* is the first task after data collection to look for obvious or subtle issues. Visualization of raw data, such as using various charts, is a good way for data inspection. We should have the predefined rules for data inspection. Data inspection also relies on researcher's experience.
- *Data cleaning* is another pretreatment process to remove incomplete, erroneous, and duplicated data, which can be important for following analysis. In addition, for a large amount of data, it is often a good practice to organize or condense the data to make them more manageable for an effective analysis.

- *Data transformation* is a process of converting data from one format, type, or structure into another one. We may consider this process part of either data preparation or an early phase of data analysis. In many cases, we transform qualitative data to quantitative data for further analysis.

Depending on the objective of research and the types of data, the data preparation may be either straightforward or complex and various software and tools are available to aid in the efforts.

Some research publications show the data preparation, for example,

> *"Using a data grid to automate data preparation pipelines required for regional-scale hydrologic modeling"* (Billah et al. 2016)
> *"Application of Chebyshev theorem to data preparation in landslide susceptibility mapping studies: an example from Yenice (Karabük, Turkey) region"* (Ercanoglu et al. 2016)
> *"Data preparation using data quality matrices for classification mining"* (Davidson and Tayi 2009)

4.1.3.2 Overall Data Analysis

Data analysis is a systematic process. The main objective of data analysis is to understand or discover the messages contained in the data by extracting and summarizing their main characteristics. There are several factors to select methods for data analysis. One factor is about the nature of the data, i.e. quantitative, comparative, or qualitative. The other is the assumptions about the data, such as distribution, independence, sample size, and statistical significance. Therefore, data analysis and analysis methods can be case and discipline dependent.

In some cases, data analysis may mean simple calculation results, such as an average and a trend. In such situations, we may present the results of data analysis in visual formats, such as graphics and tables, to illustrate certain data patterns. Based on the impression from the graphics and charts, we may draw a conclusion and/or consider a further analysis.

Data analysis software, including popular electronic spreadsheet MS Excel, has powerful functions. For example, to reveal the hiding patterns in raw data, we can sort the data. Professional data analysis software, such as Minitab, has more analytical and graphical functions and options.

Using a new method or adopting a method rarely used in a particular field may be deemed innovative and have a good potential to find something new. Therefore, trying new methods in research are encouraged. Due to the technical challenges and risks, however, some professionals warn that research should follow acceptable norms for disciplines. One statement is (NIU, n.d.):

> *"If one uses unconventional norms, it is crucial to clearly state this is being done, and to show how this new and possibly unaccepted method of analysis is being used, as well as how it differs from other more traditional methods."*

Furthermore, if the purpose of data analysis is to discover new knowledge for predictive purposes rather than primarily descriptive purposes, the data analysis is called data mining. Data mining, an active research subject related to various engineering and technology disciplines often involves new modeling, algorithm, and method development rather than use existing analysis techniques.

4.2 Types of Data

4.2.1 Basic Types of Data

4.2.1.1 Primary Data

The types of data to use and the approaches to collect the data are key focuses for researchers. Data sources can be either primary or secondary or combined, as shown in Figure 4.4. Primary data are the first-hand data that are collected by the original researchers. Such examples include the data from experiments, through observations, and interviews. Otherwise, the data are secondary. If shared with other others or to public, the data are also called open data, which are a type of secondary data in nature.

In engineering and technical research studies, we often rely on primary data. When available and appropriate, we may take advantage of using both primary and secondary data.

The primary data are also specific to our needs for the defined objectives of a study. We should have a data collection plan (e.g. where, when, and how to collect data) for our research. Primary data should be the most up-to-date form of data. The data collected for defined purposes may be challenged for the accuracy and representativeness. The conditions of primary data collection, such as experiment settings and parameters, affect the data quality and sometimes may be biased or inaccurate.

Understandably, the collection of primary data can be costly and time-consuming. They are often limited due to certain physical, financial, and resources constraints and are not practical in some cases.

4.2.1.2 Secondary Data

Secondary data are the data that already exist in the reports or databases generated by other parties. The common sources of secondary data are handbooks, journals, government databases, and commercial databases, and the like. One type of secondary data is called administrative data, which are collected routinely as part

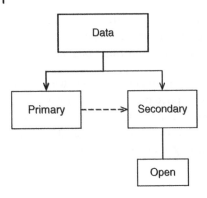

Figure 4.4 Basic types of data.

of the day-to-day operations of an organization or government agency. Secondary data may be derived from primary data as well. For instance, written sources that interpret or record primary data are secondary.

Thus, secondary data are normally inexpensive and easy to retrieve making them useful when collecting primary data are difficult or impractical for some researchers. In addition, secondary data often have large samples because the data collection is comprehensive, routine, and sometimes conducted over a long period.

However, there are a few areas of concerns when using secondary data. For example, they may include unnecessary information for a study. So, additional pre-analysis data preparation treatments, such as filtering and sorting, is necessary. We may also need to assess them for authentication, representativeness, collection conditions, etc.

In addition, secondary data may not be fully suitable for the new, specific research goals as the reliability and accuracy of some secondary data are unknown. There are possibly unidentified circumstances and assumptions in secondary data. For example, secondary data may come from or mix with other data sources rather than directly from a primary data source. If feasible, comparing secondary data from various sources may improve the reliability and increase the value of research. Besides, depending on the accessibility and currency of secondary data, they may have been already used in other studies and the research conclusions using the data again may be challenged about their novelty.

Using secondary data in research reports, we need to clearly define, assess, and disclose the conditions of derivation and conclusions based on secondary data. For example,

"Determining a Cut-Off Point for Scores of the Breastfeeding Self-Efficacy Scale–Short Form: Secondary Data Analysis of an Intervention Study in Japan" (Nanishi et al. 2015)

"Robust space time processing based on bi-iterative scheme of secondary data selection and PSWF method" (Du et al. 2016)

"The State and Prospects for Development of Railway Transport Infrastructure in Eastern Poland – Secondary Data Analysis" (Jarocka and Glińska 2017)

4.2.1.3 Open Data

Traditionally, the data used in applied research and R&D are not open to others. However, some professionals consider that closed data hinders testing the validity and reliability of research results. A trend is to share or "open" research data to external parties. Open data can be available to everyone to use without restrictions from copyright, patents, or other mechanisms of control, which is similar to those of other "open" movements, such as open sources and open access.

Therefore, using open data for a specific research topic may be limited and challenged. A recent study indicated that it is important to learn how to better integrate and take advantage of open data sources in research projects (Ruiz and Maier 2017). In addition, we need new methods to manage the datasets from different sources.

For basic research, data sharing or open is often encouraged. For example, the NSF data sharing policy states,

> *"Investigators are expected to share with other researchers, at no more than incremental cost and within a reasonable time, the primary data, samples, physical collections and other supporting materials created or gathered in the course of work under NSF grants. Grantees are expected to encourage and facilitate such sharing."* (NSF n.d.)

One important open data source is government data websites, a few listed in Table 4.2.

In addition, some commercial databases are available to the public but require a paid-subscription or purchasing of particular data files. For example, automotive companies use third-party data to study their rival's vehicles. Such information are available from the benchmarking consulting firms, such as A2Mac1 (https://portal.a2mac1.com/) and MarkLines (https://www.marklines.com), who conduct complete vehicle teardown analyses and provide various data services.

Table 4.2 Some government data websites.

Country	Website address
Canada	https://open.canada.ca
China	data.stats.gov.cn
France	www.insee.fr
Germany	www.destatis.de
Italy	https://www.dati.gov.it
Japan	https://www.data.go.jp
Russia	https://data.gov.ru
UK	www.natcen.ac.uk
US	https://www.data.gov

4.2.2 Quantitative vs. Qualitative Data

Data can be either quantitative (numerical) or qualitative (descriptive). In technical research, we mostly use quantitative data because we can analyze them using various statistical and/or numerical techniques. Other research projects, e.g. in social sciences, are often qualitative in nature.

4.2.2.1 Numerical or Non-numerical Data

Quantitative data are presented by numbers. They may be in either a continuous or discrete format. Continuous data are an infinite number of possible values in a range. For example, any number in a range between a and b; where, a and b are real numbers and $a \neq b$. It is clear that continuous data are uncountable.

On the other hand, discrete data are either finite or countably infinite. Discrete data can be given in certain types of integers, such as nonnegative integers, positive integers, or only 0 and 1. In addition, discrete data can also be categorical, like red or blue. Observation and measurement of a variable, say time, may be either continuous or discrete depending on the purpose and applications.

In contrast to quantitative data, qualitative data do not measure the attributes, characteristics, properties of a phenomenon but measure its types. Qualitative data are not expressed numerically but may be described and expressed in descriptive words, such as name, symbol, or a number code. The information from interviews, surveys, field notes, documents, electronic media, etc., is often qualitative. In many complex studies, using qualitative data and methods can make the research results more explicable.

Table 4.3 shows some relative strengths and weaknesses of qualitative and quantitative data. More comparison on their analysis is in Section 4.4.2.

Table 4.3   Characteristics of qualitative and quantitative data.

	Qualitative	Quantitative
Strength	• Detailed and in depth (rich) • Providing a nuanced understanding	• Clear and reliable • Easy to handle and analyze
Weakness	• Often small sample size • Subjective • Time for handling and analysis	• Often superficial • Not comprehensive to complex situations

4.2.2.2   Quality of Quantitative Data

The three key factors for the quality of quantitative data are their accuracy, repeatability, and reproducibility.

- *Data Accuracy.* Data accuracy is about the closeness of a measured or collected value to a true value. Data accuracy is normally addressed through the calibration of a measurement system.
- *Data Repeatability.* It is defined that the same object is measured by the same people using a single instrument on different measurement occasions.
- *Data Reproducibility.* It is measured by two or more human individuals on the same product or performance using identical measurement instruments.

All three factors are critical to research results. The combination of repeatability and reproducibility is about the variation of data. We can measure the data precision by doing gauge repeatability and reproducibility (GR&R) tests. Good quality data can ensure that other researchers will be able to replicate the same data to validate research findings.

4.2.2.3   Reliability of Qualitative Data

It may be difficult to measure qualitative data accurately. Our interpretation based on qualitative data may be more challengeable than that from quantitative data. However, this does not mean that the qualitative data and the associated findings are less valuable. The observations are normally experience-, attention-, and/or perspective-based. They may be informative and nuanced that lead to great insights into human society.

For qualitative studies, we sometimes use a term "reliability" (or other terms, such as replicability, consistency, or dependability) to describe data quality, similar to the repeatability of quantitative data. In other words, if qualitative data were

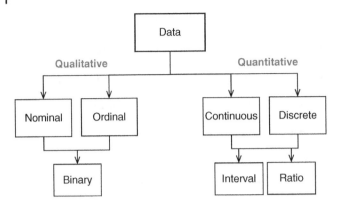

Figure 4.5 Scales of data.

to be collected a second time, they should support the same research result. Qualitative data on insubstantial phenomena tend to be less reliable than those designed to measure physical (substantial) phenomena quantitatively.

The Cronbach's alpha (α) is an overall assessment of the internal consistency (or scale reliability) of a set of test items as a group. The resulting α ranges from 0 to 1: If all of the scale items are entirely independent from one another, or not correlated at all ($\alpha = 0$). If all of the items are highly covariant, then α approaches to 1. Calculating the Cronbach's α is easy using statistical software. The Cronbach's alpha is commonly used in social studies, and occasionally in engineering applications as well, such as on a risk analysis in the integrated design and construction project (Wu et al. 2019) and the key criteria for building performance (Adamy and Bakar 2018).

Due to the significant differences between qualitative and quantitative data, some researchers assessed that the concept of reliability does not apply to qualitative data because the research findings will depend on the context (Silverman 2005).

4.2.3 Scales of Data

Data scale specifies the categories of measurements and data. Based on a measurement scale, we may consider data nominal, ordinal, etc., refer to Figure 4.5. The scale of data also determines the measurement procedure and following data analysis.

4.2.3.1 Nominal Data

Nominal data are separate, non-ranked data. Sometimes, they are called categorical data. Each data uniquely belongs to a specific category, which may be coded by a number that has no real numeric meaning. For example, data collected by color are nominal data.

Such nominal data are sorted and classified based on their names or labels, and thus can be compared. Accordingly, nominal data may be also called labeled data. Nominal data are normally analyzed using simple graphic and statistical techniques, such as bar charts and pie charts. Nominal data and combining with other types of data are used engineering disciplines. For example,

> "*Calmness of partially perturbed linear systems with an application to the central path*" (Cánovas et al. 2019)
>
> "*Fast and efficient prediction of finned-tube heat exchanger performance using wet-dry transformation method with nominal data*" (Zhou et al. 2018)
>
> "*Corporate failure prediction in the European energy sector: A multicriteria approach and the effect of country characteristics*" (Doumpos et al. 2017)

4.2.3.2 Ordinal Data

Different from (or maybe better than) nominal data, ordinal data can be an order, in terms of relative significance or priority, in either an increasing or a decreasing order. The data fall into categories, and the numbers for the categories may have physical meanings. For example, we may have a rating on a scale from 1 (lowest) to 5 (highest) for an ordinal dataset. We may use ordinal data to measure qualitative concepts.

Even with a certain sequence of importance or other types of ranking criteria, this does not mean that we can always interpret the differences between each category. We may or may not interpret the interval between categories in ordinal measures. Here are two examples of using ordinal data in engineering research:

> "*Weighted kappa loss function for multi-class classification of ordinal data in deep learning*" (Torre et al. 2018)
>
> "*Analyzing ordinal data from a split-plot design in the presence of a random block effect*" (Arnouts and Goos 2017)

4.2.3.3 Binary Data

Binary data are in two possible states, traditionally labeled as the combination of "0" and "1". In computer science and engineering, we call a binary digit a bit.

We may convert descriptive words of qualitative data into two states: 1 (yes) or 0 (no). For instance, the outcome of an experimental task is "success" or "failure" and the truth of a proposition is "true" or "false," etc. The main advantage of using binary data is that they are directly executed by a computer system. Binary data may be in form of either nominal or ordinal. There is plentiful engineering research using binary data.

4.2.3.4 Interval and Ratio Data

Interval and ratio data are quantitative. Interval data are sorted numeric scales, meaning there are not only the order of data but also the equal difference between the individual values. In interval data, the distance between attributes does have a meaning, which can be useful in carrying out more sophisticated statistical analysis. Two simple examples can be time and temperature data.

However, the data in interval scales have no "true zero." The zero point of interval data may be arbitrarily established as a reference. For example, there is no such thing as "no time." Zero time may be set up as a starting point. For temperature, zero degrees Fahrenheit and zero degrees Celsius are not the same.

The interval scale with a natural origin is called a ratio scale. A length measurement is an example of such a ratio scale. Different from interval data, ratio data have a true zero point. Ratio data are in relation to a zero value (e.g. a distance). Both differences and ratios have real meanings and are interpretable. For instance, zero inch and zero centimeter are exactly the same thing. The advantage of ratio data is that they can express values in terms of multiples of fractional parts and have a wealth of possibilities for various analyses.

4.3 Data Collection

The collection of data and information for research is a set of activities (e.g. observation, measurement, and generation). To obtain reliable and repeatable data, the collection conditions must be specifically defined and controlled. In some cases, data can be time dependent, or the data collected at a point in time may not be true or the same at another point in time.

The process and methods of data collection, such as sampling approaches, are critical to research outcomes and conclusions as they may highly rely on data collection. On the other hand, the types and characteristics of data play a determinative role for the process and methods of data collection.

4.3.1 Data Collection Sampling

4.3.1.1 Purpose of Data Sampling

In most cases of conducting research, it can be either impractical or uneconomical to get all the data of a target situation. Thus, a research project is a study of the samples of data collected from the entire population or situation.

Referring to Figure 4.6, we can know that the actual data samples are a subset of the accessible or feasible data set, which in turn is a subset of the entire data set. Note the word "population" or "data set" in research is a collective term used to

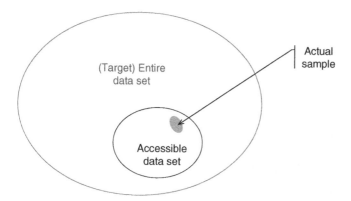

Figure 4.6 Entire data, accessible data, and samples.

describe the total quantity of things (or information) of the type. Therefore, data samples collected are always smaller than the entire population or data set.

A research process can be a type of inference (or from sample to population). We extend the conclusions based on the sample data to the entire population. We can successfully do that if and only if the data collected are truly representative to the population.

It is very possible that the samples do not accurately represent the status or condition of the entire population. Such an inaccuracy may be called sampling bias. The issue here is about external validity because the conclusion of a particular research may not be true for the entire population until the conclusion is validated by other professionals under the same or similar conditions.

4.3.1.2 General Considerations for Sampling

The process of selecting and collecting a small group of data from a population is a sampling plan. Data sampling methodology (including its plan and execution) is a key factor for research data collection, analysis, and conclusions. Basic requirements for data sampling include the representation, meaningfulness, reliability, and credibility of sampled data.

An appropriate sample size is very important to all research-based experimental and empirical studies. Sample size should be determined based on the research objective and requirements. Small samples can undermine the internal and external validity of a study. With an adequate sample size, study results may become statistically significant. Based on statistics, sampling error can be estimated. On the other hand, obtaining large samples requests more resources.

If the bias or inaccuracy of the data collected is associated with a measurement instrument, such as its settings, calibration, parameters, usage, and environment

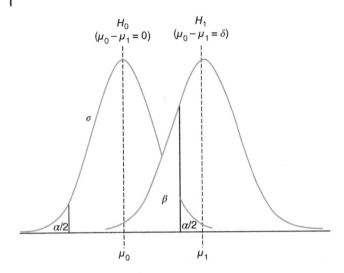

Figure 4.7 An illustration of types of I and II errors.

factors, then the bias may be called instrumentation bias. Using different measurement devices or methods on the same objects are helpful to identify possible instrument bias.

4.3.1.3 To Determine Sample Size

The question of how many samples should be collected from a population is an interesting topic. Generally, the larger the sample size is, the more accurately the samples represent the whole population. Here are some considerations about determining a sample size:

- For a statistical analysis based on quantitative data, the minimum sample size should be 25. With the minimum sample size, we may claim the analysis results are statistically significant in many cases.
- For many statistical analyses, there are recommendations for sample size requirements and calculation.
- For the research based on a hypothesis, an appropriate sample size may be determined according to an operating characteristic curve or formula for a type II error (β). For example, a two-tailed test (Figure 4.7) has

$$\beta = \Phi\left(Z_{\alpha/2} - \frac{\delta\sqrt{n}}{\sigma}\right) - \Phi\left(-Z_{\alpha/2} - \frac{\delta\sqrt{n}}{\sigma}\right)$$

where, Φ – standard normal cumulative distribution function, Z – test statistic, α – type I error, δ – mean deviation, n – sample size, and σ – standard deviation.

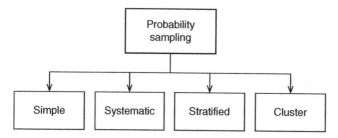

Figure 4.8 Types of probability sampling.

- For data collection, such as for a survey, we can decide the sample size based on the entire population, the confidence level (normally set at 95%), confidence interval (or margin of error, e.g. ±3%). For example, if a population is small, say fewer than 100, it is recommended that data be collected from the entire population. If a population is around 500, the sample size may be around 50%. In general, the larger the population, the smaller the sampling percentage can be, considering the feasibility and cost of a study.

4.3.2 Probability Sampling Methods

A common sampling method is probability or random sampling. Using this method, we assume that the chance of each sample is the same and that the samples statistically approximate to the characteristics of the total population. Therefore, probability sampling can give a reliable representation of the whole population. Probability sampling has four basic types, refer to Figure 4.8. There are additional types, such as unequal probability sampling and multi-stage sampling. Readers may refer to dedicated books for more information when needed.

4.3.2.1 Simple Sampling

In a simple random sampling, all elements of a population are considered and have an equal chance of being selected at any stage during the sampling process. Randomness is built into a sampling design so that the properties of the population can be assessed probabilistically. Hence, we may conclude that a simple random sampling is unbiased.

Individual samples may be determined by chance using computer software. In a simple sampling, a common practice is to avoid choosing any member of the population more than once, or sampling without replacement. This practice is important when a population is small.

Simple random sampling is easy to use with minimum knowledge of the population. If the information is available about the population, other types of sampling may be more efficient.

4.3.2.2 Systematic Sampling

We often use systematic sampling for a large list to select elements from an ordered sampling frame. We pick the every kth one from a complete set with total N elements in sampling. Thus, the sample size is $\frac{N}{k}$. In other words, the k, a fixed interval, is determined by $k = \frac{N}{n}$ if N is known and n is decided. An example of systematic sampling is in a study of "Automated Pre-Seizure Detection for Epileptic Patients Using Machine Learning Methods" (GÜL et al. 2017).

Simple random sampling plan is often preferred over systematic sampling plans because random sampling helps avoid subjective selection of samples. In fact, systematic sampling has little subjectivity if samples are determined in advance. Systematic sampling is better than simple sampling if n is large because of the more uniform coverage of an entire population.

4.3.2.3 Stratified and Cluster Sampling

Stratification sampling is the process of dividing members of a population into homogeneous subgroups before sampling. For a large population, stratified random sampling is an alternative to systematic sampling. We divide a population into mutually exclusive groups (called strata) and then use simple sampling to collect samples from each group equally. For example, this method was used in the research of "Refined Stratified Sampling for efficient Monte Carlo based uncertainty quantification" (Shields 2015).

Cluster sampling is similar to stratified sampling. Sometimes, researchers divide a population into natural groups based on their existing quantities. The sample size may be different. For example, the sampling probability may be proportional to size. The groups are called clusters in a cluster sampling. For example, an electrical study on "An I/O Efficient Distributed Approximation Framework Using Cluster Sampling" (Zhang et al. 2019) used the method.

Table 4.4 shows a comparison between stratified sampling and cluster sampling. A population may or may not be uniform or homogeneous so care is needed when using these two sampling methods. Corresponding discussion and statements are needed regarding the possibility of data not being accurately representative.

4.3.3 Non-probability Sampling Methods

4.3.3.1 Types of Non-probability Sampling

In some situations, it is impossible to know the sampling probability, and we may have to use non-probability sampling methods, which do not follow the

Table 4.4 Stratified sampling vs. cluster sampling.

	Stratified sampling	**Cluster sampling**
Groups in population	Population divided into groups	Naturally occurring groups
Sampling	Individually from all the strata	Collectively from selected group (clusters)
Homogeneity	Between groups	Within group
Advantages	Precision and representation	Cost and efficiency

random or statistical requirements. However, it may be still useful to select samples from a population. The common types of nonrandom sampling include (Figure 4.9):

- *Convenience Sampling (or accidental sampling).* Using this approach, we only select readily available samples on an opportunity. As a result, the extent to which the sample is representative of the target population is unknown. The analysis results are unlikely accurate for the target population. We should avoid using this type of sampling method if possible but we may use it as a preliminary study only.
- *Quota Sampling.* In this way, we conveniently select samples in a subgroup found in the general population, but not in a random fashion. Thus, quota sampling is a type of the convenience sampling. The results may be true for the subgroup but probably not for the entire population. There are very limited examples of using quota sampling in technical fields. An example is "Geographic Mapping of Tube Wells and Assessment of Saltwater Intrusion in the Coastal Areas of La Union" (Ngilangil et al. 2018).
- *Purposive Sampling.* We select samples for a particular objective based on our knowledge and professional judgment. This type of sampling may be acceptable only for special situations. In such cases, we should explain why the particular samples are selected. We may use this sampling method to measure a difficult-to-reach population. An example of using purposive sampling is "Designing a local Flexible Model for Electronic Systems Acquisition Based on Systems Engineering, Case Study: Electronic high-tech Industrial" (Karbasian et al. 2016). In general, purposive sampling can introduce or increase research(er) bias.
- *Expert Sampling.* Using this method, we identify the domain experts in the field of study and draw samples from them. Thus, expert sampling (or judgment sampling) is essentially a type of purposive sampling. However, the data and

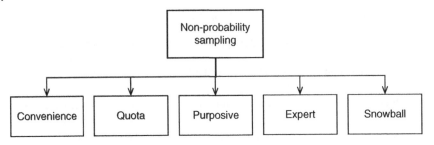

Figure 4.9 Types of non-probability sampling.

conclusion from such sampling and study may sound authoritative, which tends to be more convincing than our own viewpoint and effort defending our conclusions. The first step of using expert sampling is to define and identify the expert, which may be subjective as well. In addition, even a true expert can be wrong. A study that used the sampling method stated, "The definition of an expert is clearly open to interpretation as an 'expert' may very well be in the eye of the beholder, and an improper interpretation on the part of researchers may lead to a biased sample of participants that fails to adequately represent a population" (German and Rhodes 2017).

- *Snowball Sampling.* It is also called chain sampling or referral sampling. It begins with a few cases and spreads out on the basis of links or referrals to the initial cases. Using this approach, we choose samples from the new acquaintances of the existing study subjects. This method may be used if the desired sample characteristic is rare or cost prohibitive to find. This method is sometimes used in qualitative research. For instance, the snowball sampling is used to study what makes research software sustainable (Souza et al. 2019).

4.3.3.2 Characteristics of Non-probability Sampling

Any type of sampling method has some appropriate utility under certain circumstances. Due to the convenience and data availability, we use nonrandom sampling methods in many cases either explicitly or implicitly. The research results based on nonrandom sampling may have a good reference value but are much limited to certain and special conditions. In other words, the selection of a sampling method for a research project can be case dependent and need to study the characteristics of sampling methods with consideration of practical issues.

In general, non-probability sample techniques cannot produce a general conclusion about the whole population. We cannot guarantee whether the sample data represent the population well because some characteristics of the population have little or no chance of being included in the studies. Therefore, the disadvantages of non-probability sampling approaches outweigh the advantages because of limited value to general situations.

4.4 Method Selection

There are many research methods. Knowing most of them and selecting appropriate ones for a specific research will be beneficial. Most researchers agree that no particular method is privileged over another one for many cases.

4.4.1 Selection Factors

The method chosen will affect the research execution process and results. Selection of appropriate methods is associated with several factors such as research objective, available data, research type, and knowledge (see Figure 4.10). In addition, inventing a new method, including a new technology, algorithm, process, and model, is based on these factors and the existing methods.

4.4.1.1 Objective Driven

The purpose and expected accomplishment of a research project is the first question for us to think about when selecting a method. In many cases, objectives of research play a determinative role to select methods, assuming related data are available.

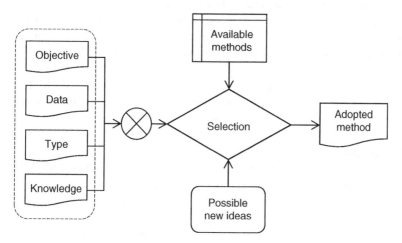

Figure 4.10 Factors of research method selection.

Table 4.5 Research objectives and methods.

Objective	Research type	Common method
(1) To generate understanding or principle	Basic (Scientific)	Qualitative
(2) To advance understanding or principle	Basic and Applied	Qualitative (as well as quantitative)
(3) To apply understanding or principle	Applied	Quantitative
(4) To develop new product or process	R&D	Qualitative (direction) and quantitative (details)

Guided by objectives, the nature of research governs the nature of the depth and breadth of the research process. Research objectives may be to explore a new discovery, explain a certain phenomenon, develop a new research method, and so on. For example, Industrial Engineering (IE) leans more toward developing, improving, and implementing integrated systems to offer solutions and optimization for problems that emerge within operations of industries and services. Accordingly, research methods, such as mathematical modeling, survey, theoretical, simulation, field study, case study, and laboratory experimentation, are often used.

However, there is probably no universal answer what methods should be selected for a certain research project. A major reason is that research can have various, mixed objectives (e.g. advance understanding and apply it to real-world problems).

The objectives, types of research, and method used are often associated with the type of data. Table 4.5 provides a general reference. Engineering research projects frequently aim at (2), (3), and (4) objectives.

4.4.1.2 Data Based

The data play a crucial role in the method selection, as a method is to use for data collection and analysis. In other words, data and methods can go hand in hand in engineering and technical studies. For example, we may use a linear regression and other types of statistical analysis for continuous data. Sometimes, we need to work on their relations and determine the best pair of data and method.

The type of data also decides what instrument we should use to collect the data. For example, transmission engineers are concerned with studying vehicle transmission controls and clutch pressures that are measured with transducers or strain gages.

Interestingly, the research objectives, data, and method are associated with each other and intertwined. For the particularly data, multiple methods are often applicable. For example, in the operation management of manufacturing, researchers analyzed over 8600 abstracts of the papers published in International Journal of Production Research for 55 years (Manikas et al. 2017) and they found that the six types of data generation were simulation, primary/secondary data, case study, mathematical modeling, experiment, and review. They also found the six types of data analysis methods: statistical method, meta-heuristic, data mining, optimization, comparative method, and typology. In recent years, mathematical modeling has been a leading data generation method. While, meta-heuristic and optimization are the main methods, about 50% and 40%, respectively.

4.4.1.3 Various Process Steps

We discuss the overall process of research in Chapter 1 (refer to Figure 1.2). The detailed steps and tasks of a research process can have a significant variance. Here, we use a process called action research for a research process discussion.

The unique characteristics of action research are to go through the simultaneous process of taking informed actions and doing research to seek transformative change. From this point, action research is a kind of comparative investigation on the conditions and effects of various forms of actions and research.

The corresponding effects and results can be different with changing actions or practices. Therefore, the process of action research is a loop with multiple cycles or a spiral of steps due to the feedback from improved understanding and results. The data analysis in action research is iterative and a step of data analysis may lead to new or further questions and actions. Figure 4.11 shows the process cycle of action research. The main difference is the embedded tasks in the process compared with Figure 1.2 in Chapter 1.

Action research has been used in engineering and technical research projects. Here are a few examples.

"Using Inertial Measurement Units and Electromyography to Quantify Movement during Action Research Arm Test Execution" (Repnik et al. 2018)
"A Maturity Assessment Model for Manufacturing Systems" (Vivares et al. 2018)
"Condition Monitoring for Airport Baggage Handling in The Era of Industry 4.0" (Koenig et al. 2019)

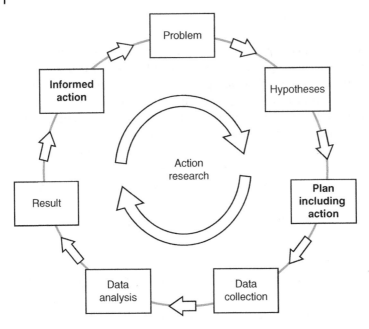

Figure 4.11 A process of action research.

"Digital Manufacturing Applicability of a Laser Sintered Component for Automotive Industry: A Case Study" (Ituarte et al. 2018)

If working in the operations in industries, researchers may be familiar with a common process of study called Plan–Do–Check–Act (PDCA). The PDCA has been widely used for the continuous improvement of processes and products and industry R&D. The process is simple but effective to explore new understanding and solve difficult problems. The general research process introduced in Chapter 1, the action research discussed above, and PDCA all follow a similar iterative route.

4.4.2 Qualitative and Quantitative

Generally, we categorize research methods as qualitative, quantitative, and mixed based on data types. First, let us check the differences between qualitative and quantitative methods based on the review of qualitative and quantitative data in Section 4.2.2.

4.4.2.1 Qualitative vs. Quantitative

Selecting analysis methods is based on the type of data. Quantitative research analysis is often concerned with the relationship between variables and explains what

Table 4.6 Characteristics of qualitative and quantitative analyses.

Characteristics	Qualitative	Quantitative
Origin	Art and social science	Natural science
Researcher's knowledge	Rough idea	Clear understanding
Question	What, why, how	How many, when, where
Data sampling	Purposeful	Probabilistic/random
Format	Words, image, etc.	Numerical data
Reasoning (epistemology)	Empiricism/induction	Rationalism/deduction
Results	Interpretation	Detailed and structured
Extension	Non-generalizable	Maybe generalizable
Nature	More subjective	More objective
Strength	Observational	Statistical

are observed with counting and classifying features, such as using statistics. On the other hand, qualitative analysis seeks to understand phenomena and explanation in depth. If both qualitative and quantitative data are available, we have more options for method selection and more avenues for research to explore.

There are some perspectives regarding the characteristics of both methods, refer to the following characteristic comparison (Table 4.6). Some characteristics may have exceptions and not be agreed by all professionals.

Qualitative and quantitative methods are not mutually exclusive and can be complementary to one another. We may use qualitative methods to understand or improve our understanding the meaning of the numbers obtained by quantitative methods and to provide in depth explanations to quantitative research questions.

The researchers in different areas, such as science, engineering, medical, or social sciences, may have different mature methods and practices. Even though quantitative methods are generally preferred for engineering and technical professionals, qualitative methods also play an important role, particularly for large research projects. Qualitative research can serve as a guide to quantitative research.

Some types of studies are normally conducted as qualitative, like survey studies. However, depending on particular approach, they may also be conducted as quantitative. When choosing analysis methods based on the available types of data, we may also think if other types of data and corresponding methods can be helpful to pursue our research goals. Many projects use both types, which are called mixed-method research and will be discussed in depth in Chapter 6.

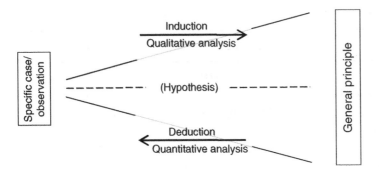

Figure 4.12 A diagram of induction vs. deduction.

4.4.2.2 Induction vs. Deduction

There are two basic approaches to acquiring knowledge. One is called empiricism, in which human beings gain knowledge by experience and using inductive reasoning. The other approach is rationalism: gaining knowledge by using deductive reasoning based on existing knowledge and new observations.

Inductive reasoning starts from specific data and develops a general conclusion, such as a new theory and model. From a problem, we may induct to a systematic observation. This approach is exploratory rather than confirmatory. Using this approach, we may develop a theory or principle rather than use it. In other words, we induct from specific observations to generalize how that thing works.

While using a deductive reasoning, we start with a theory and general ideas (called premises) and apply them to a specific situation to get a conclusion through logical argument. Often, we predict an outcome of a special case.

Induction reasoning is generally associated with qualitative research, while a deductive approach is commonly associated with quantitative analysis. Figure 4.12 shows a diagram of induction and deduction in research. Furthermore, it is possible that we jointly use both types of data analysis method and reasoning in a research project. For example, "Towards understanding requirements engineering in IT ecosystems" (Knauss et al. 2012). "Cladding materials in non-residential construction: choice criteria for stakeholder in the Province of Quebec" (Guy-Plourde et al. 2018).

In engineering applied research and R&D, deductive approaches are widely used with the known general principles and theories. While when approaching a topic with little background on the subject, induction is more likely to be used. In a word, which reasoning approach to use depends on research objective. Readers may refer to relevant literatures, for instant, *How to Get a Ph.D.: Methods and Practical Hints* (Mämmelä 2009) that provides a summary.

Table 4.7 Similar criteria of method evaluation.

Qualitative	Quantitative
Transferability	Generalizability
Credibility	Validity
Reflexivity	Objectivity
Dependability	Reliability

It is important to know that the conclusion using deductive reasoning is certain only when the premises or propositions are true. It is possible to reach a logical conclusion when the premises are not true. In other words, when the premises are wrong, the conclusion may be logical, but could be false.

4.4.2.3 Method Evaluation

Based on the characteristics of data, the two types of methods can be compared. Table 4.7 lists the pairs of similar characteristics. Some characteristics have similar meanings, such as transferability and generalizability. It might not be necessary to identify their subtle differences, but it is recommended to use the common terminology in proposal and in manuscript for publication.

Moreover, Figure 4.13 shows the main types of research methods and their characteristics in terms of qualitative and quantitative spectrums in general. Using both types of methods, or mixed-methods, in a project may be a bit more complex, which will be discussed more in Chapter 6.

4.4.3 Other Considerations

4.4.3.1 Knowledge and Preference

In addition to research objectives and data availability, other factors, such as the personal knowledge and experiences of researchers and the audience are important as well (Creswell 2002). It is true that many researchers choose a path they are most comfortable with and prefers.

The personal experiences and preference of researchers can be either an advantage or a disadvantage to research innovation. For students, they can take their advisor's suggestions as a starting point to have a good likelihood of student's initial research success.

However, a familiar method would not be necessarily the best method for a new research project. Always using the existing, common methods can limit the innovation and critical thinking of students and new researchers.

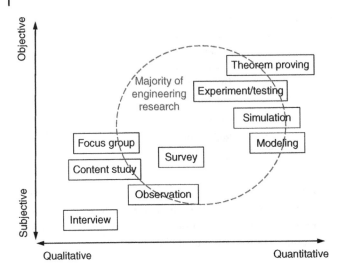

Figure 4.13 Characteristics of research methods.

As mentioned before, most engineering and technical researchers and audience generally favor quantitative methods and experimental approaches due to their education background. One study showed a trend towards the use of more quantitative methods (Borrego et al. 2009). If using keywords ("quantitative analysis" engineering) and ("qualitative analysis" engineering), Google Scholar showed up 2 040 000 and 371 000 publications, respectively, on June 4, 2019, indicating that engineering quantitative analysis are about 5.4 times more common than those based on qualitative analysis.

4.4.3.2 Possibility of Different Methods

Many research projects are not built from the ground up but rather evolved continuously over the course of a series of projects by numerous researchers. As mentioned before, it may be a good idea to try a different method even on the same research topics. Readers should feel encouraged to investigate alternate research methods used in other well-established research areas, such as medicine research, to engineering and technical studies.

Applying different methods across disciplines has good potentials for significant benefits. The methods or principles that are widely implemented in one field may be applicable to another field. Some techniques, methods, and procedures in one field may be considered for applications in different technical fields. For example, we may consider a project whose objective is to apply method X, which has been successful in field A, to solve problem Y in the field B. Here are a couple of examples:

"We adopt a method which combines both types, aiming to expand the human thinking spaces through the support of computers. Thus, humans could generate ideas they did not previously have, thereby improving their ability of imagination." (Yamada et al. 2019)

"Thus, in addition to the mesh-based perspective (homography) transformation, we adopt a method for diffeomorphic registration/warping [23]. The diffeomorphic registration reduces the alignment errors caused by large parallax, as in Fig. 1(d)." (Jacob and Das 2018)

4.4.3.3 Purposeful Data Selection

When developing research proposals, we have objectives and often have predicted results. Having a clear aim is very important to proposal as it serves a guide in the course of research execution. However, the data collected and the results from the research executions may not always meet the expectations. It can be challenging to accept unexpected results and report them.

When trying to prove something to be true but the data suggest otherwise, it is possible but unethical to interfere in data selection or a test in a way that would skew the results, and thus sway the outcomes in an expected favor.

For example, to get desired outcomes, researchers may intentionally select a specific data set for a study. Another possibility is that researchers may keep running additional tests until obtaining desired results, instead of completing the test and drawing a conclusion as planned. For such situations, the researchers intentionally try to have and only report on the "positive" results. Even through the report is still based on the factual results, it is incomplete and possibly misleading if without reporting and discussion on the failed tests. This type of "cherry-picking" practice in data selection and result reporting, sometimes called research bias or probably researcher bias, is harmful and may encounter ethical issues because of its intentionality, which will be discussed more in Chapter 7.

4.4.3.4 Non-data-related Research

Most of research in engineering and technical fields is data driven or data related, and research method selection is closely related to the data. However, there are a few exceptions – some types of research may be conducted not based on data or with no data analysis at all.

Research on theory, methodology, and procedure is to create new ones, which may be from innovative insight or theoretical derivation but with no data. Here are two examples.

"The Second-Order Adjoint Sensitivity Analysis Methodology for Nonlinear Systems—I: Theory" (Cacuci 2016) The author *"presents the second-order adjoint sensitivity analysis methodology (2nd-ASAM) for nonlinear systems,*

which yields exactly and efficiently the second-order functional derivatives of physical system responses to the system's model parameters."

"Derivation and tuning of a solvable and compact differential–algebraic equations model for LiFePO$_4$–graphite Li–ion batteries" (Lee et al. 2018) The authors *"presents a procedure for deriving and tuning a solvable and compact differential–algebraic equation (DAE) model for the LiFePO4–graphite lithium–ion battery cell."*

However, it is likely that additional work with data is needed to validate or prove the new methodology later on. Researchers, original inventors, or other professionals can apply the new approach by using existing information to show the advantage of using the new technique. For instance, after talking about the structure of the proposed methodology in details, the authors used an example (type of data) to illustrate the use of the methodology and did empirical evaluation (Rodriguez-Martinez et al. 2019). For another example, the authors presented "a new two-steps pre-heat treatment procedure of carbonate precursors." and used the procedure to cases (Zhang et al. 2015). In other words, data was used to support the new procedure presented. Therefore, regardless the type of data roles in engineering and technical research, they are an integral part of research.

Summary

Data in Research

1. Data play a critical role in research. Most research is data driven and research outcomes need data to verify and validate.
2. Data management includes the creation, administration, storage and security, publication, and maintenance of data.
3. Data science is based on applied statistics. It can be a research subject and more often the method for other research subjects.
4. There are several data distributions, such as normal, exponential, Weibull, etc., commonly used in engineering research.
5. Considerations for research data include their source, elusiveness, conditions, and ephemerality.
6. Main steps of data handling in research are preparation, analysis, and interpretation.

Types of Data

7. *There are two types of data in terms of data sources*: primary (to collect new) data and secondary (to use existing) data. Open data is a type of secondary data, which have various sources.
8. The data can be either quantitative (numerical) or qualitative (in formats of words, etc.)
9. The quality of quantitative data can be measured on their accuracy, repeatability, and reproducibility.
10. The scale of data can be categorized into normal, ordinal, binary, interval, and ratio.

Data Collection

11. Collected data samples are the subset of the accessible data set, which is a subset of the entire data set.
12. There are several methods to determine sample size.
13. Probability (or random) data sampling methods should be used when feasible, which include simple, systematic, stratified, and cluster methods.
14. Non-probability data sampling methods include convenience, quota, purposive, expert, and snowball methods.
15. Non-probability data sampling has some advantages but is lack of representation of the entire population.

Method Selection

16. Data type is a determinative factor for research method selection. Another factor is the type of research (basic, applied, or R&D).
17. Knowledge, experience, personal preference, and available resources are other factors for research method selection.
18. Based on data types, research methods are categorized as qualitative, quantitative, and mixed methods.
19. Quantitative research analysis is often concerned with the relationship between variables. While qualitative analysis seeks to understand phenomena and explanation.
20. Two basic types of reasoning approaches are induction and deduction. The former is to obtain general principle from special cases primarily based on

qualitative methods; the latter is to use general principles for special cases using quantitative analysis.

Exercises

Review Questions

1 Dr. W. Edwards Deming said, without data you are just another person with an opinion. Discuss the importance of data in a research project.
2 Find an example of the application of data science in your field.
3 Discuss the factors, such as availability, cost, and permission needed, for data collection in your project.
4 Explain the data management of a research proposal with an example.
5 Discuss the considerations for data collection and usage.
6 Use an example to discuss the application of data distribution.
7 Compare the pros and cons of primary data and secondary data for research with an example.
8 Compare quantitative data and qualitative data and discuss which type of data more often used in your field.
9 How to evaluate the quality of data?
10 Based on the characteristics of different types of probability sampling, which type is likely to be used in your research project.
11 Discuss the considerations to determine sample size.
12 Talk about sampling bias, instrumentation bias, or research bias with an example.
13 List the considerations or influencing factors for data sampling.
14 Discuss potential issues in research when using a non-probability sampling method.
15 If a random sampling is not practical, which type of non-probability sampling method you like to use in your research project.
16 Discuss the meaning of Type I and Type II errors. Which one do you think is the more serious to a special case?
17 There are key factors (such as research objective, qualitative vs. quantitative, method characteristics, etc.) for research method selection. Which one is the most important and why for your research project.
18 Some data are solely applicable to qualitative or quantitative methods. Can some data be used in the other type research method?
19 List a few of pros and cons of selecting non-conventional methods in your research.

20 Explain inductive and deductive reasoning approaches with examples.

21 Find an example of research work (or paper) without using data.

Mini-project Topics

1 Think about the data in your new project on
   (a) Data availability and accessibility
   (b) Plan of data collection, including recourses
   (c) Data sampling considerations
   (d) Consideration of data reliability

2 Search one paper using secondary data and review for the appropriateness of using such data.

3 Search one paper using non-probability sampling and review for the appropriateness of using such sampling method.

4 One states that sampling bias is virtually unavoidable. If it is true, how we consider and handle the sampling bias in our research. Please explain with an example of a published research paper.

5 Select a potential research subject and search recent publications on the data distributions used for studying such a subject.

6 One states that research objective, data to collect, and methods to use are inextricably intertwined. Analyzing a research paper or project to support such a statement with explanations.

7 Search a recent scholarly paper and analyze its:
   (a) Data type – quantitative or qualitative
   (b) Data analysis to the research conclusion
   (c) Data validity and reliability

8 In your discipline, search recent scholarly papers and find out the popular and uncommon pairs between research method and data, refer to Table 4.8.

Table 4.8 Relation analysis between research data and methods.

Method	Data				
	Experiment	Simulation	Samples	Secondary	... ...
Statistics					
Optimization					
Modeling					
Data mining					
... ...					

9 Use "research method" in quotation marks as one of your keywords to search the literature in recent five years in your discipline. Select five interesting papers and summarize the methods they used.

10 Find a research paper that is on methodology development rather than being based on data. Assess the role of data in such a study.

11 Selecting a research method can be based on personal experience, skill, and/or preference. Can the bias emerge in the form of method selection?

12 In your research field, what is the most popular method used? Do you think the method is the best effective? Provide your insights with justification.

References

Adamy, A. and Bakar, A.H.A. (2018). Key criteria for post-reconstruction hospital building performance. *IOP Conference Series: Materials Science and Engineering* 469: 012072. https://doi.org/10.1088/1757-899X/469/1/012072.

Agostino, I., Sousa, S., Frota, P. et al. (2018). Modeling and simulation of operations: a case study in a port terminal of vale S/A. In: *New Global Perspectives on Industrial Engineering and Management. Lecture Notes in Management and Industrial Engineering* (eds. J. Mula, R. Barbastefano, M. Díaz-Madroñero and R. Poler). Cham: Springer https://doi.org/10.1007/978-3-319-93488-4_11.

Aririguzo, J. and Ekwe, E. (2019). Weibull distribution analysis of wind energy prospect for Umudike, Nigeria for power generation. *Robotics and Computer-Integrated Manufacturing* 55: 160–163. https://doi.org/10.1016/j.rcim.2018.01.001.

Arnouts, H. and Goos, P. (2017). Analyzing ordinal data from a split-plot design in the presence of a random block effect. *Journal of Quality Engineering* 29 (4) https://doi.org/10.1080/08982112.2017.1303069.

Billah, M.M., Goodall, J.L., Narayan, U. et al. (2016). Using a data grid to automate data preparation pipelines required for regional-scale hydrologic modeling. *Environmental Modelling and Software* 78: 31–39. https://doi.org/10.1016/j.envsoft.2015.12.010.

Borrego, M., Douglas, E.P., and Amelink, C.T. (2009). Quantitative, qualitative, and mixed research methods in engineering education. *Journal of Engineering Education* 98 (1): 53–66.

Cacuci, D.G. (2016). The second-order adjoint sensitivity analysis methodology for nonlinear systems—I: theory. *Nuclear Science and Engineering* 184 (1): 16–30. https://doi.org/10.13182/NSE16-16.

Cánovas, M.J., Hall, J.A.J., López, M.A. et al. (2019). Calmness of partially perturbed linear systems with an application to the central path. *Optimization* 68 (2–3): 465–483. https://doi.org/10.1080/02331934.2018.1523403.

Creswell, J.W. (2002). *Research Design: Qualitative, Quantitative, and Mixed Methods Approaches*. New York: Sage Publications.

Davidson, I. and Tayi, D. (2009). Data preparation using data quality matrices for classification mining. *European Journal of Operational Research* 197 (2): 764–772. https://doi.org/10.1016/j.ejor.2008.07.019.

Diky, V., Bazyleva, A., Paulechka, E. et al. (2019). Validation of thermophysical data for scientific and engineering applications. *The Journal of Chemical Thermodynamics* 133: 208–222. https://doi.org/10.1016/j.jct.2019.01.029.

Donoho, D. (2017). 50 years of data science. *Journal of Computational and Graphical Statistics* 26 (4): 745–766. https://doi.org/10.1080/10618600.2017.1384734.

Doumpos, M., Andriosopoulos, K., Galariotis, E. et al. (2017). Corporate failure prediction in the European energy sector: a multicriteria approach and the effect of country characteristics. *European Journal of Operational Research* 262 (1): 347–360. https://doi.org/10.1016/j.ejor.2017.04.024.

Du, W., Liao, G., Yang, Z. et al. (2016). Robust space time processing based on bi-iterative scheme of secondary data selection and PSWF method. *Digital Signal Processing* 52: 64–71. https://doi.org/10.1016/j.dsp.2016.01.016.

Ercanoglu, M., Dağdelenler, G., Özsayin, E. et al. (2016). Application of Chebyshev theorem to data preparation in landslide susceptibility mapping studies: an example from Yenice (Karabük, Turkey) region. *Journal of Mountain Science* 13 (11): 1923–1940. https://doi.org/10.1007/s11629-016-3880-z.

Franch, X, Ralyté, J., Perini, A. et al. (2018). A situational approach for the definition and tailoring of a data-driven software evolution method. International Conference on Advanced Information Systems Engineering CAiSE 2018: Advanced Information Systems Engineering, Vol. 10816, 603–618, doi: 10.1007/978-3-319-91563-0_37

Gadde, S., Rosaiah, K., and Prasad, S. (2019). Bootstrap confidence intervals of CNpk for type-II generalized log-logistic distribution. *Journal of Industrial Engineering International* 15 (Suppl 1): S87–S94. https://doi.org/10.1007/s40092-019-0320-z.

German, E.S. and Rhodes, D.H. (2017). Model-centric decision-making: exploring decision-maker trust and perception of models. 15th Annual Conference on Systems Engineering Research Disciplinary Convergence: Implications for Systems Engineering Research, Redondo Beach, CA (23–25 March 2017) doi: 10.1007/978-3-319-62217-0_57.

GÜL, S., UÇAR, M.K., ÇETINEL, G. et al. (2017). Automated pre-seizure detection for epileptic patients using machine learning methods. *International Journal of Image, Graphics and Signal Processing* 9 (7): 1–9. https://doi.org/10.5815/ijigsp.2017.07.01.

Guy-Plourde, S., Blanchet, P., Blois, M. et al. (2018). Cladding materials in non-residential construction: choice criteria for stakeholder in the Province of Quebec. *Journal of Facade Design and Engineering* 6 (1): 1–18. https://doi.org/10 .7480/jfde.2018.1.1811.

Ituarte, I.F., Chekurov, S., Tuomi, J. et al. (2018). Digital manufacturing applicability of a laser sintered component for automotive industry: a case study. *Rapid Prototyping Journal* 24 (7): 1203–1211. https://doi.org/10.1108/RPJ-11-2017-0238.

Jacob G.M. and Das S. (2018). Large parallax image stitching using an edge-preserving diffeomorphic warping process. Advanced Concepts for Intelligent Vision Systems, ACIVS 2018 International Conference on Advanced Concepts for Intelligent Vision Systems, Vol. 11182, 521–533, https://doi: 10.1007/978-3-030-01449-0_44.

Jarocka, M. and Glińska, E. (2017). The state and prospects for development of railway transport infrastructure in eastern Poland – secondary data analysis. *Procedia Engineering* 182: 299–305.

Jeyadurga, P., Usha Mahalingam, U., and Balamurali, S. (2018). Modified chain sampling plan for assuring percentile life under Weibull distribution and generalized exponential distribution. *International Journal of Quality and Reliability Management* 35 (9): 1989–2005. https://doi.org/10.1108/IJQRM-02-2018-0044.

Joppen, R., Enzberg, S., Kühn, A. et al. (2019). Data map – method for the specification of data flows within production. *Procedia CIRP* 79: 461–465. https:// doi.org/10.1016/j.procir.2019.02.127.

Karbasian, M., Kashani, M., Khayambashi, B. et al. (2016). Designing a local flexible model for electronic systems acquisition based on systems engineering, case study: electronic high-tech industrial. *International Journal of Industrial Engineering and Production Research* 27 (2): 149–166.

Khan, M.A.A., Butt, J., Mebrahtu, H. et al. (2018). Data-driven process reengineering and optimization using a simulation and verification technique. *Designs* 2: 42. https://doi.org/10.3390/designs2040042.

Kim, J. and NO, H.C. (2019). Model development for fragment-size distribution based on upper-limit log-normal distribution. *Nuclear Engineering and Design* 349: 86–91. https://doi.org/10.1016/j.nucengdes.2019.04.029.

Knauss, A., Borici, A., Knauss, E. et al, (2012). Towards understanding requirements engineering in IT ecosystems. Second IEEE International Workshop on Empirical Requirements Engineering (EmpiRE), Chicago, IL, 33–36, doi: 10.1109/EmpiRE.2012.6347679.

Koenig, F., Found, P.A., Kumar, M. et al. (2019). Condition monitoring for airport baggage handling in the era of industry 4.0. *Journal of Quality in Maintenance Engineering* 25 (3): 435–451. https://doi.org/10.1108/JQME-03-2018-0014.

Lee, C.W., Hong, Y., Hayrapetyan, M. et al. (2018). Derivation and tuning of a solvable and compact differential–algebraic equations model for LiFePO$_4$–graphite Li–ion batteries. *Journal of Applied Electrochemistry* 48: 365–377. https://doi.org/10.1007/s10800-018-1164-8.

Li, Y., Roy, U., and Saltz, J.S. (2019). Towards an integrated process model for new product development with data-driven features (NPD[3]). *Research in Engineering Design* 30 (2): 271–289. https://doi.org/10.1007/s00163-019-00308-6.

Mämmelä, A. (2009). How to get a Ph.D.: methods and practical hints. II International Interdisciplinary Technical Conference of Young Scientists, (20–22 May 2009), Poznań, Poland.

Manikas, A., Boyd, L., Pang, Q. et al. (2017). An analysis of research methods in IJPR since inception. *International Journal of Production Research* https://doi.org/10.1080/00207543.2017.1362122.

Mazloum, A., Oddone, V., Reich, S. et al. (2019). Connection between strength and thermal conductivity of metal matrix composites with uniform distribution of graphite flakes. *International Journal of Engineering Science* 139: 70–82. https://doi.org/10.1016/j.ijengsci.2019.01.008.

Nanishi, L., Green, J., Taguri, M. et al. (2015). Determining a cut-off point for scores of the breastfeeding self-efficacy scale–short form: secondary data analysis of an intervention study in Japan. *PLoS One* 10 (6): e0129698. https://doi.org/10.1371/journal.pone.0129698.

Ngilangil, L.E., Vilar, D.A., Andrada, J.C. et al. (2018). Geographic mapping of tube wells and assessment of saltwater intrusion in the coastal areas of La union. *Chemical Engineering Transactions* 63: 55–60. https://doi.org/10.3303/CET1863010.

NIU (n.d.). Data analysis, Northern Illinois University, https://ori.hhs.gov/education/products/n_illinois_u/datamanagement/datopic.html (accessed April 2019).

NSF (n.d.). Dissemination and sharing of research results, proposal & award policies & procedures guide (PAPPG) chapter XI.D.4. https://www.nsf.gov/bfa/dias/policy/dmp.jsp (accessed June 2019).

NSF, a (n.d.). ENG guidance on data management plans, https://nsf.gov/eng/general/ENG_DMP_Policy.pdf (accessed April 2019).

Prasojo, S. and Prasetyoputra, P. (2019). PSO-KS algorithm for fitting lognormal distribution: simulation and empirical implementation to women's age at first marriage data. *IOP Conference Series: Materials Science and Engineering* 546 https://doi.org/10.1088/1757-899X/546/5/052052.

Qahtani, H.A., El–Hefnawy, A., El–Ashram, M.M. et al. (2019). A goal programming approach to multichoice multiobjective stochastic transportation problems with extreme value distribution. *Advances in Operations Research* 2019: 6. https://doi.org/10.1155/2019/9714137.

Repnik, E., Puh, U., Goljar, N. et al. (2018). Using inertial measurement units and electromyography to quantify movement during action research arm test execution. *Sensors* 18 (9): 2767. https://doi.org/10.3390/s18092767.

Rodriguez-Martinez, L.C., Duran-Limon, H.A., Mora, M. et al. (2019). SOCA-DSEM: a well-structured SOCA development systems engineering methodology. *Computer Science and Information Systems* 16 (1): 19–44. https://doi.org/10.2298/CSIS170703035R.

Ruiz, P. and Maier, A. (2017). Data-driven engineering design research: opportunities using open data. Proceedings of the 21st International Conference on Engineering Design (ICED17), Design Theory and Research Methodology, Vol. 7, 41–51.

Shields, M.D. (2015). Refined stratified sampling for efficient Monte Carlo based uncertainty quantification. *Reliability Engineering and System Safety* 142: 310–325. https://doi.org/10.1016/j.ress.2015.05.023.

Silverman, D. (2005). *Doing Qualitative Research: A Practical Handbook*, 2ee. London: Sage Publications.

Souza, M.R., Haines, R., Vigo, M. et al. (2019). What makes research software sustainable? An interview study with research software engineers, Cornell University, arXiv:1903.06039 (accessed April 2019).

Torre, J., Puig, D., and Valls, A. (2018). Weighted kappa loss function for multi-class classification of ordinal data in deep learning. *Pattern Recognition Letters* 105: 144–154. https://doi.org/10.1016/j.patrec.2017.05.018.

Vivares, J.A., Sarache, W., Hurtado, J. et al. (2018). A maturity assessment model for manufacturing systems. *Journal of Manufacturing Technology Management* 29 (5): 746–767. https://doi.org/10.1108/JMTM-07-2017-0142.

Wu, P., Xu, Y., Jin, R. et al. (2019). Perceptions towards risks involved in off-site construction in the integrated design & construction project delivery. *Journal of Cleaner Production* 213: 899–914. https://doi.org/10.1016/j.jclepro.2018.12.226.

Yamada, K., Ito, S., and Taura, T. (2019). A method for designing complicated emotional three-dimensional geometrical shapes through mathematical extrapolation. *Emotional Engineering* 7: 137–147. https://doi.org/10.1007/978-3-030-02209-9_9.

Zhang, Y., Hou, P., Zhou, E. et al. (2015). Pre-heat treatment of carbonate precursor firstly in nitrogen and then oxygen atmospheres: a new procedure to improve tap density of high-performance cathode material $Li_{1.167}(Ni_{0.139}Co_{0.139}Mn_{0.556})O_2$ for lithium ion batteries. *Journal of Power Sources* 292: 58–65. https://doi.org/10.1016/j.jpowsour.2015.05.036.

Zhang, X., Wang, J., Ji, S. et al. (2019). An I/O efficient distributed approximation framework using cluster sampling. *IEEE Transactions on Parallel and Distributed Systems* https://doi.org/10.1109/TPDS.2019.2892765.

Zhao, Y., Peng, X., Yang, B. et al. (2019). Dual phase nano-particulate AlN composite — A kind of ceramics with high strength and ductility. *Ceramics International* 45 (16): 19845–19855. https://doi.org/10.1016/j.ceramint.2019.06.239.

Zhou, G., Ye, Y., Zuo, W. et al. (2018). Fast and efficient prediction of finned-tube heat exchanger performance using wet-dry transformation method with nominal data. *Applied Thermal Engineering* 145: 133–146. https://doi.org/10.1016/j.applthermaleng.2018.09.020.

5

Quantitative Methods and Experimental Research

5.1 Statistical Analyses

The objectives of technical research focus on solving real-world problems, such as finding a causational relationship, explaining a phenomenon, and predicting general cases for a larger population. For example, many researchers seek to understand how certain factors affect the outcomes of a physical system. In such research, numerical data are normally available and thus quantitative methods are often used.

Among many quantitative methods, we often do calculation and analysis based on statistical principles. Bear in mind when we discuss a statistical analysis, there are other types of quantitative analysis in engineering and technology, such as a mathematical model, algorithm, differential equations, and dynamic neural network. In addition, we may jointly use multiple quantitative methods in a study.

5.1.1 Descriptive Statistical Analysis

5.1.1.1 Overview of Statistical Analysis

The basic statistics and analysis methods are in college undergraduate curricula for all technical majors and most majors of social studies. Statistical analysis is a computational procedure that enables us to explore the meanings of the data collected. Many user-friendly computer software packages may also help us perform such an analysis. With a good understanding of the principles, we can rely on software for a detailed data analysis.

There are two major types of statistical analysis: descriptive statistics and inferential statistics, refer to Figure 5.1. Briefly, descriptive statistics is to summarize and reveal the information about a given data set, present the data in a meaningful way, and come with interpretations. The conclusions from descriptive statistics stay within the data or only applicable to the sample data set.

Engineering Research: Design, Methods, and Publication, First Edition. Herman Tang.
© 2021 John Wiley & Sons, Inc. Published 2021 by John Wiley & Sons, Inc.

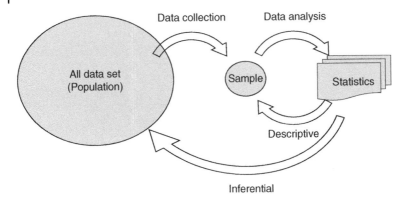

Figure 5.1 Descriptive and inferential statistical analysis.

On the other hand, inferential statistics uses the statistics to analyze a sample of data obtained from a population to make inferences about the population. The conclusions based on inferential statistics can be interesting and useful, which may better fit many research goals.

Descriptive statistics can play a supportive role to a research project that primarily uses inferential statistics. Some researchers use both types of analysis in their projects. For example,

> *"Dependency of human target detection performance on clutter and quality of supporting image analysis algorithms in a video surveillance task"* (Huber et al. 2017)
> *"Evaluation and comparison of a lean production system by using SAE J4000 standard: a case study on the automotive industry in the state of Mexico"* (Tabares et al. 2017)
> *"Measuring business performance in a SCN using Six Sigma methodology–a case study"* (Mishra and Sharma 2017)

5.1.1.2 Purposes of Descriptive Analysis

A descriptive study is to reveal the characteristics of the data collected. In general, a descriptive study does not address any relationships between variables. This approach may be useful in the cases when a topic is not well known.

With descriptive statistics, there is no uncertainty because we describe only the data measured. So we can verify the descriptive statements from the information provided. However, we should not use the statements to generalize and apply for any other groups or the entire population.

The descriptive statistics include the central tendency and variability in terms of means, standard deviations, median, mode, range, percentage, Chi-square test, z-score, correlation coefficients, etc. Some of them are briefly summarized below; readers can refer to statistical analysis books for the more and detailed information.

5.1.1.3 Central Tendency

We may use a central tendency to describe and compare data sets. Analysis of data central tendency is to find a "central" point of a data set (x_i, $i = 1$, ..., n) and is measured and represented by one or more of the following parameters:

- *Mode* is the single number that occurs most frequently in a data set.
- *Median* is the "middle" number of a set of an ordered data set. If a data set is an odd number, the median is the middle value. For an even number, the median of a data set is the average of the two middle values.
- *Mean* ($\bar{x}$) is usually referred to an arithmetic average ($\bar{x} = \frac{x_1 + x_2 + \cdots + x_n}{n} = \frac{\sum x_i}{n}$) of a data set with n samples. For an entire population, we often use a Greek letter μ for the mean parameter.
- A geometric mean ($\bar{x}_g = \sqrt[n]{x_1 \times x_2 \times \cdots \times x_n} = \sqrt[n]{\prod x_i}$) is used for some situations, such as numbers that are in different ranges and the differences among data points that are logarithmic.

5.1.1.4 Variability

Data variability shows that data points cluster around the point of central tendency and comes from data sources and measurements. Some parts of a variability are the nature of random noise since all measurement instruments and procedures are subject to random noise.

We may measure the data variability using the following:

- *Range (r)* is a simple way to reflect the spread of a data set from its lowest value to its highest value. That is, $r = \max\{x_i\} - \min\{x_i\}$, where, $i = 1$, ..., n.
- *Standard deviation (S)* is the most common way to measure of the variability of a data set. The standard deviation is $S = \sqrt{\frac{\sum (x_i - \mu)^2}{n-1}}$. For an entire data population, we normally use Greek letter σ to present the variance parameter.
- *Average deviation* is also called the mean absolute deviation. It is the average of differences of each value and the mean average for a sample data set is $\frac{\sum |x_i - \mu|}{n}$.

In addition, we may use graphic displays, such as a histogram, pie chart, and standardization error bar chart to show the variability.

5.1.2 Inferential Statistical Analysis

5.1.2.1 Characteristics of Inferential Analyses

In inferential statistics, we use similar calculations as in descriptive analysis, for example, the mean and standard deviation. However, for inferential analysis, we aim to generalize the parameters and characteristics along with a predefined degree of confidence. That is, we use a sample data set to make a generalization about the parameters of a given population.

To have a good inference, we must make sure the samples truly represent the population. This question also goes back to Chapter 4 where we discussed sample size.

We may be able to not verify an inferred statistical statement based on the sample information. However, we can evaluate the prediction with a level of confidence. The interval of numbers is an estimated range of values calculated from the sample data. A confidence interval is expressed in terms of an interval and the degree of confidence that the parameters are within the interval. For example, an interval of a mean value with 95% confidence is 55 ± 3. For the parameters of a data set (e.g. its mean and standard deviation), there are estimation formulas under different assumptions.

5.1.2.2 Data Association

In many cases, we need to examine whether variables are interrelated and have a possible causational relationship. A simple step of checking the possible relationship between two variables is by using scatter plots, which can graphically show if there is any overall pattern.

Correlation analysis is an important approach for the analysis of data association. For example, we may do a correlation analysis to identify a linear relationship between two variables. The result of a correlation analysis is a single number called correlation coefficient (r). The coefficient r indicates mathematical correlation direction varying from -1 to $+1$, as shown in Figure 5.2. The two variables are r^2% related with each other mathematically. For example, if r is $|0.6|$, then the linear correlation is a small, as $0.6^2 = 36\%$, between the two data sets. Sometimes, there is a possible nonlinear interrelationship for a low r, which may need an additional analysis.

Importantly, it is essential to note that data correlation itself does not necessarily mean a causational relationship. The variables that are mathematically correlated to each other but have no meaningful physical relationship is called a spurious relationship. Thus, with a high correlation coefficient, we may need to investigate in depth and interpret the possible causational relationship. We may create a new understanding or a prediction if we can reveal a causational relationship.

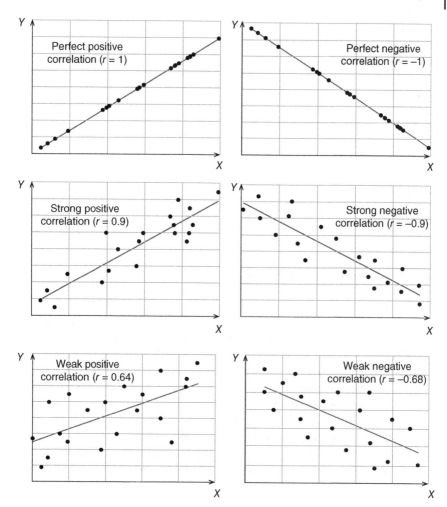

Figure 5.2 Correlations between two sets of data.

The correlation analysis is often used combined with other analyses in engineering research, for example, "Automated modal parameter estimation using correlation analysis and bootstrap sampling" (Yaghoub et al. 2018).

5.1.2.3 Analysis of Variance (ANOVA)

To analyze the differences between the data collected under two (or more) different conditions, we can use analysis of variance (ANOVA), which is an effective analysis tool based on the linearity, independence, and normality of the data

```
Method

Null hypothesis          All means are equal
Alternative hypothesis   At least one mean is different
Significance level       α = 0.05

Equal variances were assumed for the analysis.


Factor Information

Factor   Levels   Values
Factor        5   15.00%, 20.00%, 25.00%, 30.00%, 35.00%


Analysis of Variance

Source   DF   Adj SS   Adj MS   F-Value   P-Value
Factor    4    475.8   118.940    14.76     0.000
Error    20    161.2     8.060
Total    24    637.0
```

Figure 5.3   An example of ANOVA result.

collected. We can use ANOVA to test whether the means of several groups are equal. Figure 5.3 shows an example of an analysis output using Minitab software.

To determine whether the difference between two means is significant, we compare the p-value to the predetermined significance level that is often 0.05. A linear model is statistically significant only when the p-value is smaller than the significance level. More discussion on p-value is in the next subsection.

Using ANOVA, it is preferred that there are the same numbers of observations for all factors in an experimental design. Many researchers use ANOVA method, for instance, "*Modeling hydrogen storage on Mg–H$_2$ and LiNH$_2$ under variable temperature using multiple regression analysis with respect to ANOVA*" (Al-Hadeethi et al. 2017).

5.1.2.4 Regression Analyses
There are other types of statistical analysis for the interrelationship among variables. For example, a regression analysis is used to estimate both linear and nonlinear relationships among variables for a prediction purpose.

For example, a linear regression function is $y = \beta_1 x + \beta_0 + \varepsilon$, where, y represents an outcome variable, x represents its corresponding predictor variable, ε is the error, and β_1 and β_0 are parameters. Figure 5.4 shows an example of regression analysis in a form of $y = \beta_1 x^2 + \beta_2 x + \beta_0$, where x is a quadratic variable. It is still called a linear regression as expressed in a linear combination of the βs.

In some cases, a linear model does not fit data well. We may try nonlinear models $y \sim f(x, \beta)$, which are more complex. There are various forms of mathematical

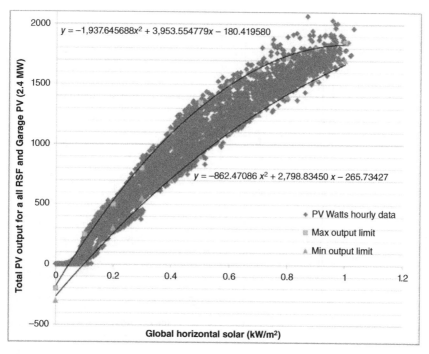

Figure 5.4 An example of nonlinear regression analysis. Source: Henze et al. (2014).

functions $f(x, \beta)$, such as exponential $e^{(x,\beta)}$, logarithmic $\log(x,\beta)$ or $\ln(x,\beta)$, trigonometric $\sin(x,\beta)$, etc., and their combinations. In such complex cases, we may break a variable in an entire course into several segments. Then we may perform a linear regression for each segment if appropriate.

Meta-Analysis

A meta-analysis is a statistical analysis of previous studies. It is an analysis of a collection of many other analyses to summarize their results into a single estimate. It is *not* a summary or systematic literature review, as a meta-analysis uses statistical approaches on the accumulated data from the results of other studies. Thus, a meta-analysis may be called the analysis of analyses.

Accordingly, we need to do an extensive search for relevant studies, identify appropriate studies, do a comparison, and convert all the previous results into statistical results. For example, a meta-analysis study was conducted on adopting global virtual engineering teams in architecture, engineering, and construction projects (Hosseini et al. 2015). Another example is on software engineering (Hayes 1999). Figure 5.5 shows an example of meta-analysis results on automation reliability.

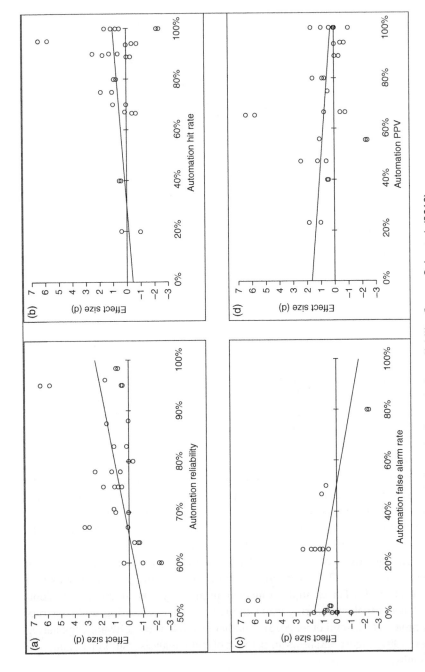

Figure 5.5 An example of meta-analysis result of automation reliability. Source: Rein et al. (2012).

Meta-analysis may be a good research as long as the outcomes improve the current knowledge and understanding. If the outcomes stay the current level of understanding, then it might be called "informational papers or research discussions." Such type of work can have a good reference value as well. Meta-analysis study is infrequent in engineering disciplines.

5.1.3 Interpretation of Analysis Results

After obtaining analysis results, we need to understand and interpret them. We try to explain the results based on the existing concepts, theories, and other studies from a literature review. If findings are within the expectations of the original research plan, the understanding and interpretation may be straightforward. However, if some results are unexpected, interpreting them can be challenging and interesting. Sometimes, we need to do additional tests and analysis.

5.1.3.1 Hypothesis Testing Process

A common format of basic research is a hypothesis-based testing. Doing this type of study, we normally have good data availability and a predetermined level (α) of confidence on research results. A common procedure of such studies is to use sample data to make an inference about the population from which the sample is drawn. Figure 5.6 shows a basic analysis flow.

Here are the key elements of hypothesis analysis.

- Establish the null hypotheses H_0 and alternative hypotheses H_1, where H_0 is often set at the status quo, based on the problem to solve and observation
- Select a required level of significance (α)
- Randomly collect data and analyze the data to get a p-value
- Determine to stay on or reject H_0 based on the p-value
- Interpret and discuss the analysis results
- Draw the conclusions for a new understanding, including the condition and limitations

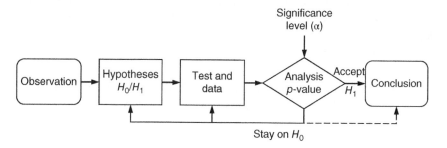

Figure 5.6 A process flow of hypothesis driven research.

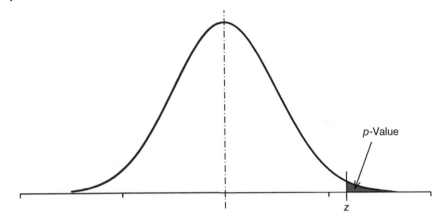

Figure 5.7 An illustration of *p*-value.

5.1.3.2 Hypothesis Test Results

We often use a statistic *p*-value in hypothesis-based research. The *p*-value is the calculated probability of finding the observed results when the null hypothesis (H_0) of a research question is true. Figure 5.7 shows as an example of one-side *p*-value. In the figure, the *p*-value is the outside area of the normal distribution region defined by the sample test statistic z $(= \frac{x-\mu}{\sigma})$.

In a study, we compare the calculated *p*-value with the predetermined significance level α. If the $p > \alpha$, we stay with H_0, which can be a new foundation for further studies. If we reject H_0 and accept H_1, then H_1 becomes a new understanding for the particular case. It is also possible that we may not disprove all (but maybe only parts) of the hypotheses.

5.1.3.3 Outlier Detection and Exclusion

Some methods of statistical analysis are sensitive to the presence of data outliers. There are several approaches for outlier detection (Kriegel et al. 2010; Yang et al. 2019; Bankar and Metre 2019).

For a univariate (one-variable) data set with an approximately normal distribution, we may use the Grubbs's test for an outlier. The Grubbs's test statistic (G) is defined as follows:

$$G = \frac{\max|x_i - \bar{x}|}{S}$$

where, $\bar{x}$ and S denote the sample mean and standard deviation, respectively. With the given significant level α and sample size n, for a two-sided test, the hypothesis of no outliers is rejected if

$$G > \frac{n-1}{\sqrt{n}} \sqrt{\frac{t^2_{\alpha/2n,n-2}}{n-2+t^2_{\alpha/2n,n-2}}}$$

where, $t^2_{\alpha/2n,n-2}$ is the critical value of the t distribution with $(n-2)$ degrees of freedom and a significance level $\frac{\alpha}{2n}$. We can do one-sided Grubbs's tests in a similar way. Please note that Grubbs's test is good for detecting one outlier. It is not recommended using Grubbs's test for the second time after the first outlier is removed. If two or more outliers are possible, other methods, such as Tietjen–Moore test (Karagöz and Aktaş 2018), may be used. We can also use graphical tools, such as normal probability plot, histogram, and box plot charts, to visualize the data distribution and identify possible outliers.

After detecting an outlier, it can be challenging to understand the root causes of the outlier. For research, we should exclude an outlier from subsequent data analysis only if the reason of the outlier is explored and its removal has a team consensus.

5.1.3.4 Identifying Limitations of Study

Since research outcomes are based on the samples rather than the entire data set (or population), some risks of making an erroneous inference using statistical analysis are inevitable. There are two types of errors of hypothesis analyses (refer to Table 5.1):

- Type I (α): True H_0 is incorrectly rejected when the hypothesis is actually true.
- Type II (β): False H_0 is incorrectly accepted or failing to reject the null hypothesis when it is actually false.

These two types of errors exist and are associated with each other. If the probability of making one type of error is decreased, the other type of error will be increased at the same time. In other words, if we want a very small type I error α, unfortunately, then the risk of making a type II error β will be high. Normally, α is set to a small value (0.05 or 5%), which means no more than a 5% chance of incorrectly rejecting H_0 when it is actually true.

The only way to reduce both types of errors at the same time is to increase sample size. That is, using as a large sample set as possible can increase the validity and reliability of research.

Table 5.1 Types I and II errors.

Truth	Decision	
	Accept	**Reject**
True H_0	Correct decision	Type I (α)
False H_0	Type II (β)	Correct decision

5.2 Quantitative Research

Quantitative analysis is often known as statistical analysis in many cases, and there are numerous quantitative methods in engineering R&D. In this section, we discuss three types: mathematical modeling, optimization, and computer simulation.

5.2.1 Mathematical Modeling

5.2.1.1 Concept of Modeling

Modeling is an approach and a process of using various scientific principles and terms to present real-world situations, as illustrated in Figure 5.8. Researchers using modeling techniques to describe the different aspects of the real world, advance scientific understandings, and solve problems.

A simple mathematical model may be a set of equations. For an example of a linear regression model discussed earlier, it may be in the form of $y = \beta_1 x + \beta_0 + \varepsilon$. The regression model determines the specific relationship between the input variable x and the outcome variable y.

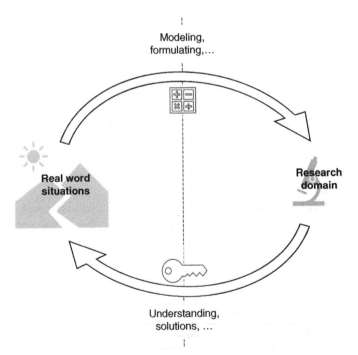

Modeling,
formulating,...

Real word
situations

Research
domain

Understanding,
solutions, ...

Figure 5.8 Modeling to bridge division between real word and research.

Modeling must have some assumptions and constraints. For example, both *y* and *x* of the linear regression are in their defined ranges and follow a certain distribution. Additionally, we often combine mathematical modeling along with other such approaches, such as finite element analysis (FEA), in research projects.

5.2.1.2 Applications of Math Modeling

A lot of research is based on mathematical modeling. With advanced computer technology, much mathematical modeling work is on algorithms for various engineering tasks, such as design, analysis, implementation, and optimization.

Table 5.2 shows a few examples and the widespread applications of mathematical modeling in engineering research. Most modeling methods, including the ones listed in the table, are suitable for diverse applications. Readers can find more examples online related to your research subjects.

5.2.2 Optimization

5.2.2.1 Concept of Optimization

In theory, optimization is based on the modeling of scientific principles. The concept of optimization started from mathematics. For a general case, a mathematical function may be expressed by:

$$y = f(x_1, x_2, x_3, \ldots)$$

where, *y* is the output of a function $f(\bullet)$; x_1, x_2, and x_3, etc., are input variables and the optimization task is about either maximizing or minimizing the output *y* by methodologically choosing input values of x_1, x_2, and x_3, etc., using various optimization algorithms and considering certain constraints. In other words, optimization aims to find the best solution when considering inputs and constraints.

There are numerous optimization theories and methods; new ones are being proposed and developed almost every day. Due to the nature of optimization, it is deemed as applied research rather than basic research. When applying basic research grants, we may avoid using the word "optimization" and alike in proposals as it implies the research likely for applications. Unquestionably, there are exceptions.

The concept of continuous improvement (CI) is different from the idea of optimization. An optimization process's goal is to seek the best status (may be theoretically ideal) of an object under certain constraints, while CI is a gradual and continuous process of improving an output. The theories and procedures of the two methodologies are different as well. Therefore, optimization is generally conducted by specialists and researchers; while CI can be practiced by anyone.

Table 5.2 Examples of mathematical modeling in engineering research.

Engineering	Method	Example topic
Mechanical	Dimensionless governing equations	The development of mathematical modeling for nanofluid as a porous media in heat transfer technology (Tongkratoke et al. 2016)
Computer	Extensible network service model	The modeling and analysis of the extensible network service model (Ji et al. 2018)
Electrical	State-space modeling	State-space modeling and reachability analysis for a DC microgrid (Ghanbari et al. 2019)
Industrial	Energy consumption of stereolithography	Energy consumption modeling of stereolithography-based additive manufacturing toward environmental sustainability (Yang et al. 2017)
Material	Synchronous generator model in Park–Gorev's equations	The synthesis of precise rotating machine mathematical model, operating natural signals and virtual data (Zhilenkov and Kapitonov 2017)
Civil	Mualem–Van Genuchten method	Mathematical modeling of hydrophysical properties of soils in engineering and reclamation surveys (Terleev et al. 2016)
Manufacturing	Economic manufacturing quantity model	Mathematical modeling for exploring the effects of overtime option, rework, and discontinuous inventory issuing policy on EMQ model (Chiu et al. 2017)
Chemical	Math models based on binary gas separation	Mathematical modeling and investigation on the temperature and pressure dependency of permeation and membrane separation performance for natural gas treatment (Hosseini et al. 2016)
Nuclear	Nondimensional model	Mathematical modeling of orifice downstream flow under flow-accelerated corrosion (Sanama and Sharifpur 2018)

5.2.2.2 Optimization Applications

Engineering optimization is to target a specific process, condition, and design and uses optimization techniques to achieve the best performance and status based on engineering principles. Many optimization research efforts are to transfer practical solutions to optimal solutions.

Optimization is a very active area of research. A few examples are listed in Table 5.3 to illustrate the variety of methods in different engineering fields. Again, the specific optimization methods listed may be applicable to other subjects and fields, at least in principle.

Table 5.3 Examples of engineering optimization research.

Engineering	Method	Example topic
Mechanical	Empirical model	Optimization of mechanical properties of printed acrylonitrile butadiene styrene using RSM design (Abid et al. 2019)
Computer	Process regression	Gaussian process regression tuned by Bayesian optimization for seawater intrusion prediction (Kopsiaftis et al. 2019)
Electrical	Surrogate-assisted robust algorithm	A new surrogate-assisted robust multiobjective optimization algorithm for an electrical machine design (Lim and Woo 2019)
Industrial	Multiobjective integer nonlinear programming model	Optimization of multiperiod three-echelon citrus supply chain problem (Sahebjamnia et al. 2019)
Material	Finite element analysis and neural networks	Development of an elastic material model for bcc lattice cell structures using finite element analysis and neural networks approaches (Alwattar and Mian 2019)
Civil	Teaching learning based	Time-cost trade-off optimization of construction projects using teaching learning-based optimization (Toğan and Eirgash 2019)
Manufacturing	3D kinematic	Optimization of centerless through-feed grinding using 3D kinematic simulation (Otaghvar et al. 2019)
Chemical	Multi-objective	Multi-objective optimization method for enhancing chemical reaction process (Cao et al. 2018)
Nuclear	Genetic algorithm	Module layout optimization using a genetic algorithm in light water modular nuclear reactor power plants (Wrigley et al. 2019)

5.2.3 Computer Simulation

5.2.3.1 Concept of Simulation

Physical systems in the real world are extremely complex, which challenges analytical solutions and computing resources. In some cases, analytically or experimentally studying a system or a situation in the real world is yet either impractical or unable to get good results due to limited resources and/or human knowledge.

Computer simulation on the other hand can be sued to solve those problems. Simulation is a research method and process to produce the functions, behaviors, and outcomes of a physical system based on a certain algorithm and using a computer technology in a digital and virtual environment. As a type of virtual experimental study, the information and conditions in a physical world are modeled into a virtual world.

We may simulate physical systems of any type, such as electrical, mechanical, operational, human actions, and so on. Modeling is a central part of computer simulation, which is based on scientific and engineering principles. Most simulation work uses commercial software and sometimes may need additional programming efforts.

To make a simulation study feasible, a simplification of real-world situations is necessary; some assumptions and parameters (including input data) are determined based on the assumptions. Therefore, a simulation may not be 100% representative to the corresponding real world, but hopefully close to. We should check the output of a simulation for its validity against the real situations if possible. In many cases, we can also perform sensitivity analyses for the parameters to improve the accuracy of the simulation results.

5.2.3.2 Common Types of Simulation

There are a few types of computer simulation, which may be categorized in different ways with significant overlaps between some types. Figure 5.9 shows the common types of computer simulation.

Based on a deterministic model, a computer simulation can generate the output that is fully determined by the parameter and the initial conditions. Due to the complexity of the real world, deterministic models may be used as an approximation of reality with simplified inputs and assumptions. For example, a deterministic model was used to study the electron transport for electron probe microanalysis (Bünger et al. 2018).

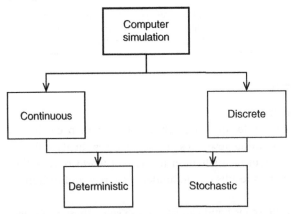

Figure 5.9 Types of computer simulation.

In contrast, the outputs of a stochastic simulation are different with the same parameter values and initial conditions. That is, the behaviors of a stochastic model cannot be entirely predictable. Thus, a stochastic simulation may be used to trace the evolution of variables that can change randomly. Comparatively, a stochastic simulation can be more complicated and closer to the real world than a deterministic simulation. For instance, stochastic method was used to solve a complex problem in production (Moheb-Alizadeh and Handfield 2017).

FEA, which is based on the principles of engineering, physics, mathematics, and statistics, can be either deterministic or stochastic depending on the principles used (Roirand et al. 2017). FEA has been widely used in various engineering studies. Originally started on the problems of mechanical structural analysis, FEA has successfully been applied on other areas, such as heat transfer, fluid flow, acoustics, electromagnetic fields, and electrical-chemical process. Figure 5.10 shows two FEA models (NASA 2017).

Continuous simulation and discrete event simulation are distinctive in terms the state variables change either in a continuous way or at a countable number of points in time. Therefore, using whether continuous and discrete event simulation depends on the nature and states of physical systems and the data types. Both continuous and discrete event simulation may be either deterministic or stochastic.

We often use the term of dynamic simulation, which is to model the time varying behaviors, characteristics, and outputs of a physical system. Dynamic simulation can be viewed consisting of two major parts: simulation calculation over time and real-time graphic animation, which are integrated into powerful software packages. The simulation method itself is based on any number types of computer calculation, such as mathematical equations and numerical analysis. Dynamic simulation is widely used in industrial applications, such as vehicle performance modeling and movements of complex equipment.

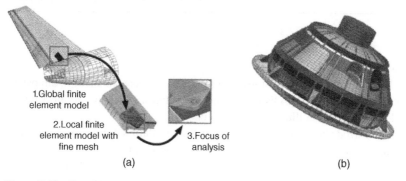

Figure 5.10 Two FEA examples. (a) Airplane empennage. (b) Orion crew module mounted in heatshield. Source: NASA (2017).

Table 5.4 Examples of simulation in engineering research.

Engineering	Method	Example topic
Mechanical	FEA and computational fluid dynamics	Simulated transcatheter aortic valve flow: implications of elliptical deployment and underexpansion at the aortic annulus (Sirois et al. 2018)
Computer	Annealing algorithm	Training ANFIS structure using simulated annealing algorithm for dynamic systems identification (Haznedar and Kalinli 2018)
Electrical	Adaptive electro-kinematical model	Electrokinematical simulation for flexible energetic studies of railway systems (Mayet et al. 2018)
Industrial	Agent-based, discrete event, and system dynamics	Centralizing the admission process in a German hospital (Reuter-Oppermann et al. 2019)
Material	FEA	Simulation and experimental tests of ballistic impact on composite laminate armor (Soydan et al. 2018)
Civil	Agent-based	Advances in probabilistic and parallel agent-based simulation: Modeling climate change adaptation in agriculture (Troost and Berger 2016)
Manufacturing	Discrete event simulation	Effects analysis of internal buffers in serial manufacturing systems for optimal throughput (Imseitif and Tang 2019)
Chemical	Linear viscoelasticity	Micromechanics and rheology of colloidal gels via dynamic simulation (Johnson 2018)
Nuclear	Heat transfer model	Coupled thermochemical, isotopic evolution and heat transfer simulations in highly irradiated UO_2 nuclear fuel (Piro et al. 2016)

5.2.3.3 Examples of Simulation

Regardless how to categorize simulation methods, we may use them in almost all engineering fields. Table 5.4 lists a few more examples, which are not necessarily typical ones, just for an illustration purpose.

As emphasized above, we need to validate the simulation outcomes. A common way is to use physical experiments under the same settings and parameters. The simulation results have a good validity if they are in agreement with the physical experimental observations.

5.2.4 New Technologies

Inventing a new approach and technology in engineering fields can be more challenging and exciting than applying existing methods. There is no limit for new technology advances. For example of FEA, there are several research advances, such as in nonlinear and new types of materials or in a new algorithm and new technology in FEA. Another example is on additive manufacturing – more and more new types of metals and new technologies are introduced for additive manufacturing processes.

One remarkable characteristic of new technologies is that they are cross-disciplinary. Most research conducted in such emerging areas is related to new approaches, modeling, and computer simulation. The most popular research subjects include

- Virtual reality (VR) and augmented reality (AR)
- Industry 4.0 (cyber-physical systems, etc.)
- Blockchain (sensing and digital ledger)
- Autonomous vehicles
- Cybersecurity engineering
- 3D metal printing

These new subjects themselves are huge research areas. Studying the sub-subjects of those may mean good opportunities to get grants as well as non-monetary supports and playing a leading role in technology advancement and breakthrough.

It is interesting to note that the topic, approach, and direction of new technologies may not be entirely new. Based on an existing technology, many initiatives suggest a gradual evolution and advance. Table 5.5 lists a few examples and readers may search more in their own disciplines. Understanding these characteristics and examples may provoke and generate new ideas, which lead to new methods, products, materials, processes, etc.

5.3 Experimental Studies

Sometimes, it can be difficult to characterize a subject theoretically or in a mathematical model, so one possible way to study them is through experiments. In many fields, scientists and researchers start their new investigations with experiments. In addition, experimental studies may be more cost-effective compared with real operations out in the field, since it may be easier to control the environment of the experimental runs to obtain required data.

Table 5.5 Examples of new technologies in engineering research.

Engineering	Type	Example topic
Mechanical	Process	New principle for aerodynamic heating (Sun and Oran 2018)
Computer	Method	Utilizing ECG waveform features as new biometric authentication method (Shdefat et al. 2018)
Electrical	Method	A new demand response algorithm for solar PV intermittency management (Sivaneasan et al. 2018)
Industrial	Method	Demand driven MRP: assessment of a new approach to materials management (Miclo et al. 2019)
Material	Process	Sandwich manufacturing with foam core and aluminum face sheets – a new process without rolling (Hohlfeld et al. 2018)
Civil	Method	A new method for estimation of aerostatic stability safety factors of cable-stayed bridges (Dong and Cheng 2019)
Manufacturing	Method	A new reverse engineering method to combine FEM and CFD simulation three-dimensional insight into the chipping zone during the drilling of Inconel 718 with internal cooling (Kaya and Kaya 2019)
Chemical	Process	A new, sustainable process for synthesis of ethylene glycol (Gong 2018)
Nuclear	Method	Constraint annealing method for solution of multiconstrained nuclear fuel cycle optimization problems (Kropaczek and Walden 2019)

5.3.1 Overview of Experimental Studies

5.3.1.1 Basic Elements of Experiments

We may have various objectives to conduct experimental studies to predict the outputs, find causational relationships, and validate analysis results.

An experiment design plays a key role in experimental research's success. When designing an experimental plan, we must consider several aspects of a study. We must select a proper experimental approach; for example, using a simple comparative design vs. a factorial experimental design. It is also important we have reliable instruments and means to measure the input factors as well as response variables in an experimental study. For a large experiment, we should use project management approaches for experiment planning and execution phases.

During an experimental study, we must carefully monitor and control experiment parameters and conditions, then observe the experiment output when the parameters and conditions are manipulated. The results from an experiment study are often compared against the expectations or the hypotheses.

We conduct most experimental studies in laboratories because of the well-controlled variables of experiments, although the results from laboratories may or may not precisely reflect real-world situations.

5.3.1.2 Influencing Factors

Many known and unknown factors may influence a particular condition or phenomenon of an experiment. Thus, we should evaluate these factors during the design and planning phase of an experimental study as much as possible. We may also try to optimize or narrow down parameters by investigating a phenomenon before trying with a large scale of experiments.

A design of an experimental study includes two parts about its factors:

1. Having a good control for all known factors.
2. Estimating influence of unknown factors and holding them fixed to ensure that the extraneous conditions are not influencing the response to be measured.

The number of factors involved in an experimental study is important in order to select an appropriate experimental approach. If there is only one factor with two varying levels, we may have a treatment group and a control group that has no treatment, which is a comparative experiment. If there are multiple factors with multiple levels of each factor, then the response may be affected by not only these factors themselves but also their interactions. Therefore, a factorial experiment design should be used for such cases, which is discussed in Section 5.4.

5.3.1.3 Other Considerations

The randomness of a data collection in an experiment design is critical for the validity of an experiment. For example, when doing a comparative experiment with two groups (a treatment and a control), we must assign the treatments by a random process to eliminate potential biases. There are various ways to obtain randomness, including but not limited to randomization tables and computerized random number generators.

Replication is another essential factor. The repeated experimental results may have a significant within-treatment variation. In addition, during an experimental study, we may achieve a better estimation or prediction through replication. However, replication may not be always practical or economical.

Due to physical constraints, unknown factors, and researcher's capabilities, the results are only true under certain circumstances and assumptions and may not always represent the overall state of affairs. Considering and acknowledging these limits and possible bias are a good practice to assure the proper interpretation of experimental results.

We should record the conditions and variables of all experiments, regardless of the levels of scale and complexity, for tracking pertinent information related to each experimental run. The conditions and variables include input data, process status, environmental conditions, special situations, time stamps, personnel involved, temporary and final results, as well as any anomalies observed during the experiment. Sometimes, these record log files are required by legal regulations.

5.3.2 Comparative Studies

5.3.2.1 Concept of Comparative Studies

A comparison study is a process of comparing two or more things to discover their differences or prove the same about them. In general, there are four types of comparative designs (see Figure 5.11).

Essentially, an experiment has two types of variables, independent x and dependent y. By designing and conducting an experiment, we study their relationship $y = f(x)$. When the objective is to find a causation relationship, the independent and dependent variables are called cause and effect, respectively. For example, we may try to know how using different methods (an independent variable) affect the research effectiveness (a dependent variable) on a subject.

Furthermore, extra variables may enter into the equation that is not accounted for, which are called confounding variables. For example, the researcher's experience can be a variable to the method study above. Figure 5.12 illustrates the relationship among independent, dependent, and confounding variables with the example. As an outside influencer, a confounding variable affects the dependent and independent variables themselves and their relationship in an experiment. In real world cases, there are often multiple confounding variables, which may or may not be recognized or understood.

Understanding the concept of confounding variables and their influences is important to have reliable study conclusions. There are a variety of ways to control

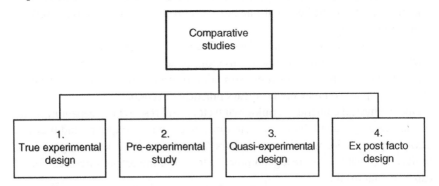

Figure 5.11 Four types of comparative experiment studies.

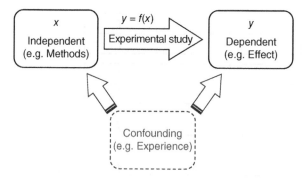

Figure 5.12 Independent, dependent, and confounding variables.

the confounding variables. Below are some helpful strategies to set up experiment settings, either individually or combined:

- Keep certain things or parameters constant
- Include a control group with a placebo treatment or without a treatment
- Randomly assign participants or members to groups
- Increase diversity to reduce the likelihood of systematic confound
- Assess equivalence before the treatment with one or more pretests
- Expose participants to all experimental conditions
- Use "within-subjects" design (also known as repeated-measures)

Comparative research is widely used in social and medical sciences, while engineering researchers use these methods occasionally. Here a few examples:

"Numerical investigation on ballistic resistance of aluminium multi-layered panels impacted by improvised projectiles" (Szymczyk et al. 2018)
"Influence of the homogenization scheme on the bending response of functionally graded plates" (Srividhya et al. 2018)
"A New Multistep Technique Based on the Nonuniform Rational Basis Spline Curves for Nonlinear Transient Heat Transfer Analysis of Functionally Graded Truncated Cone" (Heydarpour and Malekzadeh 2019)

5.3.2.2 True Experimental Design

Some comparative experimental studies do not include a control group, which can be a problem if two treatments yield similar results. A true experimental design is to compare two (or more) groups or situations, with a control group or situation. The true experimental designs are characterized by the random selection of all the cases to be tested and the use of a control group parallel to the experimental group(s). Normally, we select only one variable to manipulate and test. The statistical analysis of multiple variable experiments tends to be cumbersome.

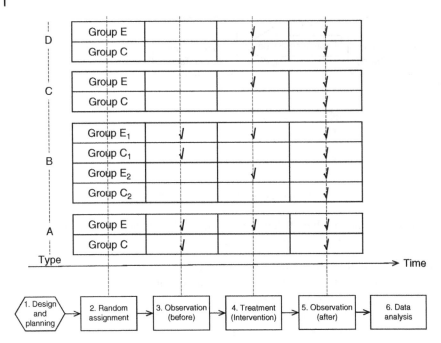

Figure 5.13 Four types of true experimental studies and overall process.

Figure 5.13 shows the common types of true experimental designs, where Group E has experimental treatments, but Group C is a control group without treatment. The process of true experimental studies has six stages. Each of the observations often has multiple measurements.

Type A is classical and most often used, called a pretest-posttest group design. The measurements or observations are taken before and after a treatment for both groups (Group E and Group C). So the design has a good internal validity.

Type B is called Solomon four-group design, without pretest for the additional two groups (E_2 and C_2). This type of design can enhance external validity as we may know the possible influence of confounding variables and extraneous factors. Gathering the necessary samples of each group may be a challenging part in the planning for this type of study.

Type C is a simplified design with posttests only. Since no pretest, there may be no interaction effect between the pretest and the treatment.

Type D is another type of posttest-only design, which may be used to focus on whether the two groups are different or not to assess a possible cause–effect relationship.

Table 5.6 An example of *F*-test result.

Source of variation	SS	DF	MS	F	P-value	F crit
Between groups	10.04	3	3.25	1.30	0.28	2.72
Within group	196.03	76	2.58			
Total	206.07	79				

In data analysis, we compare the means and variance of the two data sets (E and C). The data analyses, such as *t*-test and *F*-test, are well established and have been widely used. For example,

- $$\begin{cases} H_0 : \mu_E = \mu_C \\ H_1 : \mu_E \neq \mu_C \end{cases}$$

- $$\begin{cases} H_0 : \sigma_E = \sigma_C \\ H_1 : \sigma_E \neq \sigma_C \end{cases}$$

To compare with the mean difference, we use a *t*-test. If the data sets come from the same population, we conduct a paired *t*-test. Otherwise, use an unpaired *t*-test. To compare the variances of the two sets of data, we should use an *F*-test, which is the basis of ANOVA. As long as the requirements of the experiment and data are met, software can take care of the data analysis. Table 5.6 shows an example of an *F*-test result using Microsoft Excel.

5.3.2.3 Other Comparative Designs

There are several other types of comparative studies, which are similar to the true experimental designs.

Pre-experimental Study. This type of study is the simplest form. In such a study, either a single group or multiple groups are observed presuming a treatment causes the change. The experimental and control groups in this type of study are not composed of equivalent or randomly selected members or there may be no control group. Pre-experimental study should be considered as a merely test and should be followed by more controlled experiments.

Quasi-experimental Design. This type of experiment is similar to regular true experiment designs, except its randomness is not practical or ensured. In such experiments, two groups (a control group and an experimental treatment group) are still matched. We apply some criterion other than random assignments in the experiments. e.g. planning your experimental runs based on one factor that is hard to change over by running all the experiments where that factor is at one setting, then the remaining experiments with that factor at another setting. In this way, it becomes more feasible to set up than true experimental designs.

Due to the lack of randomness, the statistical analysis, results, and conclusions from a quasiexperiment do not have as strong internal validity. Quasi experiments are often used for causal relationships with two or more variables. Quasi-experimental studies are not often used for engineering and technology research.

Ex Post Facto Design. This is a type of quasi-experimental study but is based on the existing (secondary) data. In such studies, there is no direct control or manipulation of the independent variables. Using this kind of study, we may reveal the presumed cause that has already occurred for certain effects. Therefore, we are to collect existing data to investigate a possible relationship between the factors and subsequent characteristics. There is a chance of searching in the wrong areas and getting incorrect conclusions.

5.4 Factorial Design of Experiment (DOE)

5.4.1 Introduction to DOE

5.4.1.1 Concept of DOE

Design of experiments (DOEs) is a methodology that has various methods, design techniques, and processes for engineering study, research, and optimization. DOE is powerful and practical for the cases with multiple factors. The applications of DOE involve experimental planning, conducting experiments, fitting models to the outputs, and so on.

We often use DOE to find the relationship between the controlled variation of input factors and the variation of a system outcome under certain conditions. Establishing the relationship between the input and output, we may predict a "what-if" outcome or confirm a suspected outcome by introducing changes to the input factors under the same conditions.

Researchers also combine DOE and another method for their research to address multiple input variables. Table 5.7 lists a few examples.

5.4.1.2 Advantage of DOE

Conventional comparative experiments are sometimes called OFAT (One-Factor-at-a-Time). If multiple input factors are considered, a conventional experiment design has to hold all factors constant except one factor to vary at a time during experiments. Obviously, OFAT is inefficient and ineffective to study multiple factors.

Table 5.7 Examples of DOE in engineering research.

Engineering	Combined with	Example topic
Mechanical	Computational fluid dynamics	Effects of piston geometry and injection strategy on the capacity improvement of waste heat recovery from RCCI engines utilizing DOE method (Nazemian et al. 2019)
Computer	Genetic algorithm	A genetic algorithm for the hybrid flow shop scheduling with unrelated machines and machine eligibility (Yu et al. 2018)
Electrical	Finite element analysis	Flashover process analysis of non-uniformly polluted insulation surface using experimental design methodology and finite element method (Terrab et al. 2018)
Industrial	Optimization design method	Optimization of rotor shaft shrink fit method for motor using "Robust design" (Toma 2018)
Material	Thermodynamic modeling	Prediction of steel transformation temperatures using thermodynamic modeling and design of experiments (DOE) (Loucif et al. 2018)
Civil	Finite element analysis and artificial neural networks	Reliability assessment of buckling strength for imperfect stiffened panels under axial compression (Mouhat et al. 2015)
Manufacturing	Discrete event simulation	Throughput analysis of manufacturing systems with buffers considering reliability and cycle time using DES and DOE (Imseitif et al. 2019)
Chemical	Uncertainty modeling	Model-based design of experiments in the presence of structural model uncertainty: an extended information matrix approach (Quaglio et al. 2018)
Nuclear	Surrogate modeling	Efficient global sensitivity analysis for flow-induced vibration of a nuclear reactor assembly using Kriging surrogates (Banyay et al. 2019)

One strength of DOE is about handling multifactors in an experimental design. Figure 5.14 shows a diagram that DOE can be used for studying the relationship between three input factors and one response of a process. The input factors are called "treatments" in DOE, and they are assignable at different levels.

Another issue of conventional comparative experiments for multiple input factors is that they cannot find the possible interactions among the factors, which can be significant. Therefore, the most significant advantage of DOE is that it allows for multiple input factors to be manipulated and determining their effects, including

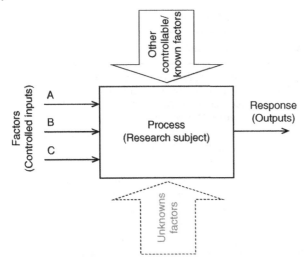

Figure 5.14 A process with inputs and output for DOE.

their interactions, on the output (or the response of the experiment), which is often the case in engineering and technical projects.

5.4.1.3 Two-level Factorial Design

A two-level factorial design means that each input factor has two values, i.e. low and high. A design of full two-level factors is the foundation of DOE and is widely used in practice because it is simple and efficient to design and execute. If there are k factors, the full two-level factorial design consists of 2^k treatment combinations. The simplest design is 2^2, meaning two levels for each of the two input factors.

The 2^2 design of an experiment needs to run at least four times. In addition, under each combination of levels and factors, we may have multiple runs (replicates). Replicates provide an estimate of the pure trial-to-trial "noise" in an experiment and allow for more precise estimates of the effects. In other words, with two replicates, a 2^2 design needs to run eight times (see Table 5.8). We normally generate a DOE design matrix is using software, such as Minitab.

Similarly, three-level factors have three levels of values: low, normal, high presented by 3^k. Three-level factorial design can yield more information, particularly possible nonlinear relationship. However, four-level factors are uncommon due to the complexity of experiments.

5.4.2 Process of DOE Applications

5.4.2.1 Overall Procedure

The requirements, design, and analysis of DOE are different from those of conventional experimental studies. Figure 5.15 shows a process flow of DOE, which has three stages: design and planning, execution, and analysis.

Table 5.8 An example of 2^2 DOE with two replicates.

Trial			Factors or treatments		Response
Standard order	Run order	Blocks	A	B	Y
1	5	1	−1	−1	y_1
2	8	1	1	−1	y_2
3	1	1	−1	1	y_3
4	6	1	1	1	y_4
5	4	2	−1	−1	y_5
6	2	2	1	−1	y_6
7	7	2	−1	1	y_7
8	3	2	1	1	y_8

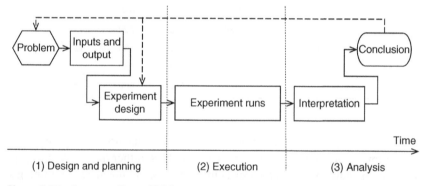

Figure 5.15 A process flow of DOE.

(1) Design and Planning

The first stage is to design and plan an experiment. With an identified problem, we define the input factors and output response. Then we determine the reasonable ranges of input factors and assign them specified levels based on the real situations and the purpose of the investigation.

With the input factors and output response defined, we can establish an experiment plan or a design matrix. Thanks to computer software, we can generate such a matrix quickly. Table 5.9 shows an example of 2^4 experiment in the laser welding experiment (data source: Graf et al. 2013). In general, we need to determine the required number of experimental runs based on technical feasibility and cost considerations.

For this 2^4 example, it has four input factors (laser power – P, spot diameter – d, welding speed – v, and powder mass flow – m); and each factor has two levels. This experiment needs to run total 16 times without a replicate.

Table 5.9 An example of 2^4 DOE.

Standard order	Run order	P	d	v	m
2	1	1700	1.2	320	5
10	2	1700	1.2	320	11
3	3	800	1.8	320	5
15	4	800	1.8	680	11
8	5	1700	1.8	680	5
4	6	1700	1.8	320	5
7	7	800	1.8	680	5
12	8	1700	1.8	320	11
1	9	800	1.2	320	5
11	10	800	1.8	320	11
16	11	1700	1.8	680	11
6	12	1700	1.2	680	5
9	13	800	1.2	320	11
14	14	1700	1.2	680	11
13	15	800	1.2	680	11
5	16	800	1.2	680	5

For most engineering and technical projects, input factors and the output in an experiment are quantitative (numerical) variables, rather than attributes (e.g. pass/fail). In addition, even if the exact output may be unknown before running a DOE, the experimenter should estimate the range of the output and appropriately prepare its measurements.

(2) Execution

Following the experimental design, we need to run the experiments based on the design (or at the levels of each factor) and record the output responses. There is no physical difference between conducting DOE tests and doing any other types of experiments. For example, we must record the inputs, experiment conditions, and output. We should follow the running order based on randomized combinations of factors and levels, which is included in the design matrix.

(3) Analysis

Many textbooks have relevant mathematic equations and calculation examples. However, manual analysis of DOE results are tedious. Using computer software, the analysis is quick and easy, as we just need to input the experiment data and select the correct functions of software. The output of software analysis is often in

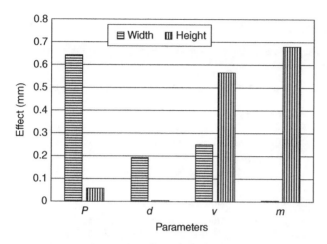

Figure 5.16 An example of 2^4 DOE analysis results.

the form of ANOVA and various charts. Figure 5.16 shows the results of process parameters on the bead size of the laser welds from the 2^4 DOE example (data source: Graf et al. 2013).

Understanding and interpreting the analysis results may be a major task for researchers. The correct and appropriate explanation heavily depends on the knowledge and experiences of researchers.

5.4.3 Considerations in DOE

5.4.3.1 Basic Requirements

In a DOE planning, beyond defined input factors, there may be other known or unknown factors influencing the experiment output. We can minimize the effects of uncontrolled variability and random variation (noise) in an experiment using randomization. Randomizing experimental runs also satisfy the statistical assumptions of independently and randomly distributed observation.

Replication, as mentioned, means a test under each factor combination runs multiple times. Running DOE with replications is an effective way to reduce and estimate the errors associated with an experiment. Generally, the more an experiment is replicated, the more reliable the results can be. When testing replication is technically and financially feasible, it is recommended that we run the test under each condition three times.

Experiments have some factors that likely affect the output but are not of primary concern to the research objectives. These factors are called nuisance factors if they are controllable. To control these factors, we may use a blocking approach, or assigning some tests into groups, in an experiment design. Using a blocking

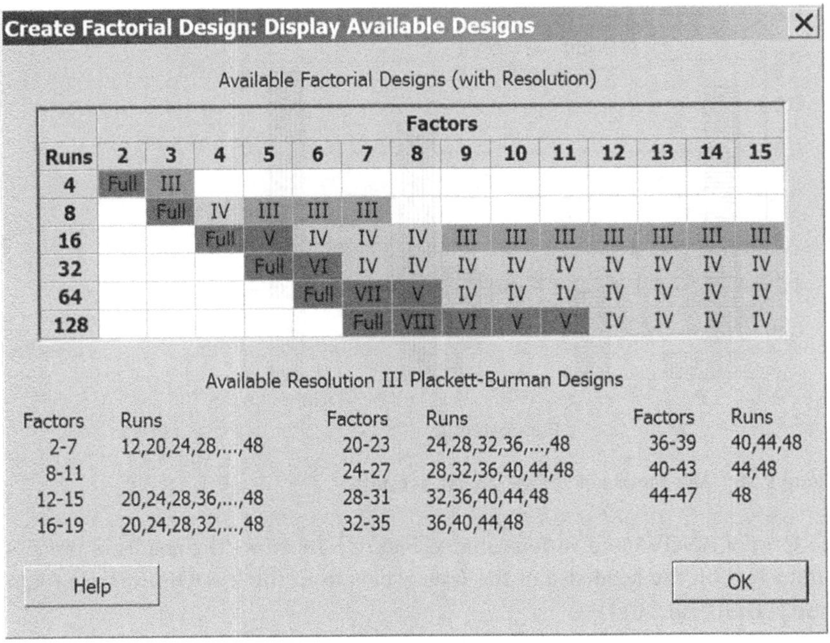

Create Factorial Design: Display Available Designs ✕

Available Factorial Designs (with Resolution)

Runs	2	3	4	5	6	7	8	9	10	11	12	13	14	15
4	Full	III												
8		Full	IV	III	III	III								
16			Full	V	IV	IV	IV	III	III	III	III	III	III	III
32				Full	VI	IV	IV	IV	IV	IV	IV	IV	IV	IV
64					Full	VII	V	IV	IV	IV	IV	IV	IV	IV
128						Full	VIII	VI	V	V	IV	IV	IV	IV

Available Resolution III Plackett-Burman Designs

Factors	Runs	Factors	Runs	Factors	Runs
2-7	12,20,24,28,…,48	20-23	24,28,32,36,…,48	36-39	40,44,48
8-11		24-27	28,32,36,40,44,48	40-43	44,48
12-15	20,24,28,36,…,48	28-31	32,36,40,44,48	44-47	48
16-19	20,24,28,32,…,48	32-35	36,40,44,48		

Help   OK

Figure 5.17 Options of fractional factorial designs. Source: Minitab.

approach, we can increase the probability of revealing the true differences in identification of the main effects influencing the output from a DOE test. DOE software packages have blocking functions to use in experiment designs.

5.4.3.2 Fractional Factorial Designs
Full factorial designs may not be practical when there are more than four factors because the number of tests doubles sequentially with an additional factor. For example, nonreplicated runs need $2^5 = 32$ and $2^6 = 64$ times for five and six factors with two levels, respectively.

If there are several influential factors, a DOE study can be in two steps. The first step, called screening, is to run a low-resolution experiment, or fractional factorial 2^{k-p} to identify the main effects, where p is smaller than k. For example, if there are seven initial factors, the options of fractional factorial designs are 2^{7-1}, 2^{7-2}, 2^{7-3}, and 2^{7-4}. If using Minitab, we have the options of fractional factorial designs (refer to Figure 5.17). They have resolutions called VII, IV, IV, and III and need to run 64, 32, 18, and 8 times without replicates, respectively.

After successfully screening and finding significant factors, which may be three or four, we can perform the second step – a sequential full factorial design on the significant factors to have an in-depth study to understand the effects and interactions. If there are four significant factors, then we may run a full DOE design with two levels: $2^4 = 16$. Comparatively, for a full factorial design for all seven factors 2^7, we would have to run 128 times without replicates.

DOE itself is a science based on statistics. However, successfully conducting DOE experiments in engineering and technical research is more like an art and is experience-related. It is recommended that novice researchers consult with experienced professionals for their first DOE study.

Summary

Statistical Analyses

1. Two major types of statistical analysis are descriptive statistics (to reveal and summarize the information about a data set) and inferential statistics (to analyze sample data to make inferences about the population).

2. A data set is often analyzed and described by its central tendency (mode, median, and/or mean) and variability (range, standard deviation, and/or average deviation).

3. The association of two data sets is often analyzed for their correlation. The correlation coefficient r indicates mathematical correlation, but does not necessarily present a causational relationship between the two data sets.

4. ANOVA and regression analysis are often used for relationship between data sets.

5. A meta-analysis is a statistical analysis of previous studies.

6. A hypothesis testing is a process following certain format and steps and often concludes based on the calculated p-value. The analysis has types I (α) and II (β) errors.

7. There are several methods to detect outliers in a data set.

Engineering Quantitative Research

8. There are various mathematical approaches to model real-world situations or cases.

9. Optimization has wide applications in applied research and R&D, good research potentials, and significant benefits.

10. CI in real world is different from optimization.

11. Computer simulation is a research method and process to produce the functions, behaviors, and outcomes of a physical system. Simulation has growing applications for complex real-world situations.

12. Research can be the development of new technologies, methods, and processes.

Experimental Studies

13. Experimental study is a common research approach, including comparative studies and factorial DOEs.
14. Comparative studies, simple and effective, include true experimental, pre-experimental, quasiexperimental, and ex post facto design. However, they are not suitable for the studies with interactive and confounding variables.
15. When randomness is not designed into a comparative study, it is called quasi-experimental design. If using existing data, a comparative study is called an ex post facto design.

Factorial Design of Experiment (DOE)

16. DOE is an approach and a process that has various methods and design techniques for research.
17. DOE is more complex than the comparative studies but can handle multiple variables and reveal possible interactions between variables.
18. DOE has certain processes and considerations and often use software for the designs, experimental plan, and consequent data analysis.

Exercises

Review Questions

1 One states that a typical objective of quantitative research is to find the causational relationship, explain a phenomenon, or predict the findings onto general cases or a larger population. Do you agree? Why?
2 Compare between the descriptive and inferential statistics with an example.
3 Find an example of confidence interval for the estimated mean of a data set.
4 Discuss a strong correlation and possible corresponding causational relationship.
5 What factor could be a spurious variable between your stipulated independent and dependent factors? Explain with an example.
6 Explain the type of meta-analysis with an example.
7 When analyzing data, how to determine outliers?
8 Discuss the types I and II errors in hypothesis-based research with examples.
9 Can you find an example of quantitative study that is not directly related to statistical analysis?
10 Explain mathematical modeling applications in research with an example.
11 Discuss the basic concept of optimization with an example.

12 Compare between discrete event simulation and continuous simulation with examples.

13 Compare analytical and experimental studies on their characteristics.

14 Discuss one of the four types of true experiment studies.

15 Discuss whether researchers should have a system viewpoint in addition to detailed research work.

16 Think about a simple experiment to discuss its independent, dependent, and confounding variables.

17 Explain the basic concept of DOE.

18 Compare the differences between a comparative experiment and DOE.

19 Discuss the overall DOE process with a screening process.

20 Review DOE design considerations on input factors, input levels, run replicates, etc.

Mini-project Topics

1 In your study discipline, find and review a paper that uses statistical analysis.
   (a) Summarize its method, assumptions, data analysis, and results
   (b) Discuss the type of analysis (e.g. descriptive or inferential) and interpretation

2 Review a paper that uses a modeling as a research method in your field, identify and discuss the model whether it uses an existing one or proposed a new one in the paper.

3 In your familiar subject, find and review a recent scholarly paper using optimization.
   (a) The concept of optimization method
   (b) Justification of the method application
   (c) Research results interpretation

4 Review a paper that uses computer simulation is as a research method in your field and discuss the optimization method.

5 In your study discipline, find and review a paper that uses comparative analysis.
   (a) Summarize its method, assumption, data analysis, and results
   (b) Explain the type of experiment (e.g. true experimental design) and its process
   (c) Discuss the results and their reliability or limitations.

6 Select one emerging technology in your field and discuss the research opportunities associated with the technology.

7 Define the problem, input factors, and output, create a DOE design matrix, and explain your process and considerations.

References

Abid, S., Messadi, R., Hassine, T. et al. (2019). Optimization of mechanical properties of printed acrylonitrile butadiene styrene using RSM design. *The International Journal of Advanced Manufacturing Technology* 100 (5–8): 1363–1372. https://doi .org/10.1007/s00170-018-2710-6.

Al-Hadeethi, F., Haddad, N., Said, A. et al. (2017). Modeling hydrogen storage on $Mg-H_2$ and $LiNH_2$ under variable temperature using multiple regression analysis with respect to ANOVA. *International Journal of Hydrogen Energy* 42 (40): 25558–25564. https://doi.org/10.1016/j.ijhydene.2017.05.101.

Alwattar, T. and Mian, A. (2019). Development of an elastic material model for BCC lattice cell structures using finite element analysis and neural networks approaches. *Journal of Composites Science* 3 (2): 33. https://doi.org/10.3390/ jcs3020033.

Bankar, M.R. and Metre, K.V. (2019). A review on outlier detection approaches. *International Journal for Research in Applied Science and Engineering Technology* 7 (3): 1854–1857. https://doi.org/10.22214/ijraset.2019.3345.

Banyay, G.A., Shields, M.D., Brigham, J.C. et al. (2019). Efficient global sensitivity analysis for flow-induced vibration of a nuclear reactor assembly using Kriging surrogates. *Nuclear Engineering and Design* 341: 1–15. https://doi.org/10.1016/j .nucengdes.2018.10.013.

Bünger, J., Richter, S., and Torrilhon, M. (2018). A deterministic model of electron transport for electron probe microanalysis. *IOP Conference Series, Material Science and Engineering* 304: 012004. https://doi.org/10.1088/1757-899X/304/1/012004.

Cao, X., Jia, S., Luo, Y. et al. (2018). Multi-objective optimization method for enhancing chemical reaction process. *Chemical Engineering Science* 195: 494–506. https://doi.org/10.1016/j.ces.2018.09.048.

Chiu, S.W., Lin, H., Chou, C. et al. (2017). Mathematical modeling for exploring the effects of overtime option, rework, and discontinuous inventory issuing policy on EMQ model. *International Journal of Industrial Engineering Computations* 9 (4): 479–490. https://doi.org/10.5267/j.ijiec.2017.11.004.

Dong, F. and Cheng, J. (2019). A new method for estimation of aerostatic stability safety factors of cable-stayed bridges. *Proceedings of the Institution of Civil Engineers–Structures and Buildings* 172 (1): 17–29. https://doi.org/10.1680/jstbu.17 .00083.

Ghanbari, N., Shabestari, P.M., Mehrizi-Sani, A. et al. (2019). State-space modeling and reachability analysis for a DC microgrid. 2019 IEEE Applied Power Electronics Conference and Exposition. Anaheim, CA (17–21 March 2019). https://doi.org/10 .1109/APEC.2019.8721914.

Gong, J. (2018). A new, sustainable process for synthesis of ethylene glycol. *Journal of Energy Chemistry* 27 (4): 949–950. https://doi.org/10.1016/j.jechem.2018.04.005.

Graf, B., Ammer, S., Gumenyuk, A. et al. (2013). Design of experiments for laser metal deposition in maintenance, repair and overhaul applications. Second International Through-life Engineering Services Conference, Procedia CIRP. Vol. 11, 245–248. https://doi.org/10.1016/j.procir.2013.07.031.

Hayes, W. (1999). Research synthesis in software engineering: a case for meta-analysis. Proceedings of the Sixth International Software Metrics Symposium (Cat. No. PR00403), Boca Raton, FL, 143–151. https://doi.org/10.1109/METRIC .1999.809735

Haznedar, B. and Kalinli, A. (2018). Training ANFIS structure using simulated annealing algorithm for dynamic systems identification. *Neurocomputing* 302: 66–74. https://doi.org/10.1016/j.neucom.2018.04.006.

Henze, G.P., Pless, S., Petersen, A. et al. (2014). Control Limits for Building Energy End Use Based on Engineering Judgment, Frequency Analysis, and Quantile Regression. *Tech. Rep., Natl. Renewable Energy Lab., NREL/TP-5500-60020.* https:// www.nrel.gov/docs/fy14osti/60020.pdf (accessed June 2018).

Heydarpour, Y. and Malekzadeh, P. (2019). Dynamic stability of cylindrical nanoshells under combined static and periodic axial loads. *Journal of the Brazilian Society of Mechanical Sciences and Engineering* 41 (4): 184. https://doi.org/10.1007/ s40430-019-1675-1.

Hohlfeld, J., Hipke, T., Schuller, F. et al. (2018). Sandwich manufacturing with foam core and aluminum face sheets–a new process without rolling. *Materials Science Forum* 933: 3–10. https://doi.org/10.4028/www.scientific.net/MSF.933.3.

Hosseini, M.R., Chileshe, N., Zuo, J. et al. (2015). Adopting global virtual engineering teams in AEC projects: a qualitative meta-analysis of innovation diffusion studies. *Construction Innovation* 15 (2): 151–179. https://doi.org/10.1108/CI-12-2013-0058.

Hosseini, S.S., Aminian-Dehkordi, J., Kundu, P. et al. (2016). Mathematical modeling and investigation on the temperature and pressure dependency of permeation and membrane separation performance for natural gas treatment. *Chemical Product and Process Modeling* 11 (1): 7–10.

Huber, S., Dunau, P., Wellig, P. et al. (2017). Dependency of human target detection performance on clutter and quality of supporting image analysis algorithms in a video surveillance task. Proceedings of the SPIE 10432, Target and Background Signatures III, https://doi.org/10.1117/12.2278342.

Imseitif, J. and Tang, H. (2019). Effects analysis of internal buffers in serial manufacturing systems for optimal throughput. MSEC2019-2912, Proceedings of the ASME 2019 14th International Manufacturing Science and Engineering Conference, Erie, PA, USA (10–14 June 2019).

Imseitif, J., Tang, H. and Smith, M. (2019). Throughput analysis of manufacturing systems with buffers considering reliability and cycle time using DES and DOE. 25th International Conference on Production Research Manufacturing Innovation:

Cyber Physical Manufacturing, Chicago, Illinois (12–15 August 2019). https://doi.org/10.1016/j.promfg.2020.01.423.

Ji, Z., Shen, J., Ding, D. et al. (2018). The modeling and analysis of the extensible network service model. *IEEE Access* 6: 7301–7309. https://doi.org/10.1109/ACCESS .2017.2789178.

Johnson, L. (2018). Micromechanics and rheology of colloidal gels via dynamic simulation. PhD Dissertation in Chemical Engineering, Cornell University. https:// doi.org/10.7298/9601-8472

Karagöz, D. and Aktaş, S. (2018). Generalized Tietjen–Moore test to detect outliers. *Mathematical Sciences* 12 (7): 1–15. https://doi.org/10.1007/s40096-017-0239-8.

Kaya, E. and Kaya, İ. (2019). A new reverse engineering method to combine FEM and CFD simulation three–dimensional insight into the chipping zone during the drilling of Inconel 718 with internal cooling. *The International Journal of Advanced Manufacturing Technology* 100 (5–8): 2045–2087. https://doi.org/10.1080/10910344 .2017.1415933.

Kopsiaftis, G., Protopapadakis, E., Voulodimos, A. et al. (2019). Gaussian process regression tuned by Bayesian optimization for seawater intrusion prediction. *Computational Intelligence and Neuroscience* 2019: 12. https://doi.org/10.1155/ 2019/2859429.

Kriegel, H. , Kröger, P. and Zimek, A. (2010). Outlier detection techniques. The 2010 SIAM International Conference on Data Mining, Columbus, Ohio (20 April–1 May 2010). https://archive.siam.org/meetings/sdm10/tutorial3.pdf (accessed January 2020).

Kropaczek, D.J. and Walden, R. (2019). Constraint annealing method for solution of multiconstrained nuclear fuel cycle optimization problems. *Journal Nuclear Science and Engineering* 193 (5): 506–522. https://doi.org/10.1080/00295639.2018.1554173.

Lim, D.K. and Woo, D.K. (2019). A new surrogate-assisted robust multi-objective optimization algorithm for an electrical machine design. *Journal of Electrical Engineering and Technology* 14 (3): 1247–1254. https://doi.org/10.1007/s42835-019- 00120-1.

Loucif, A., Touazine, H., and Jahazi, M. (2018). Prediction of steel transformation temperatures using thermodynamic modeling and design of experiments (DOE). *Materials Science Forum* 941: 2284–2289. https://doi.org/10.4028/www.scientific .net/MSF.941.2284.

Mayet, C., Bouscayrol, A., Delarue, P. et al. (2018). Electrokinematical simulation for flexible energetic studies of railway systems. *IEEE Transactions on Industrial Electronics* 65 (4): 3592–3600. https://doi.org/10.1109/TIE.2017.2750632.

Miclo, R., Lauras, M., Fontanili, F. et al. (2019). Demand Driven MRP: assessment of a new approach to materials management. *International Journal of Production Research* 57 (1): 166–181. https://doi.org/10.1080/00207543.2018.1464230.

Mishra, P. and Sharma, R. (2017). Measuring business performance in a SCN using Six Sigma methodology – a case study. *International Journal of Industrial and Systems Engineering* 25 (1): 76–109.

Moheb-Alizadeh, H. and Handfield, R. (2017). An integrated chance-constrained stochastic model for efficient and sustainable supplier selection and order allocation. *International Journal of Production Research* 56 (21): 6890–6916. https://doi.org/10.1080/00207543.2017.1413258.

Mouhat, O., Khamlichi, A., and Limamc, A. (2015). Reliability assessment of buckling strength for imperfect stiffened panels under axial compression. *Canadian Journal of Civil Engineering* 42 (12): 1040–1048. https://doi.org/10.1139/cjce-2014-0401.

NASA (2017). Finite element analyses – not all beautiful color plots are precise or accurate. Last Updated: 5 October 2017, Editor: Daniel Hoffpauir. https://www.nasa.gov/offices/nesc/articles/finite-element-analyses (accessed August 2019).

Nazemian, M., Neshat, E., and Saray, R.K. (2019). Effects of piston geometry and injection strategy on the capacity improvement of waste heat recovery from RCCI engines utilizing DOE method. *Applied Thermal Engineering* 152: 52–66. https://doi.org/10.1016/j.applthermaleng.2019.02.055.

Otaghvar, M.H., Hahn, B., Werner, H. et al. (2019). Optimization of centerless through-feed grinding using 3D kinematic simulation. *Procedia CIRP* 79: 308–312. https://doi.org/10.1016/j.procir.2019.02.072.

Piro, M.H.A., Banfield, J., Clarno, K. et al. (2016). Coupled thermochemical, isotopic evolution and heat transfer simulations in highly irradiated UO_2 nuclear fuel. *Journal of Nuclear Materials* 478: 375–377. https://doi.org/10.1016/j.jnucmat.2016.06.030.

Quaglio, M., Fraga, E.S., and Galvanin, F. (2018). Model-based design of experiments in the presence of structural model uncertainty: an extended information matrix approach. *Chemical Engineering Research and Design* 136: 129–143. https://doi.org/10.1016/j.cherd.2018.04.041.

Rein, J., Masalonis, A., Messina, J. et al. (2012). Separation management: automation reliability meta-analysis and conflict probe reliability analysis. U.S. Department of Transportation, DOT/FAA/TC-TN-12/65, 30. hf.tc.faa.gov/publications/2012-11-separation-management/full_text.pdf (accessed March 2019).

Reuter-Oppermann, M, Kienzle, L., Zander, A. et al. (2019). Centralising the admission process in a German hospital. Proceedings of the 52nd Hawaii International Conference on System Sciences, 4098–4106. hdl.handle.net/10125/59847

Roirand, Q., Missoum-Benziane, D., Thionnet, A. et al. (2017). Finite element modelling of woven composite failure modes at the mesoscopic scale: deterministic versus stochastic approaches. *Continuum Mechanics and Thermodynamics* 29 (5): 1081–1092. https://doi.org/10.1007/s00161-017-0553-2.

Sahebjamnia, N., Goodarzian, F., Hajiaghaei-Keshteli, M. et al. (2019). Optimization of multi-period three-echelon citrus supply chain problem. *Journal of Optimization in Industrial Engineering* 12: 41–50. https://doi.org/10.22094/joie.2017.728.1463.

Sanama, C. and Sharifpur, M. (2018). Mathematical modeling of orifice downstream flow under flow-accelerated corrosion. *Nuclear Engineering and Design* 326: 285–289. https://doi.org/10.1016/j.nucengdes.2017.11.031.

Shdefat, A.Y., Joo, M., Choi, S. et al. (2018). Utilizing ECG waveform features as new biometric authentication method. *International Journal of Electrical and Computer Engineering* 8 (2): 658–665. https://doi.org/10.11591/ijece.v8i2.pp658-665.

Sirois, E., Mao, W., Li, W. et al. (2018). Simulated transcatheter aortic valve flow: implications of elliptical deployment and under-expansion at the aortic annulus. *International Center for Artificial Organs and Transplantation* 42 (7): E141–E152. https://doi.org/10.1111/aor.13107.

Sivaneasan, B., Kandasamy, N.K., Lim, M.L. et al. (2018). A new demand response algorithm for solar PV intermittency management. *Applied Energy* 218: 36–45. https://doi.org/10.1016/j.apenergy.2018.02.147.

Soydan, A.M., Tunaboylu, B., Elsabagh, A.G. et al. (2018). Simulation and experimental tests of ballistic impact on composite laminate armor. *Advances in Materials Science and Engineering* 2018: 1–12. https://doi.org/10.1155/2018/4696143.

Srividhya, S., Basant, K., Gupta, R.K. et al. (2018). Influence of the homogenization scheme on the bending response of functionally graded plates. *ACTA Mechanica* 229 (10): 4071–4089. https://doi.org/10.1007/s00707-018-2223-2.

Sun, B. and Oran, E.S. (2018). New principle for aerodynamic heating. *National Science Review* 5 (5): 606–607. https://doi.org/10.1093/nsr/nwy035.

Szymczyk, M., Sumelka, W., Łodygowski, T. et al. (2018). Numerical investigation on ballistic resistance of aluminium multi-layered panels impacted by improvised projectiles. *Archive of Applied Mechanics* 88 (1–2): 51–63. https://doi.org/10.1007/s00419-017-1247-8.

Tabares, L., Robles-Cárdenas, M., Romainville, F. et al. (2017). Evaluation and comparison of a lean production system by using SAE J4000 standard: a case study on the automotive industry in the state of Mexico. *Brazilian Journal of Operations and Production Management* 14 (4): 461–468. https://doi.org/10.14488/BJOPM .2017.v14.n4.a3.

Terleev, V., Petrovskaia, E., Sokolova, N. et al. (2016). Mathematical modeling of hydrophysical properties of soils in engineering and reclamation surveys. *MATEC Web of Conferences* 53: 6. https://doi.org/10.1051/matecconf/20165301013.

Terrab, H., Boulanouar, H., and Bayadi, A. (2018). Flashover process analysis of non-uniformly polluted insulation surface using experimental design methodology and finite element method. *Electric Power Systems Research* 163 (Part B): 581–589. https://doi.org/10.1016/j.epsr.2017.12.016.

Toğan, V. and Eirgash, M.A. (2019). Time-cost trade-off optimization of construction projects using teaching learning based optimization. *KSCE Journal of Civil Engineering* 23 (1): 10–20. https://doi.org/10.1007/s12205-018-1670-6.

Toma, E. (2018). Optimization of rotor shaft shrink fit method for motor using "Robust design". *Journal of Industrial Engineering International* 14 (4): 705–717. https://doi.org/10.1007/s40092-018-0255-9.

Tongkratoke, A., Pramuanjaroenkij, A., and Kakac, S. (2016). The development of mathematical modeling for nanofluid as a porous media in heat transfer technology. *Heat Pipe Science and Technology* 7 (1–2): 17–29. https://doi.org/10.1615/HeatPipeScieTech.2016017200.

Troost, C. and Berger, T. (2016). Advances in probabilistic and parallel agent-based simulation: modelling climate change adaptation in agriculture. Eighth International Congress on Environmental Modelling and Software – Toulouse, France (July 2016). https://www.iemss.org/publications/conference/proceedings-of-the-iemss-2016-conference

Wrigley, P.A., Wood, P., Stewart, P. et al. (2019). Module layout optimization using a genetic algorithm in light water modular nuclear reactor power plants. *Nuclear Engineering and Design* 341: 100–111. https://doi.org/10.1016/j.nucengdes.2018.10.023.

Yaghoub, V., Vakilzadeh, M., and Abrahamsson, T. (2018). Automated modal parameter estimation using correlation analysis and bootstrap sampling. *Mechanical Systems and Signal Processing* 100: 289–310. https://doi.org/10.1016/j.ymssp.2017.07.004.

Yang, Y., Li, L., and Pan, Y. (2017). Energy consumption modeling of stereolithography-based additive manufacturing toward environmental sustainability. *Journal of Industrial Ecology* 21 (S1): S168–S178. https://doi.org/10.1111/jiec.12589.

Yang, J., Rahardja, S. and Fränti, P. et al. (2019). Outlier detection: how to threshold outlier scores? The International Conference on Artificial Intelligence, Information Processing and Cloud Computing, Article No. 37, Sanya, China (December 2019), 7. https://doi.org/10.1145/3371425.3371427

Yu, C., Semeraro, Q., and Matta, A. (2018). A genetic algorithm for the hybrid flow shop scheduling with unrelated machines and machine eligibility. *Computers and Operations Research* 100: 211–229. https://doi.org/10.1016/j.cor.2018.07.025.

Zhilenkov, A. and Kapitonov, A.A. (2017). The synthesis of precise rotating machine mathematical model, operating natural signals and virtual data. IOP Conference Series: Materials Science and Engineering, Vol. 221, VIII International Scientific Practical Conference "Innovative Technologies in Engineering" (18–20 May 2017), Yurga, Russian Federation. https://doi.org/10.1088/1757-899X/221/1/012003

6

Qualitative Methods and Mixed Methods

6.1 Qualitative Research

6.1.1 Qualitative Research Basics

6.1.1.1 Qualitative Methods Overview

We as human beings always attempt to understand more and better the world we live. In many cases, we do not have numerical data for our research. In such cases, we must utilize the nonnumerical information or qualitative data, which are often expressed in the form of words, such as descriptions, accounts, and opinions that are explained in Chapter 4.

Qualitative research involves the collection and analysis of nonnumerical data. That is, qualitative research is about argumentations rather than based on calculations. Using qualitative data, we can conduct a research project to solve problems or simplify information that does not require numbers, mathematical calculations, or tangle factors.

While quantitative and qualitative research differ in types of information, research methods, and objectives. The two types of research do have some overlaps (refer to Figure 6.1).

An overall purpose of conducting a qualitative study is primarily for exploration and explanation of a phenomenon, i.e. on the meaning without numbers. Qualitative research is usually applied in sciences when there is little understanding of a subject or phenomenon. To explore new things, we often start with qualitative research, which can lead to further research ideas. Qualitative research help build off a body of knowledge until the patterns are constant the subject may be studied quantitatively.

A small amount of engineering research is still qualitative based, such as engineering design (Daly et al. 2013) and software development that often has a human interface component. While the qualitative method is not used often in the engineering fields, it does not necessarily mean that is less important. As engineering

Engineering Research: Design, Methods, and Publication, First Edition. Herman Tang.
© 2021 John Wiley & Sons, Inc. Published 2021 by John Wiley & Sons, Inc.

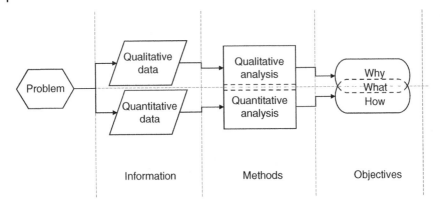

Figure 6.1 Methodology of qualitative and quantitative research.

research becomes more cross-disciplinary and novel, understanding uncommon methods to a field may lead to different ways of doing research.

Qualitative methods have good potentials in engineering and technical research. For example, a qualitative study may be exploratory to investigate a problem without knowing the exact numerical parameters of the problem. Such a qualitative research project may underpin a later quantitative study as one researcher suggested, "Engineering research should be expanded beyond its current positivist paradigm. Engineers can learn a lot about qualitative research methodologies and methods from our colleagues in the social sciences who have been using such methods for many years" (Kelly 2011).

6.1.1.2 Methods of Qualitative Analysis

Along with the nature of qualitative information and research objectives, there are many types of qualitative analysis methods. The common qualitative methods are shown in Figure 6.2.

We may consider these qualitative methods into two categories: information collection and data analysis, both in any qualitative research project. We may also use multiple analysis methods in a study. For example, we may do a case study based on a survey data and using a grounded theory from interviews along with content analysis from literature reviews.

Qualitative data processing, a part of data analysis, is about sorting and categorizing. A preliminary sort and category may reveal new findings to guide further data collection and analysis as does reviewing data within and across the categories.

6.1.1.3 Concepts and Applications

Table 6.1 lists some examples of qualitative method applications in engineering and technology fields.

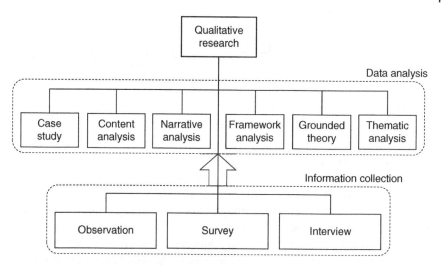

Figure 6.2 Methods of qualitative research.

Table 6.1 Basics of some qualitative analysis methods.

Method	Concept	Engineering example
Content analysis	To study documents, consisting of various formats, texts, pictures, video, etc., to examine patterns in a replicable and systematic manner	Software engineering (DeFranco and Laplante 2017)
Narrative (case) study	To study field texts, such as stories, journals, field notes, letters, conversations, and interviews, and understand the ways people create meanings in their lives	Computer science (Casebeer et al. 2018)
Framework analysis	To organize and manage research through the process of summarization, resulting in a robust and flexible matrix output for analyzing data	Manufacturing engineering (Gerritsen 2010)
Grounded theory	To gather, synthesize, analyze, and conceptualize qualitative data to develop theories	Architectural engineering (Wonoto 2017)
Thematic analysis	To pinpoint, exam, and record patterns or "themes" within data	Information technology (Wang et al. 2018)

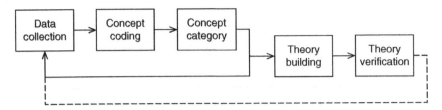

Figure 6.3 A process flow of grounded theory.

The purpose of grounded theory, its process steps shown in Figure 6.3, is to discover a theoretical framework, say a pattern that explains the process or action from the collected data. Given engineering research often leaning to applications instead of new theory development, the grounded theory may be used when there are sufficient data on a specific engineering subject for potential theoretical development, such as in software engineering (Stol et al. 2016), ergonomics (McNeese 2017), robotics (Ganesan 2017), and architectural engineering (Wonoto 2017).

6.1.1.4 Using Qualitative Methods

First, we may identify what a research problem is and what methods are applicable. One way to select a method is to develop a preliminarily plan how to conduct the research based on research objectives. Then, we exam the applicable methods and see which methods better fit our research objectives and are suitable for the anticipated outcomes. During the proposal development stage, we can review the target methods and revise the research plan.

Regarding research objectives, qualitative research is primarily used for exploratory purposes. We may apply qualitative methods to obtain a comprehensive description of a process, mechanism, or setting and the outcome can be in forms of recurrent themes or hypotheses, survey instrument measures, taxonomies, or even a new theory. In other words, we use qualitative methods to gain a new understanding, provide insights into a problem, or help develop new ideas.

Researchers often use qualitative methods on the subjects of human behaviors-related subjects to understand why something happened or not happened. For example, it may be appropriate to examining what people think, know, conceive, or perceive using interviews and surveys.

Research bias, discussed in Chapter 4, is about our influences on the results to portray a certain outcome that can significantly affect the internal validity. Even through this type of bias can happen when using all types of research methods, many people think it is more likely and severely in qualitative research than that in quantitative research. Therefore, it is important that we acknowledge and discuss the possible bias when reporting qualitative study results.

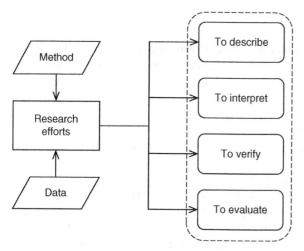

Figure 6.4 Purposes of qualitative research.

6.1.2 Discussion on Qualitative Analysis

6.1.2.1 Qualitative Data and Research

Qualitative research can have four kinds of purposes (see Figure 6.4.). For a particular study, it may focus on one purpose or have any combination of the four types.

The questions and tasks of qualitative research can be general and open-ended. For example, a qualitative study task is to answer what is the trend of autonomous vehicles in the coming 10 years. In many researchers' perspectives, the qualitative information collected may be rich because we can find not only new things but also their reasons.

Figure 6.5 shows an overall process flow of qualitative data analysis. In the process, data review is important to ensure the data pertinent to a specific study.

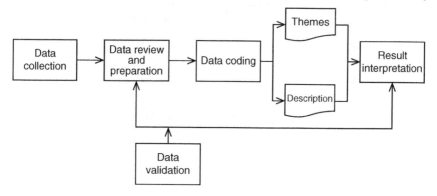

Figure 6.5 A data analysis flow in qualitative research.

Researchers suggested, "As data accumulate, the more the researcher reviews them, the easier subsequent work becomes" (Chism et al. 2008).

Interviews or observations in fieldwork are usually qualitative, which could be very subjective and thus the results of a qualitative data collection are often not repeatable. Furthermore, we often use inductive techniques to draw conclusions, so from this perspective, their validation may be challengeable.

6.1.2.2 Reflexive Thinking

Reflexivity is a process of reflecting on ourselves as researchers or on a research process. Researchers understand how a research process can influence the outcomes and findings. From this point, we may define the reflexivity as "the interpretation of interpretation" (Alvesson and Sköldberg 2009). Researchers of social sciences often use this thinking approach for the relationships between causes and effects, especially in human belief structures. Figure 6.6 illustrates a circular reflective thinking process. In this thinking process, the analysis, evaluation, and adjustment are the main efforts.

Reflexivity thinking is a type of critical thinking, which focuses specifically on the processes of analyzing and making judgments about what has happened and what will happen. For example, we may ask ourselves in the various stages of our research:

1. Why do we plan an experiment in this way?
2. Are there any theories that would help explain the results?
3. Have the data changed our results and interpretation?
4. Can the data analysis interpretation be different if using another method?

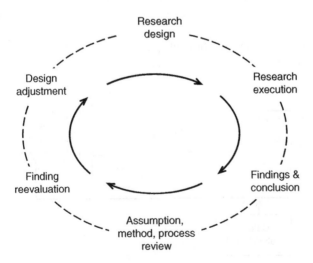

Figure 6.6 A reflexive process in research.

5. Is the interpretation on the results the only way or best way?
6. Is there evidence to show the conclusion might change if done differently?

It is likely that we have already thought this way without knowing the name of thinking reflexively. It makes us consider critically about our preconceptions and experience that can affect the research results and improves our understanding on the limits of research outcomes. Such reflective questions can be a healthy self-assessment and may reduce the possibilities of influencing by our own experiences and lead to additional research tasks or a revised research direction.

Reflexive thinking can be a good approach to examine and improve the process and results of engineering research as well. One study showed that "some engineering challenges… can be more complex because of the social contexts they inhabit." The author suggested, "All engineers could gain from integrating broader concerns into their work" (Robbins 2007). The approach of reflexive thinking has limited applications in the research of engineering and technology fields. Here are a few studies:

> "Confronting the methodological challenges of engineering practice research: A three-tiered model of reflexivity" (Sochacka et al. 2009)
> "A Case Study of a Case Study: Analysis of a Robust Qualitative Research Methodology" (Snyder 2012)
> "An Empirical Study of System Development Method Tailoring in Practice" (Fitzgerald et al. 2000)

6.1.2.3 Qualitative Analysis in Engineering

Since most engineers and technical professionals are trained primarily using quantitative methods, they may not be familiar with some norms of qualitative research.

In contrast to quantitative research, qualitative research is concerned with understanding and words, referring to the meanings, concepts, characteristics, metaphors, symbols, and/or description. For example, engineering design should be based on the requirements of customers, and we may use qualitative research to ascertain the requirements. Some cases may be good for qualitative research in engineering and technology, for example,

1. To generate knowledge or understanding of new or complex situations
2. To obtain the opinions from domain experts
3. To improve the understanding on a phenomenon

Limited engineering research has used qualitative methods without combining with a quantitative method. Here are a few examples:

> "Software development as an experiment system: a qualitative survey on the state of the practice" (Lindgren and Münch 2015)

"Qualitative research methods in spatial urban development: a methodological investigation of approaches into urban development based on centralities in the context of a medium-sized European" (Brabant 2016)
"Qualitative and kinetic analysis of torrefaction of lignocellulosic biomass using DSC-TGA-FTIR" (Acharya et al. 2015)
"A qualitative study of vortex trapping capability for lift enhancement on unconventional wing" (Salleh et al. 2018)

We may view qualitative methods complementary to commonly used quantitative research in engineering and technology research. If we master both quantitative and qualitative methods, we can be more competent to do engineering research. More discussion in the mixed methods will be in a later section of this chapter.

6.2 Questionnaire Survey

6.2.1 Basics of Survey

6.2.1.1 Survey Basics

Surveys are an important research tool when studying the information of or related to human beings (e.g. reactions, behaviors, perspectives). The objective of survey research is to learn the perspectives of a population from surveying a small group of that population. Researchers design a series of questions and other prompts for gathering information from respondents.

We may categorize survey studies into two types: descriptive and analytical. Descriptive survey is the most common type and aims to identify the frequency of a particular response. For example, multicultural human-systems engineering (Smith-Jackson 2015). Analytical survey aims to analyze a special relationship or phenomenon. Two such examples are in smart electricity meters using global system mobile (Mendiratta et al. 2018) and the distributed file systems in cloud storage (Ramesh et al. 2016).

As communication technology advances, it changes survey studies in terms of executions. For example, telephone surveys require that people pick up and stay on the line. That approach was popular but nowadays less practical with cell phones, caller identification, and blocking. Most people do not answer a phone call if the number looks not familiar with because of their feelings against tele-marketers and tele-researchers. As a telephone survey only has respondents who are willing to answer the phone, the survey respondents may not be representative that we would hope. Similarly, mail surveys in a paper-and-pencil form are outdated and their feedbacks may be biased toward older generations.

Online surveys are a common practice now. There are companies dedicating survey research, including data collection and analysis services, which are often specialized in some areas and industries. There are some Internet-based survey tools available, such as SurveyMonkey (https://www.surveymonkey.com), SurveySparrow (https://surveysparrow.com/), and Google Forms (https://www.google.com/forms/about/). In most cases, we have a relatively simple design in a survey with 20 or fewer questions.

The advantages of using electronic survey tools include

1. Improved survey efficiency
2. Low-cost delivery and submission
3. User-friendly with easy completion
4. Embedded data collection and analysis tools

6.2.1.2 Applications of Survey in Engineering

Survey method is widely used in social sciences and business disciplines while a small number of survey-based papers are published in engineering and technology. However, when studying the relationship between engineering and human beings, one of the best approaches probably is to conduct a survey. For example, several survey studies in different fields listed below:

> *"Enablers and barriers of sustainable manufacturing: results from a survey of researchers and industry professionals"* (Bhanot et al. 2015)
> *"Design structure matrix extensions and innovations: a survey and new opportunities"* (Browning 2016)
> *"Cloud manufacturing: key characteristics and applications"* (Ren et al. 2014)
> *"A Survey of networking challenges and routing protocols in smart grids"* (Sabbah et al. 2014)
> *"A study of transitions between technology push and demand-pull strategies for accomplishing sustainable development in manufacturing industries"* (Singla et al. 2018)
> *"A Compressor fouling review based on a historical survey of ASME turbo expo papers"* (Suman et al. 2017)
> *"Quality of service in cloud computing in higher education: a critical survey and innovative model"* (Upadhyaya and Ahuja 2017)

6.2.1.3 Structure of a Survey

A formal survey normally consists of four sections with key elements in each section (refer to Figure 6.7):

1. Introduction
   - Survey purpose and background
   - Information about the survey organizer
   - Statement and assurance of anonymity
   - Size of the survey and estimated time needed

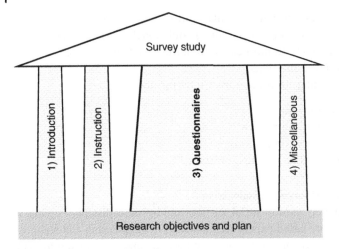

Figure 6.7 Key elements of survey study.

- Ethics and Institutional Review Board (IRB) approval, if applicable
2. Instruction
   - How to complete the survey
   - Scales and their meanings
   - Required and optional questions
   - Demographics – respondent's background (work, age, etc.)
   - Additional specifics for online, etc.
3. Main body – Questionnaires
   - Multiple-choice questions
   - True/false questions
   - Open-ended questions (often optional)
4. Miscellaneous
   - Thank-you message
   - Contact/follow-up info (optional)
   - Respondent's info for incentive drawing

6.2.1.4 Characteristics of Survey Study

One key factor is to engage participants and motivate them why to participate in a survey. The quality of these first two sections, Introduction and Instruction, as the first impression largely sets the pace and return rate for a survey. For example, the introduction should not only be concise but also sound meaningful. It should be simple enough so the participants do not get frustrated and throw it away but detailed and interesting enough to get good responses.

Table 6.2 Required sample sizes (at 95% confidence level).

Population size	Sample size (3% margin of error)	Sample size (5% margin of error)
50	48	45
100	92	80
200	169	132
300	235	169
500	341	218
1000	517	278

Sample size is one of the important factors, discussed in Chapter 4. In general, the larger the sample size, the more accurate representation the target population. To determine the sample size of a survey study, we need to first decide both the confidence level (normally set at 95%), confidence interval (or called margin of error, e.g. ±3%), as well as target population. Many survey books and online tools have sample size guidelines and calculators. Statistical analysis software, such as Minitab, has sample size estimation functions as well. Table 6.2 lists typical sample sizes for reference. Please note the calculated sample size is the necessary size for a study. The actual distribution of a survey study should be much larger considering a return rate.

Moreover, the response or return rate of a survey is often a challenging issue and can be difficult to predict. In addition to the interest in the subject to the participants, being familiar with a survey conductor is another factor. In general, internal surveys have a higher return rate than that of surveys conducted externally. For a large, external survey, the return rate is frequently between 10% and 20%. While an internal survey can have an 80% or more return rate. We will discuss how to improve the return rates in a later section.

Survey studies can be inexpensive and quick. Like a conventional paper-and-pencil way, an electronic survey is normally anonymous, which is a key assurance of a high return rate. However, survey approach has some limitations. An anonymous survey has limited respondent's information. Additionally, there is no control over participant interpretation. Responding is interests related, i.e. people with strong opinions are likely to participate in a survey. That may mean the respondent samples do not accurately represent the whole population.

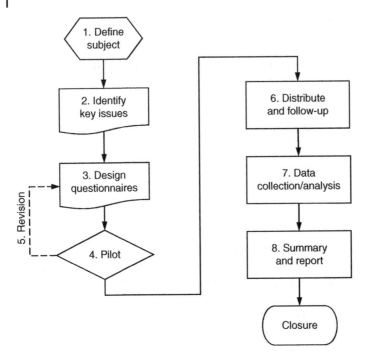

Figure 6.8 A process flow of survey study.

6.2.2 Questionnaire Development

6.2.2.1 Development Tasks

We need to develop questionnaires following a typical procedure, referring to Figure 6.8.

The main tasks of a survey research project include

1. Define the research subject and develop a work plan, including study objectives, resources, and constraints.
2. Identify the key issues and interests. As a driving force to a survey, we need to know what data to represent and how to reach the population.
3. Design questionnaires based on the objective, key issues, and interests.
4. Run a tryout or pilot test within a small group to evaluate its questionnaires and estimate the responses. Have an external review by who are not on your research team for clarity if possible.
5. Revise the questionnaires based on the tryout, mainly on the wording clarity.
6. Start the survey distribution and follow-up. To get as more responses as possible, substantial effort of follow-up is critical.

7. Conduct data analysis and interpretation. We may need to code and adjust the data received if necessary. Sometimes we can assign weights on the data to match the known population parameters.
8. Summarize the findings.

6.2.2.2 Preparing Questions

Based on the defined research subject and issues, such as behaviors, characteristics, or other entities, we develop questionnaires, as the third step of a survey research process. We may view a survey development as both an art and a science. In the development, in addition to the wording of questionnaires, we need to address the overall flow and format of questions.

A neutral position is very important for research, including survey, interview, and observational studies. For example, for a survey question design, we should avoid leading questions or having implying answers. Not give any clues and hints about preferred responses.

We may arrange the suggestions and considerations into two groups: question design and question arrangement.

For a specific question design:

1. Address one concept a question
2. Keep questions straightforward to respondents
3. Not require the participants to calculate or do extra work.
4. Not use undefined descriptive words, such as "large" and "small"
5. Not bounce back and forth between the positive and negative phrases, and no double negatives

For an overall question arrangement:

1. Keep related questions together and arrange them from general to specific
2. Design all the questions in a consistent rating order, e.g. the lowest to the highest
3. Consider crosscheck questions if needed. We may design a few questions to serve as crosschecks on the answers to important questions.
4. Offer some flexibility. Some questions are very important and we do not want any missing answers. While other questions may be less important or indirect to the research. Accordingly, the latter type of questions can be designed optional.

The total questions should be limited to 20 or 25. More questions may result in a drop in the middle before finishing. In general, a short survey can have a better chance for more responses. Without an appropriate arrangement, agreement, or compensation, a longer than 30-minute survey may be deemed impractical.

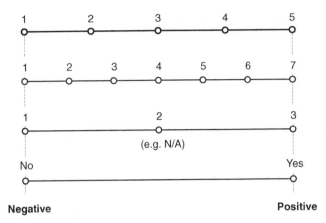

Figure 6.9 Rating scales of survey studies.

6.2.2.3 Rating Scale

Most survey questions are designed in a closed format, meaning the respondents must choose from a set of given answers. The answers may be in a two-point scale (yes and no), three-point scale (1, 2, and 3), five-point (Likert) scale (1, 2, 3, 4, and 5), or seven-point (Likert) scale, refer to Figure 6.9. A few examples in a 5-point scale are shown in Table 6.3.

The scale to use is case-dependent. If not enough response options, respondents have to choose the next best alternative, which may introduce inaccurate information. However, it is debatable that increasing the number of points in a scale can provide valuable information. Some researchers prefer a 7-point scale. However, other researchers think that too many possibilities may complicate the survey results, as the differences between choices could be more subjective or subtle.

A simple answer "yes" or "no" does not require in-depth thinking, which might encourage a better return rate. We may intentionally design some questions into a yes/no or true/false format, not allowing a neutral choice. As an example, "do you

Table 6.3 Examples of five-point Likert scale.

Rating	Example 1 (agreement)	Example 2 (occurrence)	Example 3 (importance)
1	Strongly disagree	Never	Unimportant
2	Disagree	Rarely	Little important
3	Neutral	Sometimes	Moderately important
4	Agree	Often	Important
5	Strongly agree	Always	Very important

believe that artificial intelligence replacing human beings is justified?" We might just provide two choices, "yes" or "no," not allowing a possible third answer, such as "don't know," "depends," or alike. Some researchers suggest a midpoint or N/A option. A review paper (Chyung et al. 2017) summarized the benefits and problems in several cases about the middle point as well as evidence-based strategies to employ.

The total number of questions may be a factor when deciding a point-scale. For a short survey, a 7-point scale could be fine to have detailed information. If a survey has more than 10 questions, a 5-point scale may be more suitable.

6.2.2.4 Open-Ended Questions

We often want to have additional information from respondents, which may lead to further study opportunities. When an open-ended question is available, respondents have the opportunity to further explain their responses or add more comments beyond the existing survey questions.

We often put open-ended questions at the end of a survey as seen in Figure 6.10. For a large survey with multiple sections, we may locate an open question at the end of a section.

However, it can be difficult to handle the unexpected comments as they may vary in terms of topics, contents, style, length, clarity, etc. We may try to categorize such comments into a few groups. Bear in mind, the input comments in open questions may or may not be representative because of their variation and smaller sample size.

Therefore, we should have a specific goal for an open-ended question. For example, we may provide some hints or a specific guidance to an open-ended question, instead of a blank question, say, "please identify one thing that causes you concern in this new technology."

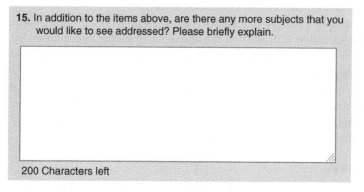

Figure 6.10 An example of open-ended question.

6.2.3 Considerations in Conducting Survey

6.2.3.1 Validation

There are two inherent issues in survey research. One is the subjectivity of the responses. Researchers and participants bring their own background, assumptions, values, perspectives, and practices into their research results. Even a survey focuses on technical; its results may be different if we conduct the same survey in different regions, on different group of people, or during different time, although such differences can be useful information as well.

The other issue of survey research is sample bias as the respondents may or may not represent the entire population. That is due to not only a small sample size but also the diverse enthusiasm and willingness of targeted participants to a survey. Some survey results were incorrect or even misleading due to inappropriate participant selection and sampling. It is difficult to know whether a survey result is fully valid.

Keeping the two potential issues in mind, we should consider the limitations of validity and strive for the balance and accuracy when developing questionnaires and planning a sampling process.

As discussed in Chapter 4, researchers sometimes use convenience sampling and purposive sampling for small surveys. The former is to collect samples that are easy to access. The latter is sample selection based on the characteristics of a population and the objective of a study. The concern about the two approaches is about their nonrandomness, which can result in incorrect conclusions. For large surveys, we should use random sampling techniques for a good validity, such as simple random sampling and stratified random sampling.

6.2.3.2 Anonymity

Anonymity and confidentiality are important for making survey participants comfortable to provide honest responses. Therefore, asking full personal identifiable information in survey questionnaires should be avoided.

We also should have a privacy statement explaining with whom the researchers will share the information with and how the research will use the information. For a possible further analysis, we may also need to get the respondents permission to use the data they provided.

One concern for anonymous surveys is that some people might be less truthful when their identities are not associated with the survey because they do not have to be held accountable for their inputs. In general, such unfavorable information to certain questions still has a reference value.

6.2.3.3 Return Rate

We should think a survey from the standpoint of survey participants. A participant may ask what in it is for me and what I will gain. If a survey makes participants care about why they participate in the survey, they would be more willing to take time for the survey. We may ask ourselves: what value are we providing for the participants?

It is often true that respondents like to access the survey results. Offering study results is an effective way to connect the target responders and increase their interests of participation. For survey results availability, we should state and explain it at the beginning of a survey.

The timing of sending a survey is also a factor to avoid holidays and typical vacation times. For internal surveys in an organization, we should have at least several weeks of interval between two surveys. There are a few additional procedural tactics to motivate survey respondents, for example:

1. Send a precontact notice
2. Write a personalizing, concise cover letter with meaningfully explanation
3. Send a reminder email to follow-up

6.2.3.4 Incentive

Many people would fill out a survey if there is an incentive. They may feel obligated to fill out and return a survey since they get "paid." If a long survey is necessary, a corresponding compensation is even more important for respondent's time.

An incentive of "a chance to win" is another way to promote a survey. However, its effectiveness may or may not be significant, particularly the odds to win is slim. Some people prefer a small and definite reward rather than a tiny chance of winning a large reward.

An incentive is not necessarily monetary and may be a "hope of change" or a "hope of no change" as well. When people feel that they are passionate about the subject, they are more likely to respond a survey. If the results of a survey can be determinative or with a clearly influence, the people with strong opinions for or against a change are likely willing to respond the survey. However, such survey results may or may not be fully representative to the entire population.

6.2.4 Data Analysis of Survey Results

6.2.4.1 Data Coding

The data analysis in qualitative studies is often called coding. It is a process of organizing data into chunks that are alike, i.e. shared properties that are meaningful for the analysis, and can help organize data and provide conceptualization as a first step.

A coding process requires reviewing, selecting, interpreting, and summarizing the information without distorting it. We should have the data analysis methods in mind during survey questionary development.

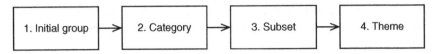

Figure 6.11 An analysis process of survey data.

The main steps of a survey data coding include (Figure 6.11):

1. *Initial Grouping.* In this first step, we group and label the raw data. The purpose is to organize the data superficially to get an overall impression with identified patterns and relationships.
2. *Category Development.* In this step, we examine the initial grouping results in details to identify preliminary categories by dividing the data meaningfully.
3. *Subset Analysis.* At this stage, we, as a team with two or three researchers, review the data independently first. Then we get the team consensus based on the research objectives. We should address the exceptions and possible outliers in this step.
4. *Thematic Analysis (Coding).* This is to refine themes. Our efforts in this step should lead to meaningful conclusions.

To assess the internal consistency (or reliability) of survey data that are made up of multiple Likert-type scales and items, we may use Cronbach's α discussed in Chapter 4. Considering Cronbach's α a measurement of scale reliability, many experts suggest that it should be above 0.70 and 0.80 or greater is preferred.

6.2.4.2 Types of Data Analysis

The common methods for the qualitative analysis for surveys are listed below:

1. Thematic Analysis. This method is to identify and report the patterns of data. We can use the method to understand the meaning of data, check the relationship between data sets, and explain the difference and relationship between data sets. An example is on modeling the implementation process of a new technology (Sepasgozar 2018).
2. Content Analysis. We use this method for the analyses of words, latent meanings, and relationships to each other to describe written, spoken, or visual communication systematically, such as an analysis is on the counting word frequencies. An application example is in an area of systems engineering (Holness 2016).
3. Discourse Analysis. We use this method to analyze the text for the type of language used, the ideas that underlie the text, and how the ideas are demonstrated through language. This type of analysis is sometimes used in computer engineering fields (Madhusudanan 2016).

4. Grounded Theory. It is a type of inductive methodology. We collect, review, and code data (may be quantitative as well) into grouped concepts, and then into categories. The process can be exhaustive. An example is a study in civil and environmental engineering research (Wonoto and Blouin 2018).

Another approach we may consider is to convert qualitative data to quantitative data (for example, a rate from very unsatisfactory to very satisfactory on a score of 1–5). Then, we can use quantitative methods, such as statistics, to analyze the converted data. The two frequently used approaches are the calculated average values and other statistics for each theme and constructed histograms to show overall data distribution.

6.3 Interviews and Observations

6.3.1 Interview Studies

6.3.1.1 Overview of Interviews

We may view an interview as a type of survey for an individual participant, in either face to face, telephone, or computer online. The purpose of an interview study is to explore individual's perceptions in detail. In other words, interviews can be a better method to collect in-depth information on people's opinions, thoughts, experiences, and feelings.

An interview study has similar characteristics with a survey study but with unique features. In design and planning stages (see Figure 6.12), we should more judiciously prepare the questions, target interviewees, moderators, and possible flexibility during interviews.

Interview studies may be more suitable for a study that requires flexible probing to obtain adequate information than questionnaire surveys. Thus, an advantage of interview studies is the flexibility and possibility to ask further and/or controversial questions. Compared with a survey, the interview approach is also more intrusive and reactive.

Interestingly, there are some interview studies in various engineering fields. For example, interview studies are in applied software engineering (Karlsson et al. 2007), product development (Münzberg et al. 2016), manufacturing (Tokola et al. 2015; Corsini and Moultrie 2017), etc.

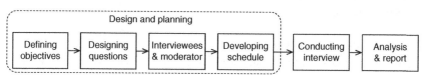

Figure 6.12 A process of interview study.

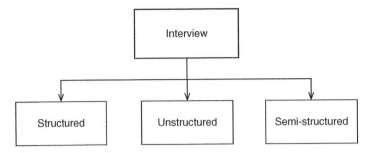

Figure 6.13 Types of interview study.

6.3.1.2 Types and Characteristics of Interviews

Figure 6.13 shows three types of interviews based on how the questions are preprepared.

An interview study with predetermined and standardized questionnaires is called a structured interview. Using this approach, we can have questions presented with the same or similar order to various interviewees to reliably aggregate the corresponding answers, which we can then do comparative analysis after interviews with a good confidence between interviewees and between different periods.

In contrast, conducting an unstructured interview means no specific question is prepared ahead of time and an interview process is "free-flowing." However, we still should have predetermined themes to cover. An unstructured interview offers us a great flexibility on the interview contents and plentiful outcomes, but they may not be fully predictable, which makes later analysis difficult.

We may also consider a semistructured style, with both predesigned and open-ended questions. In this style, we prepare a list of themes to be covered with standardized questions. During an interview, however, we may omit some questions and add new topics, depending on the flow of an interview conversation.

However, it is impractical that an interview study follows the rule of sample size for survey study. Some researchers suggest that sampling should continue until the researcher senses a saturation of knowledge on a subject, which means additional data produce little new information. A recent study concluded that the provision of sample size justifications in qualitative research is limited and is not contingent on the number of interviews (Vasileiou et al. 2018). A common practice is about 10–15 interviewees.

6.3.1.3 Considerations in Interviews

There are few practices for consideration to ensure the success of an interview. For example, we may send e-mail reminders before an interview to confirm interviewee's participation. We may also plan to limit an interview section to a half hour with about 10 questions.

We will also need to decide how to record the interview information, for example, take notes, video tape, or simply rely on our memory. Video recording or audiotaping is effective and accurate but needs the permission from interviewees beforehand. In addition, during the beginning of the interview, we may need to ensure the confidentiality and impartiality toward the interviewee.

We can elicit more in-depth responses or fill in information by helping participants to understand the questions. However, to avoid influencing or prompting a reaction from an interviewee, we should ask questions in a neutral tone of voice.

After an interview, we normally submit the transcripts and summary of interviews to the interviewees so that they may clarify matters, add a few afterthoughts, and correct misrepresentations. By doing so, we can improve interview quality by not fully relying on a quick response during interview. We may send a cordial closing and thank you letter in the following day.

6.3.1.4 Limitations of Interviews

People often welcome the opportunity to express themselves and talk about their opinions. However, there are challenging factors to a quality interview beyond the time-consumption of an interview study. The factors include

1. Analytic observation of interviewers leading to next questions
2. Possible snowball sampling in terms of questions
3. Influence of the result from the first interview on the subsequent interviews
4. Role of interviewers and their conducting consistency, if multiple interviewers

It may be difficult that interviewer keeps neutral when asking additional questions after getting answers from the basic questions. The additional questions, the tone, and interviewer's facial expression can show the interviewer's preference, which may lead to different reactions and answers.

Given the amount of time and effort required for an interview, inherently it has a limitation in the number and range of interviewees. There is an argument that an interview approach is biased in its data collection. For example, influencing factors include the social interaction between the researcher and interviewees and the researcher's own background and attitude. Such factors are difficult to be measured and fully eliminated.

6.3.2 Focus Group Studies

6.3.2.1 Objective of Focus Group

A focus group study is a guided discussion of a small group of people from a larger population. The purpose of a focus group study is to generate insights in shared experiences through group discussion. In such a study, we can concentrate in-depth on a particular theme or topic with team interaction. The development of a focus group study is similar with an interview process.

Focus group studies have been used for some engineering research, such as ecofactories (May et al. 2016), planner system (El-Sabek and McCabe 2017), and exploratory testing (Ghazi et al. 2017).

6.3.2.2 Execution of Focus Group

Working with a focus group may be in a similar way as an interview to collect opinions on a particular subject. As a common practice, we select 7–10 people into a group as a good presentation for the target population.

Based on research topics, some researchers suggest selecting and assigning group members based on their similarity. For example, we may group interviewees based on gender, position, background, etc. However, to avoid the potential conflict of interests, the group members should not have any direct family and work relationships.

During a group study, a researcher (as the group facilitator) guides a focused discussion from a general discussion to draw the main points of a research topic (see Figure 6.14). For example, a topic may be why people feel a certain way and their decision-making processes. To facilitate a group effectively, the researcher may need a dedicated note taker.

6.3.2.3 Considerations of Focus Group

The members in a group interact and influence among each other during a study. To obtain valid and meaningful results efficiently, we, as researchers, need to pay attention to the following:

1. Guide group members focusing on the topic without shutting them down
2. Encourage free discussion but be ready to intervene to resolve group problems
3. Avoid too much leading to get group member contributions
4. Not discourage unpopular or "incorrect" answers

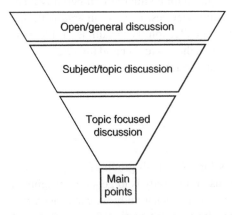

Figure 6.14 A process flow of focus group study.

While the group interaction is a strength to get additional information from the group, it also has potential downsides. For example, group dynamic stimulates conversation and reaction to discover new ideas and directions. In addition, we can find the norms and different opinions in a short time. However, a focus group study may or may not be appropriate to address some sensitive topics and some cases that group discussion could be problematic. Hence, the researcher, as a moderator, must appropriately manage and control the group discussion. For example, a moderator should eliminate or reduce the feeling of peer pressure to agree with a dominant viewpoint.

When choosing either an interview or a group study, we need to consider the cost–benefit in addition to the feasibility. Sometimes, it is easier to reach participants for individual interviews than to coordinate a group study.

6.3.3   Observational Studies

6.3.3.1   Execution of Observational Studies

Like experimental studies, the objective of an observational study is to understand the meanings of what is going on and the cause-and-effect relationships. However, different from an experimental study, we do not change the conditions and parameters during an observational study.

Some observational studies aim to develop or test the models that describe or predict a behavior. In such a study, we directly observe the subject of research, typically in a natural setting without an interference and manipulation.

During a site visit for observation, we should carefully take records, in the form of handwritten notes, photos, and/or videos. We should get as much information as possible during a site visit but use a small part of the information collected for a later analysis and study. During an observational study, we may also participate in the activity.

An observational study may involve human participants, such as production workers. If human subjects are involved, the study needs to be approved by IRB beforehand, discussed in Chapters 2 and 7. The human participants may or may not be aware of being observed. If not aware, the observation is called unobtrusive or nonreactive.

In some cases, an observational study does not have a fully defined research problem in advance. New questions likely arise during observation. For example, we may notice an unknown phenomenon and ask, "what is happening and why?"

6.3.3.2   Advantages of Observational Studies

Compared with surveys and interviews, observational studies have a greater proximity to real-world situations. Therefore, industrial professionals often use observational studies particularly for problem-solving and validation. There is some published observation-based research work:

> "*Enhancing Efficiency of Die Exchange Process through Single Minute of Exchanging Die at a Textile Manufacturing Company in Malaysia*" (Ibrahim et al. 2015)
> "*Fuzzy Cellular Automata Model for Signalized Intersections*" (Chai and Wong 2015)
> "*Empirical research methods for human-computer interaction*" (MacKenzie and Castellucci 2016)
> "*Mixed-Model Assembly Lines and Their Effect on Worker Posture and Recovery Time*" (Carrasquillo et al. 2017)

Other advantages of observational studies include that they are usually inexpensive and can generate new research initiatives or hypotheses. They may also be complementary to other research methods and simultaneously used.

6.3.3.3 Limitations of Observational Studies

Since observational studies are passive, not altering any conditions and parameters during a study can limit the depth of a study. We often use an observational approach at an early phase of a large research project.

Another concern is related to observational studies is the possibility of being subjective. The influence of observers, with their unique experience and background, can be significant because observers have a certain level of judgment and inference in their results. A study showed only 5–20% of observational studies could be replicated (Naik 2012). Using such results may inherit the invalidated results and make further research results questionable.

In order to reduce such limitations, we should get trained and practice in advance for consistent results. For example, let observers use the same rating scale to evaluate the same phenomenon independently. Then, we analyze the observational results to assess their consistency, which is similar to the reproducibility tests for quantitative data.

6.4 Mixed-Method Approaches

6.4.1 Combination of Two Types of Methods

6.4.1.1 Characteristics of Mixed Methods

When both quantitative and qualitative data available, we have an opportunity to use and analyze both of them. Qualitative research, for example, may include quantitative components and data, such as categories and frequency, and vice versa.

Using both quantitative and qualitative methods are called mixed methods or multimethods. As discussed, quantitative and qualitative analyses methods have their own characteristics (see Table 6.4). Mixed methods research may draw on potential strengths of both types of methods. Using a mixed method may, in principle, offset the weaknesses and allow both exploration and analysis in the same study. Mixed-method approaches may also provide additional evidence and support to research findings. We may contain the reduced personal biases and improved validity of results.

After understanding both types of methods, we may assess them for our research need. One study showed, "engineering studies will probably follow the contemporary trend, with respect to an increase of purely qualitative multi-method researches" (Reis et al. 2017).

Mixed methods, as a methodology, are also in debates. For example, qualitative and quantitative data can be under different assumptions and may have incommensurate differences. Some papers have detailed discussion, for example (Younas et al. 2019). Readers may like to search and study the different perspectives.

6.4.1.2 Considerations for Using Mixed Methods

We may ask ourselves whether we should use both types of methods to take the advantages of each type in one study. However, simply having both qualitative and quantitative methods in a study does not necessarily mean using an optimal methodology. In other words, both methods should have their purposes and meaningful contributions to a study. There are a few factors we may consider for what type of mixed method to be used in a study, see Table 6.5.

In general, selecting a mixed method depends on the research objectives, data availability, analysis tools, etc. If the use of a mixed method approach is justified, we need to develop a process of using the two methods. Figure 6.15 shows some additional considerations of using a mixed method (from Figure 6.1).

We may exam the research subject and data availability based on the six scenarios (Creswell and Clark 2011) to decide the mixed methodology usage and justify the appropriateness.

1. To explain initial results with another type of data
2. To use one type of data source
3. To generalize exploratory findings with a quantitative study
4. To enhance a study with a second method
5. To be required for a theoretical stance
6. To understand an objective through multiple phases

There are some methods that can use both types of data, e.g. grounded theory, which is often considered qualitative. The data used in a grounded theory study can be either quantitative or both types to construct a new theory using inductive reasoning.

Table 6.4 Characteristics of quantitative, qualitative, and mixed methods.

	Quantitative	Qualitative	Mixed
Advantages	To define and describe the facts in a quantitative way To have accurate estimations More objective	Often, relatively easy and inexpensive	Offset the weaknesses of the both methods More comprehensive understanding Better methodological flexibility Strengthened validity
Differences between Quan. and Qual.	Data collection (instruments) Deductive reasoning Focusing on individual More on theory testing Presentation (charts and tables) High generalizability Math-intensive data analysis Predetermined method	Data collection (survey, interview, etc.) Inductive reasoning Focusing on collections More on theory/understanding building Normally transcript presentation Results normally specific Simple data analysis, summary Emerging method	
Disadvantages	Analysis results maybe unexplainable Often more expensive	No accurate estimation Cost–benefit not clear More subjective Possible small, nonrepresentative samples	Complex designs More time and resources (cost) required Possible discrepancies in finding and interpretation More researcher's skills (training) needed

6.4.1.3 Applications of Mixed Method Research

Since there are many research methods, there are numerous innovative variations of combining different methods to form a new mixed method. For example, a focus group approach and back-box modeling method was jointly used in a sequential way in a study (Timma et al. 2015). In the study, authors used focus group approach

Table 6.5 Considerations for mixed method applications.

Factor	Possible option
Timing	• Sequential
	• Parallel
Importance	• Equal weight
	• Unequal weight
Data Mix	• Merge
	• Embed
	• Connect

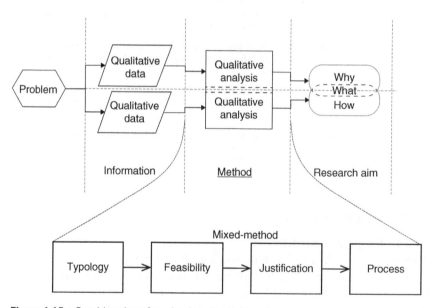

Figure 6.15 Considerations for mixed method research.

as qualitative, and the artificial neural networks and system dynamic were applied as quantitative methods. The authors stated, "This methodology combines both mixed and combined research methodology to improve the validity and reliability of study results."

To illustrate the applications of mixed methods in various engineering and technology disciplines, Table 6.6 lists some examples. They are good references for readers when you think about the applications of mixed methods for your projects.

Table 6.6 Examples of mixed method studies in engineering research.

Engineering	Method	Topic
Mechanical	Parallel	Qualitative evaluation of modeling the aramid fabric elementary cell in the piercing process with a 9 mm full metal jacket projectile (Pyka et al. 2019)
Computer	Quan–Qual	The effects of modularization in a telecommunication sector – a case study of telenor (Lindholm and Feratovic 2017)
Electrical	Qual–Quan	Ecosystem effects of the Industrial Internet of Things on manufacturing companies (Arnold and Voigt 2017)
Industrial	Quan–Qual	Business model innovation and strategy making nexus: evidence from a cross-industry mixed methods study (Cortimiglia et al. 2015)
Civil	Qual–Quan	The impact of airport performance toward construction and infrastructure expansion in Indonesia (Laksono et al. 2018)
Manufacturing	Parallel	Analysis of Overall Equipment Effectiveness (OEE) within different sectors in different Swedish industries (Cheh 2014)
Chemical	Parallel	Mixed method research: a comprehensive approach for study into the New Zealand voluntary carbon market (Birchall et al. 2016)

6.4.2 Method Integration

6.4.2.1 Integration Considerations

With two types of data and corresponding analysis methods, we need to address two considerations regarding the mixture and integration of the two types. First is how to weight quantitative and qualitative data when drawing conclusions, and the second is how to integrate quantitative and qualitative findings for interpretation. The relationship and integration of data analysis results and conclusions should ensure the validity of the overall research effort.

Overall, two types of methods may be integrated in three ways in terms of process flow in a research project (see Figure 6.16 a–c).

Normally, the different types of data are from different aspects of a research target and using different ways to collect. Therefore, it is important at an early design phase to identify and break the main research tasks down to subproblems. Every subproblem is associated with either quantitative data or qualitative data. Working on individual subproblems, we should pay attention to the independency of the data subsets since they may link each other.

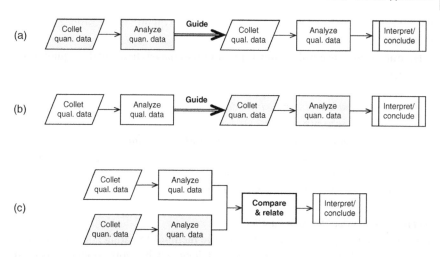

Figure 6.16 Three types of mixed method arrangements.

Qualitative then Quantitative

We may consider such a research design to explore a new phenomenon in two steps. The first step is to get a general and qualitative sense of a phenomenon. Then, we go further to do a quantitatively data analysis to quantify the results. In another case, we may use qualitative to identify and prioritize the research problems, which may not be easily obtained in quantitative.

In such a study, the qualitative research plays a directive role in the study. For example, we first conduct interviews to explore how experts prioritize main issues in a subject. Then we follow the preliminary conclusion from the interviews to further collect quantitative data and do numerical analysis. This type of research may be called an exploration design.

Here are a few of examples using a qualitative then quantitative approach:

> "*Salp Swarm Algorithm: A bio-inspired optimizer for engineering design problems*" (Mirjalili et al. 2017)
> "*Photovoltaic solar energy: Conceptual framework*" (Sampaio and González 2017)
> "*Effect of addition of micron-sized TiC particles on mechanical properties of Si_3N_4 matrix composites*" (Ye et al. 2017)

Quantitative then Qualitative

Using this type of mixed method, we may explain the quantitative results. For example, we collect quantitative data and do a statistical analysis. Then, we conduct interviews with professionals to discuss the quantitative results for a good comprehension and interpretation. The results from quantitative research lead to

form the questions and develop qualitative study on quantitative results and interpretation.

Here are a few of examples using a quantitative analysis first and then qualitative
approach:

> *"Human skeleton tracking from depth data using geodesic distances and opti-
> cal flow"* (Schwarz et al. 2012)
> *"Thermo-mechanical coupling in constitutive modeling of dissipative materi-
> als"* (Egner and Egner 2016)
> *"Effects of dual wavenumber dispersion solutions on a nonlinear monochro-
> matic wave-current field"* (Kouskoulas and Toledo 2017)

Quantitative and Qualitative in Parallel

We may also use qualitative and quantitative methods in parallel or concurrently
if appropriate. This parallel type of methodology may be more beneficial as both
methods are complementary to each other. Sometimes, this type of research
design is called triangulation or convergent, refer to Figure 6.17, which is useful
to increase the credibility and validity of research conclusions.

6.4.2.2 Discussion of Mixed Methods

Many mixed studies may not have a clear sequential order of using the two
types of methods. In the process arrangement types of A and B, one of the
qualitative and quantitative portions may play a dominative role. Even in a
parallel setting C, the two types of methods may or may not play equal roles to a
study. In such applications, the dominating method is the key for research results.

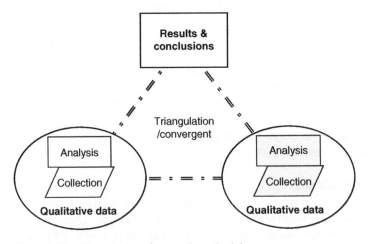

Figure 6.17 Triangulation of research methodology.

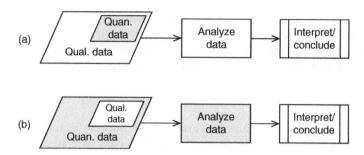

Figure 6.18   Embedded mixed method designs.

The other type plays a supplementary role. We call this situation an embedded design. Embedded designs collect both types of data but consider one of them supplemental. Figure 6.18a, b shows two possibilities of embedded designs. In most engineering research projects, quantitative methods play a dominant role.

Furthermore, we may also use both quantitative (explanatory) and qualitative (exploratory) methods into multiple iterations in a study. For example, we may go iterative steps:

1. Collect and analyze initial quantitative (measurements) data to have a baseline
2. Collect and analyze qualitative (e.g. interviews or a group study) data to interpret quantitative results
3. Collect additional data under a different setting and analyze them
4. Collect and analyze qualitative data to verify research outcomes

Mixed methods have not widely used, partially due to data availability. However, when both types of data are available, it may be beneficial to use such methods as they may bring about more opportunities to obtain research achievements.

Summary

Qualitative Research

1. Qualitative research involves the collection and analysis of nonnumerical data and information, which is often applied to when there is little understanding of a subject or phenomenon.
2. Common qualitative information comes from surveys, interviews, and observations. Qualitative methods include case study, content analysis, narrative analysis, framework analysis, grounded theory, and thematic analysis.
3. Objectives of qualitative analysis are normally on the description, interpretation, verification, or evaluation of a subject.

4. Data analysis in qualitative research includes collection, preparation, coding, analysis (themes or description), interpretation, and validation.
5. Reflexive thinking is a process of reflecting on researchers themselves or on a research process itself on assumptions, method, process, finding, etc.

Questionnaire Survey

6. There are two types of surveys: a descriptive survey to identify the frequency of a particular response and an analytical survey to analyze a special relationship or phenomena.
7. A formal survey normally consists of the four sections: introduction, instruction, main questionnaires, and miscellaneous items.
8. A survey is a study process consisting of multiple steps of design, pilot, distribution, follow-up, etc.
9. Sample size is one of the important factors to decide in survey design, which depends on the target population size.
10. Likert scales in 2-, 3-, and 5-points are often used in survey questions.
11. Open-ended questions may be used in surveys as well.
12. Return rate is a key factor to survey study success. There are a few approaches, such as various types of incentives, to improve a return rate.

Interviews and Observations

13. Interview study is a similar approach as a survey and with special characteristics, such as exploring perceptions in detail.
14. An interview study may be designed in structured, unstructured, and semistructured. The semistructured approach has a good balance of the specifics and flexibility of subjects.
15. Focus group is an approach of interviewing a small group of people, which may generate additional insights from group interactions.
16. Observational study can be used to understand to real-world situations as great proximity without any manipulate environment and conditions. However, without altering any conditions during an observational study, a study has limits on the depth of study.

Mixed Method Approaches

17. Quantitative and qualitative methods can be jointly used in a study.
18. The method combination may be in different ways in terms of order (sequential or parallel) and significance (equal or one dominative).
19. Using a mixed method can take the advantages of each method, but make the research project more complex.

Exercises

Review Questions

1 Discuss qualitative research in your field with an example.
2 Explain one type of qualitative research method and its application.
3 Compare quantitative and qualitative research for their aims.
4 Select one advantage and one disadvantage of a quantitative study over a qualitative study with an example.
5 Explain the difference between actual observations and their interpretation in a qualitative study.
6 Review the statement, "the results of a qualitative study may not be repeatable."
7 Based on your research plan discuss a step in qualitative data analysis (refer to Figure 6.5).
8 Discuss a reflexive thinking process with an example.
9 Discuss the differences between descriptive survey and analytical survey.
10 Which rating scale do you prefer? Why?
11 Discuss the pros and cons of using open-ended questions in a survey.
12 Review the subjectivity and sample bias of survey studies.
13 List a couple of tactics to improve the return rate of a survey study.
14 Discuss the effectiveness of incentives for survey studies.
15 Use examples to show the differences between the three types of interview study.
16 Review the limitation and potential issues of interview studies.
17 Compare an interview study and a focus group study and discuss how to choose one of them for a given project.
18 Discuss the uniqueness of observational studies with an example.
19 Review the advantages of mixed method research with an example.
20 Find a research paper that uses a mixed method and identify the relationship between the quantitative and qualitative methods.
21 Discuss the three types of mixed methods (Figure 6.16) for their applications.

Mini-Project Topics

1 Identify the research problem in a journal paper and discuss what type of method (quantitative, qualitative, or mixed) used.
2 Use a reflexive thinking approach to ask a couple of questions for a research project, provide possible answers, and assess the improvement opportunities.

3 Find a paper that uses survey as a research method in your discipline and review the results and conclusions on the validity and/or justification using survey method.

4 Compare two online survey tools and justify which one better fits to your study.

5 Find a survey study and review its introduction and instruction for improved return rate.

6 Find a survey and analyze its question design for possible improvement.

7 Investigate the possible sampling bias or subjectivity of a survey study.

8 Do a comparative study on the differences of methods between individual interviews and focus group studies.

9 Propose a miniature interview study with your colleague:
   a. Identify a problem around improving research project execution
   b. Craft two or three research questions for the problem
   c. Develop several interview questions to cover the research questions
   d. Perform interviews with colleagues
   e. Analyze and summarize interview results.

10 Find a paper that uses an observational approach as research method in your field and review the process and results of the observational approach and provide your comment on the effectiveness and/or justification using the approach

11 Find a paper that uses a mixed method in research in your field and review the necessity and advantages of using the mixed method.

References

Acharya, B., Pradhan, R., and Dutta, A. (2015). Qualitative and kinetic analysis of torrefaction of lignocellulosic biomass using DSC-TGA-FTIR. *AIMS Energy* 3 (4): 760–773.

Alvesson, M. and Sköldberg, K. (2009). *Reflexive Methodology: New Vistas for Qualitative Research*, 2e, 9. SAGE Publications Ltd.

Arnold, C. and Voigt, K. (2017). Ecosystem effects of the industrial internet of things on manufacturing companies. *4th International Management Information Systems Conference "Industry 4.0"* (17–20 October 2017). İstanbul, Turkey: İstanbul University.

Benner, J., Wonoto, N., and Blouin, V. (2018). Using grounded theory for the development of a structural optimization tool as a form-finding method for architectural schematic design. *Journal of Architectural Engineering Technology* 7 (1) https://doi.org/10.4172/2168-9717.1000217.

Bhanot, N., Venkateswara Rao, P., and Deshmukh, S.G. (2015). Enablers and barriers of sustainable manufacturing: results from a survey of researchers and industry professionals. *Procedia CIRP* 29: 562–567. https://doi.org/10.1016/j.procir.2015.01 .036.

Birchall, S.J., Murphy, M., and Milne, M. (2016). Mixed methods research: a comprehensive approach for study into the New Zealand voluntary carbon market. *The Qualitative Report* 21 (7) https://doi.org/10.7939/R3GF0N97Z.

Brabant, M. (2016). Qualitative research methods in spatial urban development: a methodological investigation of approaches into urban development based on centralities in the context of a medium-sized European. *Journal of Civil Engineering and Architecture* 10: 1288–1296. https://doi.org/10.17265/1934-7359/2016.11.010.

Browning, T.R. (2016). Design structure matrix extensions and innovations: a survey and new opportunities. *IEEE Transactions on Engineering Management* 63 (1): 27–52. https://doi.org/10.1109/TEM.2015.2491283.

Carrasquillo, V., Armstrong, T.J., and Hu, S.J. (2017). Mixed-model assembly lines and their effect on worker posture and recovery time. *Proceedings of the Human Factors and Ergonomics Society Annual Meeting* 61 (1): 968–968. https://doi.org/10 .1177/1541931213601723.

Casebeer, W., Ziegler, M., Kraft, A.E. et al. (2018). Human performance augmentation in context: using artificial intelligence to deal with variability—An example from narrative influence. *International Conference on Augmented Cognition.* DOI: https://doi.org/10.1007/978-3-319-91467-1_3

Chai, C. and Wong, Y.D. (2015). Fuzzy cellular automata model for signalized intersections. *Computer-Aided Civil and Infrastructure Engineering* 30: 951–964. https://doi.org/10.1111/mice.12181.

Cheh, K.M. (2014). Analysis of Overall Equipment Effectiveness (OEE) within different sectors in different Swedish industries. Master Thesis. Mälardalen University, Eskilstuna, Sweden. www.diva-portal.org/smash/get/diva2:903747/ FULLTEXT01.pdf (accessed 29 April 2019).

Chism, N.V.N., Douglas, W., and Hilson, W.J. (2008). Qualitative research basics: a guide for engineering educators, rigorous research in engineering education, NSF DUE-0341127. https://crlte.engin.umich.edu/wp-content/uploads/sites/7/2013/ 06/Chism-Douglas-Hilson-Qualitative-Research-Basics-A-Guide-for- Engineering-Educators.pdf (accessed June 2018).

Chyung, S.Y., Roberts, K., Swanson, I., and Hankinson, A. (2017). Evidence-based survey design: the use of a midpoint on the likert scale. *Performance Improvement* 56 (10): 15–23. https://doi.org/10.1002/pfi.21727.

Corsini, L. and Moultrie J. (2017). An exploratory study into the impact of new digital design and manufacturing tools on the design process. *DS 87-2 Proceedings of the 21st International Conference on Engineering Design*, Vol. 2 (21–25 August 2017). Canada: Vancouver, 21–30.

Cortimiglia, M.N., Ghezzi, A., Frank, A.G. et al. (2015). Business model innovation and strategy making nexus: evidence from a cross-industry mixed-methods study. *R&D Management* 46 (3): 414–432. https://doi.org/10.1111/radm.12113.

Creswell, J.W. and Clark, V.L.P. (2011). *Designing and Conducting Mixed Methods Research*, 2e. Los Angeles, CA: Sage.

Daly, S., McGowan, A., and Papalambros, P. (2013). Using qualitative research methods in engineering design research. *Proceedings of the International Conference on Engineering Design*, Vol.2 (19–22 August 2013). Soul, Korea, 203–212.

DeFranco, J.F. and Laplante, P.A. (2017). A content analysis process for qualitative software engineering research. *Innovations in Systems and Software Engineering* 13 (2–3): 129–141. https://doi.org/10.1007/s11334-017-0287-0.

Egner, W. and Egner, H. (2016). Thermo-mechanical coupling in constitutive modeling of dissipative materials. *International Journal of Solids and Structures* 91: 78–88. https://doi.org/10.1016/j.ijsolstr.2016.04.024.

El-Sabek, L.M. and McCabe, B.Y. (2017). Framework for managing integration challenges of last planner system in IMPs. *Journal of Construction Engineering and Management* 144 (5) https://doi.org/10.1061/(ASCE)CO.1943-7862.0001468.

Fitzgerald, B., Russo, N. and O'Kane, T. (2000). An empirical study of system development method tailoring in practice. European Conference on Information Systems. aisel.aisnet.org/ecis2000/4 (accessed June 2018).

Ganesan, R.K. (2017). Mediating human-robot collaboration through mixed reality cues. MS Thesis. Arizona State University. https://repository.asu.edu/attachments/186372/content/KalpagamGanesan_asu_0010N_16868.pdf (accessed 7 November 2019).

Gerritsen, B.H.M. (2010). Engineering frameworks: a bibliographic survey-based problem inventory. *Proceedings of the 1st IMS Summer School* (26–28 May 2010). Zürich, Switzerland, 71–106.

Ghazi, A.N., Petersen, K., Wohlin, C, and Bjarnason, E. (2017). A decision support method for recommending degrees of exploration in exploratory testing. C. https://arxiv.org/abs/1704.00994 (accessed June 2018).

Holness, K. (2016). Content analysis in systems engineering acquisition activities, SYM-AM-16-025. www.dtic.mil/dtic/tr/fulltext/u2/1016695.pdf (accessed July 2018).

Ibrahim, M.A., Mohamad, E., Arzmi, M.H. et al. (2015). Enhancing efficiency of die exchange process through single minute of exchanging die at a textile manufacturing company in Malaysia. *Journal of Applied Sciences* 15 (3): 456–464.

Karlsson, L., Dahlstedt, A.G., Regnell, B. et al. (2007). Requirements engineering challenges in market-driven software development–An interview study with practitioners. *Information and Software Technology* 49 (6): 588–604. https://doi.org/10.1016/j.infsof.2007.02.008.

Kelly, K. (2011). Qualitative research methods in engineering. *2011 ASEE Annual Conference & Exposition, Vancouver, BC* (26–29 June 2011).

Kouskoulas, D.M. and Toledo, Y. (2017). Effects of dual wavenumber dispersion solutions on a nonlinear monochromatic wave-current field. *Coastal Engineering* 130: 26–33. https://doi.org/10.1016/j.coastaleng.2017.09.016.

Laksono, T.D., Kurniasih, N., Hasyim, C. et al. (2018). The impact of airport performance towards construction and infrastructure expansion in Indonesia. *Journal of Physics: Conference Series* 954: 012015. https://doi.org/10.1088/1742-6596/954/1/012015.

Lindgren, E. and Münch, J. (2015). Software development as an experiment system: a qualitative survey on the state of the practice. *International Conference on Agile Software Development* 212: 117–128. https://doi.org/10.1007/978-3-319-18612-2_10.

Lindholm, T. and Feratovic, K. (2017). The effects of modularization in a telecommunication sector–a case study of Telenor. Master's Degree Thesis. Blekinge Institute of Technology, Karlskrona, Sweden. www.diva-portal.org/smash/get/diva2:1242710/FULLTEXT02 (accessed 29 April 2019).

MacKenzie, I.S. and Castellucci, S.J. (2016). Empirical research methods for human-computer interaction. *ACM SIGCHI Conference on Computer-Human Interaction – CHI EA 2016, San Jose, California* (7–12 May 2016), 996–999. DOI: https://doi.org/10.1145/2851581.2856671

Madhusudanan, N. (2016). Discourse analysis based segregation of relevant document segments for knowledge acquisition. *Artificial Intelligence for Engineering Design, Analysis and Manufacturing* 30: 446–465. https://doi.org/10.1017/S0890060416000408.

May, G., Stahl, G., and Taisch, M. (2016). Energy management in manufacturing: toward eco-factories of the future–A focus group study. *Applied Energy* 164: 628–638. https://doi.org/10.1016/j.apenergy.2015.11.044.

McNeese, N.J. (2017). Identification of the emplacement of improvised explosive devices by experienced mission payload operators. *Applied Ergonomics* 60: 43–51. https://doi.org/10.1016/j.apergo.2016.10.012.

Mendiratta, S., Garg, M., Jadon, J.S., and Arora, N. (2018). An analytical survey on smart electricity meter using GSM. In: *Smart Computing and Informatics. Smart Innovation, Systems and Technologies*, vol. 78 (eds. S. Satapathy, V. Bhateja and S. Das). Singapore: Springer https://doi.org/10.1007/978-981-10-5547-8_46.

Mirjalili, S., Gandomi, A.H., Mirjalili, S.Z. et al. (2017). Salp swarm algorithm: a bio-inspired optimizer for engineering design problems. *Advances in Engineering Software* 114: 163–191. https://doi.org/10.1016/j.advengsoft.2017.07.002.

Münzberg, C. Gericke, K., Oehmen, J., and Lindemann, U. (2016). An explorative interview study on crisis situations in engineering product development.

Proceedings of the International Design Conference–DESIGN '16, Dubrovnik, Croatia (16–19 May 2016). hdl.handle.net/10993/26808

Naik, G. (2012). Analytical trend troubles scientists. *The Wall Street Journal* https:// www.wsj.com/articles/SB10001424052702303916904577377841427001840.

Pyka, D., Pach, J., Bocian, M. et al. (2019). Qualitative evaluation of modeling the aramid fabric elementary cell in the piercing process with a 9 mm full metal jacket projectile. *Proceedings of the 14th International Scientific Conference: Computer Aided Engineering.* DOI: https://doi.org/10.1007/978-3-030-04975-1_67

Ramesh, D., Patidar, N., Kumar, G., and Vunnam, T. (2016). Evolution and analysis of distributed file systems in cloud storage: analytical survey. *2016 International Conference on Computing, Communication and Automation (ICCCA), Noida,* 753–758. doi: https://doi.org/10.1109/CCAA.2016.7813828.

Reis, J., Amorim, M., and Melao, N. (2017). Breaking barriers with qualitative multi-method research for engineering studies: pros, cons and issues. *Proelium* 7 (12): 275–292.

Ren, L., Zhang, l., Wang, L. et al. (2014). Cloud manufacturing: key characteristics and applications. *International Journal of Computer Integrated Manufacturing* 30 (6): 501–515. https://doi.org/10.1080/0951192X.2014.902105.

Robbins, P.T. (2007). The reflexive engineer: perceptions of integrated development. *Journal of International Development* 19: 99–110. https://doi.org/10.1002/jid.1351.

Sabbah, A.I., El-Mougy, A., and Ibnkahla, M. (2014). A survey of networking challenges and routing protocols in smart grids. *IEEE Transactions on Industrial Informatics* 10 (1): 210–221. https://doi.org/10.1109/TII.2013.2258930.

Salleh, M.B., Kamaruddin, N.M., and Mohamed-Kassim, Z. (2018). A qualitative study of vortex trapping capability for lift enhancement on unconventional wing. *IOP Conference Series: Materials Science and Engineering* 370: 012054. https://doi .org/10.1088/1757-899X/370/1/012054.

Sampaio, P.G.V. and González, M.O.A. (2017). Photovoltaic solar energy: conceptual framework. *Renewable and Sustainable Energy Reviews* 74: 590–601. https://doi .org/10.1016/j.rser.2017.02.081.

Schwarz, L.A., Mkhitaryan, A., Mateus, D., and Navab, N. (2012). Human skeleton tracking from depth data using geodesic distances and optical flow. *Image and Vision Computing* 30: 217–226. https://doi.org/10.1016/j.imavis.2011.12.001.

Sepasgozar, S.M.E. (2018). Modeling the implementation process for new construction technologies: thematic analysis based on Australian and U.S. practices. *Journal of Management in Engineering* 34 (3) https://doi.org/10.1061/ (ASCE)ME.1943-5479.0000608.

Singla, A., Sethi, A.P.S., and Ahuja, I.S. (2018). A study of transitions between technology push and demand pull strategies for accomplishing sustainable development in manufacturing industries. *World Journal of Science, Technology and Sustainable Development* https://doi.org/10.1108/WJSTSD-09-2017-0028.

Smith-Jackson, T. (2015). Research challenges in multicultural human-systems engineering. *65th Annual Conference and Expo of the Institute of Industrial Engineers*. Nashville, Tennessee, 1121–1130, ISBN: 978-1-5108-1368-7.

Snyder, C. (2012). A case study of a case study: analysis of a robust qualitative research methodology. *The Qualitative Report* 17 (13): 1–21.

Sochacka, N., Walther, J., Jolly, L. et al. (2009). Confronting the methodological challenges of engineering practice research: a three-tiered model of reflexivity. *Proceedings of the Research in Engineering Education Symposium* (20 July—23 July 2009). Queensland: Palm Cove.

Stol, K. P., Ralph, P., and Fitzgerald, B. (2016). Grounded theory in software engineering research: a critical review and guidelines. *2016 IEEE/ACM 38th International Conference on Software Engineering, Austin, TX*, 120–131. DOI: https://doi.org/10.1145/2884781.2884833

Suman, A., Morini, M., Aldi, N. et al. (2017). A compressor fouling review based on an historical survey of ASME turbo expo papers. *Journal of Turbomachinery* 139 (4): 23. https://doi.org/10.1115/1.4035070.

Timma, L., Blumberga, A., and Blumberga, D. (2015). Combined and mixed methods research in environmental engineering: when two is better than one. *International Scientific Conference Environmental and Climate Technologies–CONECT 2014, Energy Procedia*, Vol. 72. Riga, Latvia, 300–306.

Tokola, H., Jarvenpaa, E., Salonen, T. et al. (2015). Shop floor-level control of manufacturing companies: an interview study in Finland. *Management and Production Engineering Review* 6 (1): 51–58. https://doi.org/10.1515/mper-2015-0007.

Upadhyaya, J. and Ahuja, N. J. (2017). Quality of service in cloud computing in higher education: a critical survey and innovative model. *2017 International Conference on I-SMAC (IoT in Social, Mobile, Analytics and Cloud), Palladam, 2017*, 137–140. DOI: https://doi.org/10.1109/I-SMAC.2017.8058324

Vasileiou, K., Barnett, J., Thorpe, S., and Young, T. (2018). Characterising and justifying sample size sufficiency in interview-based studies: systematic analysis of qualitative health research over a 15-year period. *BMC Medical Research Methodology* 18, Article number: 148. DOI: https://doi.org/10.1186/s12874-018-0594-7.

Wang, S., Wang, H., and Khalil, N. (2018). A thematic analysis of Interdisciplinary Journal of Information, Knowledge, and Management. *Interdisciplinary Journal of Information, Knowledge, and Management* 13: 201–231. https://doi.org/10.28945/4095.

Wonoto, N. (2017). Integrating parametric structural analysis and optimization in the architectural schematic design phase. Doctoral Dissertation. Clemson University. https://tigerprints.clemson.edu/all_dissertations/1963

Ye, C., Yue, X., Ru, H. et al. (2017). Effect of addition of micron-sized TiC particles on mechanical properties of Si_3N_4 matrix composites. *Journal of Alloys and Compounds* 709: 165–171. https://doi.org/10.1016/j.jallcom.2017.03.124.

Younas, A., Pedersen, M., and Tayaben, J.L. (2019). Review of mixed methods research in nursing: methodological issues and future directions. *Nursing Research* 68 (5) https://doi.org/10.1097/NNR.0000000000000372.

Part III

Management, Writing, and Publication

7

Research Execution and Management

7.1 Basics of Project Management

As researchers, we are fully in charge of performing research activities on the technical side. Managing a research project is critical to research success and adds significant work for principle investigators (PIs). Administration related tasks, including management of operational and business aspects of a research project, need professional support, procedural review, and approval processes for large and externally funded projects. In other words, successfully preparing and submitting a research proposal, conducting an approved research project, and closing out a project require teamwork between the research team and institutional administration teams. We may discuss this teamwork starting from the understanding of research life cycle.

7.1.1 Life Cycle of Research Project

7.1.1.1 Research Life Cycle

Research life cycle is the entire process and life of a project, from inception to completion. There are four phases for a research project from a viewpoint of researchers and their organization. Figure 7.1 shows the four phases and Table 7.1 lists the main tasks of the phases for large projects.

We already discussed the tasks of the first two phases, i.e. proposal development and reviews in Chapter 2 and literature review in Chapter 3. In this chapter, we have additional discussions on the two phases regarding the collaboration between researchers and administration, and address the tasks of the last two phrases.

7.1.1.2 Main Aspects of Research Project

A research project can be complex, with the uncertainty and risks for its outcomes. Good project management and institutional administration support have better

Engineering Research: Design, Methods, and Publication, First Edition. Herman Tang.
© 2021 John Wiley & Sons, Inc. Published 2021 by John Wiley & Sons, Inc.

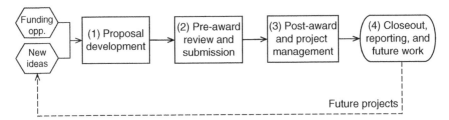

Figure 7.1 Four phases in research project life cycle.

control over on such uncertainty and risks. We, as PI and main team members, should adopt the approaches of general project management with some modified aspects due to the characteristics of research.

For a funded research project, the PI is the project director. In the entire course of research work, PI is responsible and takes care of six main aspects of a research project (Figure 7.2) on a routine basis.

Among these six aspects, project timing, cost control (budget and expenditures), and team management of research project management are similar to those of general project management. For a research project, the quality of deliveries or innovation, instead of quantity, is often more important than that of a general project. In addition, we should work closely with all internal and external stakeholders for the existing project and future opportunities.

Most experienced researchers agree that well-communicated teamwork is the key for success. The PI must take care of their team members to ensure the effective teamwork to the research goals. A PI's responsibility includes the following:

- View and work on a big picture, have good understanding on progress and road-blocks.
- Prioritize the research tasks and resources to the objectives.
- Improve the communication among team members and stakeholders.

7.1.1.3 Overall Efforts

General project management consists of four process phases: initiating, planning, executing and managing, and closing. While as discussed above, a research process has four phases. Figure 7.3 shows a direct comparison between general project management and research project management for their phases.

The proposal development of a research project is about equivalent to the initiating and planning phases combined in general project management. For research, the initiating and planning of research are often one integrated stage, while the pre-award management phase of a research project has additional tasks.

In a research proposal, we plan the research efforts to be as balanced as possible over the course of a project. However, the realistic efforts often deviate from the

Table 7.1 Tasks in four phases of research life cycle.

Phase	Tasks (ref. the book sections)	Contents
(1) Proposal development	Prepare, draft, review, revision (Sections 2.1–2.3) Literature Review (Chapter 3)	Problem statement (hypothesis) Literature review Funding sources Description of work Team build up Budget plan and justification Collaborator letter of intent (if applicable) Researchers' bio sketches Data management plan
(2) Pre-award review and submission	Review, compliance, routing approvals (Sections 2.4–2.5, and 7.3)	Administrative review and approvals Facilities and recourses review Finance review Internal coordination Collaborative planning Regulatory compliance review Submission
(3) Post-award and project management	Execution, monitoring, reporting, administrative process (Section 7.4.1)	Review, negotiation, and acceptance Project accounting set up Research tasks execution Resources and expenditures control Collaboration and sub-awards (if applicable) Cost sharing (if applicable) Award changes (if applicable) Regular reports
(4) Closeout, reporting, and future work	Reporting, administrative process (Sections 7.4.2 and 7.4.3)	Final project report Finance report Accounting closeout Equipment and facilities Records retention Dissemination Patent application and technology transfer

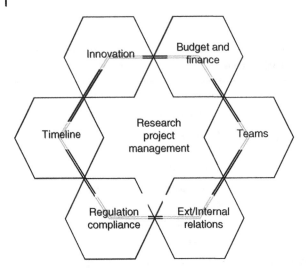

Figure 7.2 Main aspects of research project management.

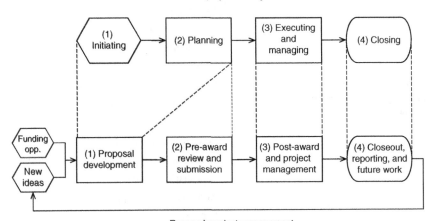

Figure 7.3 General vs. research project management.

planned, shown in Figure 7.4. For example, we may begin a project a little late and run slow at the beginning, which result in a more demanding workload later near the deadline. Please note that the total amount of the actual work can be more than that of the planned as well.

Such deviations, due to various reasons, are neither desirable nor optimal regarding the deployment of resources and team workload balance. Significant deviations from the plan can also adversely affect other work and activities. The

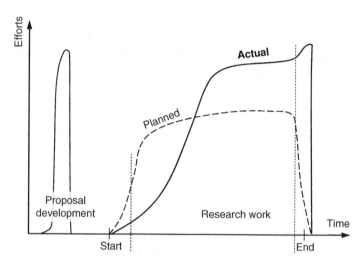

Figure 7.4 Research efforts throughout project life span.

PI should try his/her best to manage the efforts and make progress "back on the track" as planned.

7.1.1.4 Continuous Advance

As emphasized, the essential requirement for research is innovation. After a research proposal is approved, we go to the post-award stage and focus on the research execution and achievements via project management. We should keep innovative thinking during the post-award stage as well. In fact, the successful closeout of a research project heavily depends on its innovative outcomes.

We must recognize that the research life cycle never ends, in terms of new directions, activities, and research opportunities. After a project is finished, we normally move research continuously forward for new projects. A research continuation on a specific topic may be carried out by ourselves, students, and/or colleagues. In this sense, research itself can "automatically" generate new ideas, new directions, and new methodology if we are open-minded and willing to continue.

To consider a new research direction, we consider two steps of diagnostics to help finalize our research direction and plan (Czarniawska 2015):

1. Focusing on researcher himself or herself to "decide what approaches, methods and techniques would be best for you–no matter what your friends and colleagues chose."
2. Focusing on research subject "to identify the stage you have reached in your research project and make a list of actions necessary to finish the project."

Thinking about future work is a common practice in research community. Many researchers have a brief statement or discussion at the end of their reports and papers. As mentioned in Chapter 3, such a statement or discussion is a valuable reference to other researchers as well.

7.1.2 Performance of Research Project

7.1.2.1 Performance and Planning

For a research project, the three key indicators or elements of performance are outcome, cost, and timeline, as shown in Figure 7.5. We should clearly define and address all three elements in a research proposal. A viable work plan is the "blueprint" to accomplish the stated research tasks and meet the defined project objectives.

The significance of each element depends on the nature of a research project, while the outcome is often the most important and challenging part. In many cases, we consider the three key elements equally important for a research proposal and its execution. As the timeframe is normally fixed, the delivery pf an outcome on time is critical for many projects.

We measure the achievement and performance of a research project in various ways (refer to Table 7.2). For a particular project, the assessment can be on one or more aspects of the predefined goals and targets. Readers may refer to dedicated literature, such as performance measurement and reporting systems (Krugler 2008)

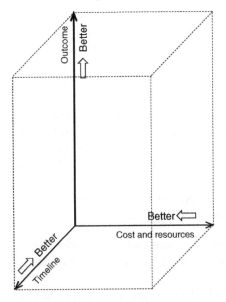

Figure 7.5   Three key elements of research proposal and execution.

Table 7.2 Evaluation considerations of research performance.

Outcome	Description	Form
Knowledge	Contribution to knowledge or establishing new understanding	Refereed publications, reports, policies, etc.
New artifacts	In forms of new products, process, software, standards, technologies, etc.	Effect data, savings, applications, etc.
Significant improvements	Impacts on environment, quality of life, safety, security, education, etc.	Changing rates, ratios, etc.

and how to measure research efficiency in higher education (Gralka et al. 2019), for more detailed discussion.

As discussed in Chapter 2, the development phase of a research proposal plays a vital role to the success of a project since a proposal shows a real plan and path of research. For external funding applications, even though the approval rate is low, we should prepare a research project assuming it will get approval.

7.1.2.2 Execution and Reporting

In the life cycle of a research project, project execution is the main effort. The detailed tasks of research execution can be project-dependent. In general, the main tasks of conducting a large research project, from a PI's standpoint, include:

- Establish a project team
- Refine and update a research plan
- Guide a team for technical tasks, such as:
  - o Retrieve information and collecting data
  - o Analyze information
  - o Obtain and reviewing results
  - o Revisit and validate
  - o Draw conclusions
- Manage resources and expenditure
- Coordinate with external partnership or suppliers, if applicable
- Work with institutional research administration (RA)
- Report to the funding sponsor

Research administrative services are integral to a project. PI's effectively using such services can improve the efficiency of research work and reporting. It is PI's responsibility to submit to the sponsor annual technical reports and final reports. Other deliverables may be required under a research agreement. More discussion on the project reporting is in Section 7.4.

7.1.2.3 Progress Monitoring

We often use a Gantt chart for a research plan during proposal development (Chapter 2). We use the same approach in the monitoring and management of research execution. Figure 7.6 shows an example of a research project Gantt chart (using Microsoft Project).

From a standpoint of project management, we should view and use a Gantt chart as a dynamic tool and update the chart on a regular, for example biweekly, basis. The research tasks, current date, expected status, and actual status should be clearly indicated in a Gantt chart. We recommend having team members update their progress to the PI in a regularly meeting, which may take a few minutes as a project task. Then, the PI discusses the issues and tasks with the team from there.

For reported major issues or roadblocks, a PI can call a dedicated meeting with the involved team members to address the issues and find remedy. We should review a major issue in its two key aspects. Figure 7.7 shows a review process flow.

1. The immediate impacts on the project timing and progress.
2. The relatively long-term, potential risk.

The impact review is to check the possible delay of research progress due to the reported issue. A common problem is the actual progress of a task is slower than the planned because of unexpected difficulties or factors. The risk assessment is proactive. Even if an issue may not have a major and immediate impact, the issue can have a potential effect in the near future. So, we should consider actions now to mitigate future risks. Therefore, if we realize a major consequence from any of these two key aspects, we must make adjustments and revision of the project plan, in terms of timing, resources, methodology, etc.

7.1.2.4 Project Adjustments

It is difficult to predict precisely the timeline of research tasks in a proposal planning phase. However, during project execution, we can monitor and detect the progress of some individual tasks not following the schedule for various reasons. Thus, it is essential that we review and consider updating the plan based on the realistic achievements up to date. One important capability of a PI is sensitivity to the progress and willingness to adjust when needed. The required adjustments may involve the project schedule, budget, personnel, techniques, etc.

Even if most adjustments are realistically feasible, they often require additional resources or money. Figure 7.8 illustrates a concept that shortening the duration of a task can result in a demand for additional resources. We often call a new, shortened schedule a "crash" schedule. It is PI's responsibility to make decisions on such adjustments to ensure the project's overall success.

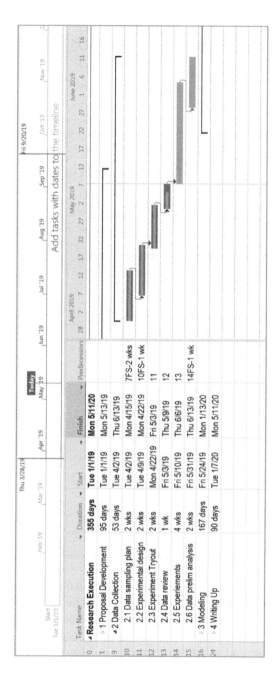

Figure 7.6 An example of Gantt chart for research project.

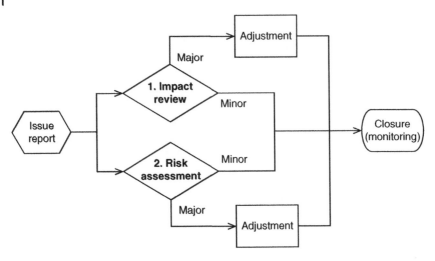

Figure 7.7 A review process for reported issues in execution.

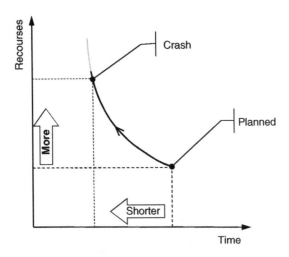

Figure 7.8 A diagram of trade-off between time and resources of a task.

For a graduate student thesis or dissertation, the student is also the project leader under the supervision of a faculty advisor. As a part or a subset of faculty research, a graduate student normally does not control the project expenditures or recourses, and requires faculty advisor's approval. Thus, students need to work closely with their advisors to update and adjust their project as necessary.

7.2 Research Administration

7.2.1 Overall Functionality

7.2.1.1 Goals of Research Administration

RA at a research organization is a main function for the institute. Overall, RA maintains the relationship of trust and cooperation between funding sponsors, the institute, and researchers, which is also a foundation of proposal approval and future awards. For example, the mission of the Office of Research, University of Michigan is "To catalyze, support and safeguard U-M research" (UM n.d.). Other examples include the research mission statement of University of California Santa Cruz, "... supports, fosters, and champions world-class research and academic excellence" (UCSC n.d.) and the University of Southern California Office of Research "... building interdisciplinary research collaborations that address societal needs and by increasing the impact and prominence of our research" (USC n.d.).

The RA is integral part of research that aids the institutional research mission and ensures all required compliance. For both external and internal projects, researchers must work with their RA team to ensure a successful project. Therefore, understanding the RA functions as well as its organization can make research project management smooth and effective.

7.2.1.2 Administration and Support

There are many functions in an RA office (Figure 7.9). The functions include establishing policies, helping researchers find funding, processing grant applications,

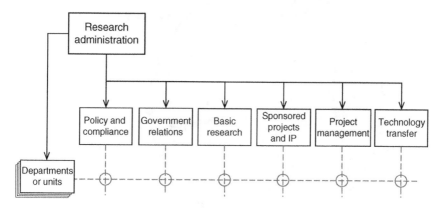

Figure 7.9 A diagram of functionality of research administration.

liaising with external funding sponsors, compiling various reports, coordinating internal funding and resources, ensuring research compliance with government regulations, arranging training, and so on. As an RA member stated, "research administration: it's complicated" (Jeracki 2018).

Depending on the size of an organization and volume of research activities, the RA functions may be performed by individual staff or a small team. Some functions may be combined in an institute. The RA's accountability is normally to the vice president of the institute.

In the different stages of approved research projects, the RA staff's work focuses may be on the pre-award and post-award phases in the life cycle of research projects. Figure 7.10 shows these phases, and the relative amounts of workload for PI and RA. Research project management (or administration) really begins at the time of proposal initiation to assure the compliance with award requirements, terms and conditions, etc.

In pre-award stages, the RA staff interacts with researchers and helps them be successful in proposal development. During post-award stages, the staff supports project execution in various ways. We will discuss both pre-award management and post-award management in the following sections.

For student research, it can be either an internal project or a part of an external project. Students often contact with the RA offices through their advisors for project-related tasks, such as proposal requirements, Institutional Review Board (IRB) approval, etc.

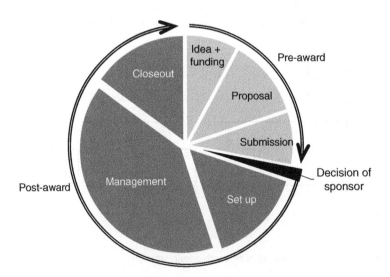

Figure 7.10 Phases of research project management.

7.2.1.3 Main Research Offices
Sponsored Projects
Sponsored projects are funded by external sources, such as the federal and state government, an industrial corporation, and charity foundations. The RA sponsored projects office helps researchers with contract terms, export control, intellectual property (IP) matters, proprietary rights, legal issues, material transfer, nondisclosure agreements, etc. This office also helps researchers negotiate contract terms and grants with sponsors.

The agreement of a sponsored project has three characteristics per Stanford University (n.d.). Without these characteristics, the support is considered as a gift or fellowship.

1. Has a detailed statement of work and commitment to a specified project plan.
2. Includes detailed financial accountability.
3. Includes the terms and conditions for the disposition of tangible and intangible properties.

Compliance
Research compliance and integrity are expected so that we carry out our work according to high ethical and professional standards. We must comply with the government's regulatory requirements relating to research, such as IRB, animal subjects, and export-control. In addition, the team of compliance and integrity in an RA office can help researchers reduce regulatory burden.

All the compliances must be satisfied with formal audits. Each federal agency has its own Office of Inspector General (OIG) and annual audit plans. For example, readers may visit the NSF's OIG website: www.nsf.gov/oig/. The highlights of audits include grants policy implementation, information security, cost principals, and 2 CFR Part 200, Subpart F (originally OMB A-133 Audit Supplement) (OMB, Office of Management and Budget 2018). The NSF publishes their audit report with findings of noncompliance every year. For example, one report states (Smith+Brown 2019),

> *"The auditors found $233,075 of inadequately supported charges, $125,458 in equipment purchases that did not benefit the award, $44,469 for unreasonable or unallowable travel and related charges, $19,208 in unreasonable materials and supplies, $2,465 in unallowable indirect costs, and $1,992 in unallowable salaries and wages."*

Technology Transfer
This function is for IP commercialization and industry relationships; some organizations call the function industry/corporate relations. This function provides the

connection between researchers and industry partners for the applications of basic and applied research results of the institute. In addition, this function is to identify new corporate partnership opportunities, match research assets, achievements to industry applications, support new sponsored research projects, and enhance research enterprise.

The technology transfer function covers the aspects of law, commerce, marketing, and financial analysis and provides services of patenting and licensing for IP. For example, "The Technology Transfer Office connects researchers with industry to bring inventions to the marketplace" (OU n.d.). This function may be called Technology Commercialization as well (WSU n.d.), "responsible for the identification, protection, marketing and licensing of intellectual property (e.g. patents, unique biological or other materials, and certain copyrights) developed by WSU faculty, staff and students."

Government Relations

Government relations is another important function of a research institute. This office is responsible for contact with Congress, the executive branch, federal agencies, research organizations, and state government to bridge the government with researchers of the institute. Government Relations is an institutional function and may have an office within the RA.

7.2.1.4 Teamwork Between PI and RA

Generally, we as researchers are a representative of the institute and as such are responsible for assuring efforts of good stewardship. The PI, by name, plays a principal role in conduct of research activities. The PI and his/her research team should work with RA and get administrative support on a regular basis. The PI may even delegate some work to RA administrators.

For externally funded projects, a research PI needs to cooperate with RA and/or his/her unit administration. The PI and team members are responsible (R) for most research tasks with the support from RA. On the other hand, RA is in charge, as a responsible (R) and/or approver (A) of some tasks for a project. For some tasks and functions, for example, PI should consult (C) or just inform (I) the RA and/or Unit Administration. Table 7.3 summarizes the general relationships in an responsible, approver, informed, and consulted (RAIC) matrix. In some of cases, approver (A) can be combined with another role (X) as (X/A).

The work relationship table also includes a research unit (departmental or central) administration. Note the functionality and requirements of a research unit may vary from one organization to another. However, we may think about these tasks based on this general example and work with the institute RA and research unit with specific requirements and responsibilities.

Table 7.3 An example of work relationship between PI, RA, and Unit.

Phase	Task	PI	RA	Unit
Proposal	Provide grant information	C	R	I
	Identify funding opportunities	R	C	C
	Provide training and advice	I	R	I
	Prepare quality proposal	R	C	I
Pre-award	Review and revise proposal	R	C/A	C
	Do regulatory compliance and certification	S	R/A	I
	Prepare and route for approvals	C	R	S
	Submit proposal	I	R	I
Post-award	Accept (and negotiate) award	C	A	S
	Set up accounting	I	R	S
	Set up team	R/A	I	S
	Monitor and assure expenditures	R	C/A	I
	Manage research efforts	R	I	S
	Do regular reporting as required	R/A	S	I
Closeout	Prepare final (tech and fiscal) reports	R/A	C	C
	Review and submit final reports	S	R/A	I
	Maintain closeout doc and records	S	R	I

7.2.2 Academic Integrity

Research institutions, via their RA, play a central role in administrating research compliance and protecting research integrity.

7.2.2.1 Research Misconduct

We must conduct research projects and activities according to high standards of academic integrity and ethics. Academic integrity and ethics should be self-policed and administrated by an RA office to avoid possible research misconduct.

There are three types of research misconduct in the proposing, executing, reviewing, and reporting of research. Their definitions by the U.S. Department of Health and Human Services are (HHS n.d.):

"(a) Fabrication is making up data or results and recording or reporting them.

(b) Falsification is manipulating research materials, equipment, or processes, or changing or omitting data or results such that the research is not accurately represented in the research record.
(c) Plagiarism is the appropriation of another person's ideas, processes, results, or words without giving appropriate credit."

Fabrication and falsification are serious issues. However, even senior researchers can make such mistakes. If research is government funded, suspicious issues can be investigated by the institute and the corresponding government OIG. For example, a Semiannual Report to Congress of the NSF (2019) reported a few instances:

"The professor allegedly altered a figure in a manuscript to show a desired result rather than the experimental result obtained by his graduate students."
"In the other instance, the professor claimed a colleague provided him the questioned data at a conference. The IC (Investigation Committee) learned the professor did not attend that conference, and that colleague did not exist."

Other examples are from the Office of Research Integrity, the U.S. Department of Health and Human Services. They investigate and publish the case summaries of research misconduct (https://ori.hhs.gov/case_summary). Here are a few examples from their case summaries.

"Respondent knowingly, intentionally, and recklessly falsified and/or fabricated Western blot data for…" (HHS 2019a)
"ORI found that Respondent engaged in research misconduct by intentionally, knowingly, or recklessly: fabricating data and analyses in a manuscript submitted to Nature, which was subsequently voluntarily withdrawn." (HHS 2019b)
"ORI found that Respondent engaged in research misconduct by knowingly and intentionally falsifying and/or fabricating data reported in the following three (3) published papers and seven (7) grant applications submitted to NIH:" (HHS 2018)

When an allegation arises about possible research misconduct associated with published papers, journals often first ask the researchers' institution to investigate. Playing a central role for such investigations, the RA has room to improve based on some research administrators and research (Gunsalus et al. 2018).

Serious research misconduct ruins researchers' reputation and career. Misconducted researchers, once an investigation is concluded, may face various types of formal penalties. The investigations may result in an administrative action, civil/criminal action, monetary recovery, and/or a questionable practice warning.

It is important to distinguish between honest errors and research misconduct, and between different opinions and research misconduct. Simplified, misconduct is more than an error or disagreement. A study suggested that for a deviation from disciplinary norms: (i) if unintentional, it is an honest error; (ii) if on a difference of opinion, it is a scientific disagreement. An intentional deviation not based on opinion may be a potential misconduct, which needs a formal inquiry (Resnik and Stewart 2012).

7.2.2.2 Plagiarism

Plagiarism means using the material of others without acknowledging its sources or gaining authorization as required, which is a relatively common issue and may be serious as well. This may happen when copying information, including other's opinion with a long phrase, a table, and a figure from literature, but without an appropriate citation. Plagiarism can be a copyright issue, which not only is a research ethics concern but also become a legal problem.

Researchers can unintentionally use other's work without an appropriate acknowledgement and citation. In other words, they may plagiarize by accident. For example, a researcher fails to give a proper credit to someone else's ideas or statements probably due to not knowing how to do it appropriately. Training on the reference citation is important for student and novice researchers.

Interestingly, there is another concern called "paraphrasing", which means one takes another person's new and specific idea and putting it in his/her own words. In such a situation, we should have an indirect citation – even with no need for quotation marks. Note, using different words for general knowledge and common ideas is not "paraphrasing."

Another issue is called "self-plagiarism" or "text-recycling," meaning use of one's own previous work in another context without citing. This topic is controversial and in an ethical gray area, which is still being studied (Moskovitz 2019). It is true that we are the owner of the work and may reuse it, where the concern is on citation or referencing. It is appropriate that we let the readers know that this is not the first use of the material when reusing it. If there is a new development based on the original work, then the reused part should be a small portion in the new work or publication.

7.2.2.3 Conflicts of Interest

A conflict of interest (COI) is a situation in which a researcher's outside financial interest has the potential to bias a research project. All researchers are required to avoid or mitigate real or perceived financial COI based on the institutional and sponsor's policies. We conduct research and official duties on behalf of the institute in such a manner consistent with statutes and regulations.

COI policies and requirements vary with research institutes and sponsors. However, the policies and requirements are similar. For example, a financial interest is deemed significant if the amount exceeds $5000 during a 12-month period. The significant financial interest (MIT, Massachusetts Institute of Technology 2017),

> *"means a financial interest that meets any of the criteria for significance set forth below and is received or held:*
> *1. by an Investigator; or*
> *2. by an Investigator and members of his or her Family; or*
> *3. solely by members of the Investigator's Family, but only if the financial interest could reasonably appear to be related to the Investigator's Institutional Responsibilities."*

The NSF COI policy is typical for basic research. The NSF states (NSF 2005),

> *"a. NSF requires each grantee institution employing more than fifty persons to maintain an appropriate written and enforced policy on conflict of interest. Guidance for such policies has been issued by university associations and scientific societies.*
> *b. An institutional conflict of interest policy should require that each investigator disclose to a responsible representative of the institution all significant financial interests of the investigator (including those of the investigator's spouse and dependent children) (i) that would reasonably appear to be affected by the research or educational activities funded or proposed for funding by NSF; or (ii) in entities whose financial interests would reasonably appear to be affected by such activities."*

More and more research reports and publications have a COI declaration, which will be discussed in the next chapter.

7.2.2.4 Export Controls

The US export controls addresses concerns of homeland security, terrorism, and leaks of technology to US economic competitors. The Export Controls Act of 2018 extends to emerging and foundational technologies by imposing restrictions on transfer to foreign persons (Braverman and Wong 2018).

The export includes the actual shipment of a controlled commodity out of the country and sharing the technical information of controlled commodities to foreign persons inside the United States. If a researcher take a laptop computer issued by an institute to travel or do research internationally, for example, it is likely that he takes a controlled device abroad. We need to consult with our institutional RA office, understand our regulations, and follow the required procedure to get a license if required. Violations can result in both civil and criminal penalties for the researcher individual and for the institution.

However, fundamental research may be excluded from the export control regulations. The US government defines fundamental research by 22 C.F.R. § 120.11(8) (CFR, Electronic Code of Federal Regulations 2019):

> *"basic and applied research in science and engineering where the resulting information is ordinarily published and shared broadly within the scientific community, as distinguished from research the results of which are restricted for proprietary reasons or specific U.S. Government access and dissemination controls. University research will not be considered fundamental research if:*
>
> *(i) The University or its researchers accept other restrictions on publication of scientific and technical information resulting from the project or activity, or*
>
> *(ii) The research is funded by the U.S. Government and specific access and dissemination controls protecting information resulting from the research are applicable."*

As a regulatory and legal issue, readers should consult with their institutional RA office. Research institutes have their own policies on how to comply with export control laws and regulations. For example, as a reference, University of California has a policy on export controls (UC, University of California 2018). The policy states,

> *"The Fundamental Research Exclusion (FRE) provides that technology or software that arises during, or results from, fundamental research and is intended to be published is excluded from the export control regulations."*
>
> *"Generally speaking, the export control regulations permit U.S. universities to allow foreign nationals (e.g., students, faculty, academic appointees, and non-employee participants in University programs) to participate in fundamental research projects without securing a license. They also permit U.S. universities to share with foreign nationals in the U.S. or abroad 'technology' or 'software' that arises during, or results from, fundamental research and is intended to be published," also without securing a license."*

7.3 Pre-Award Management

7.3.1 Tasks and Funding

7.3.1.1 Pre-Award Tasks

Pre-award research management is primarily for proposal development and completes its duties when a given proposal is approved or rejected. This RA function

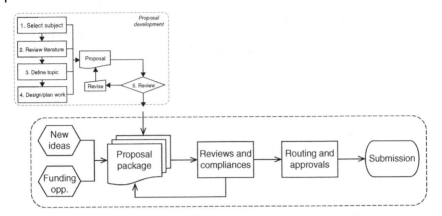

Figure 7.11 Pre-award tasks for external sponsors.

and staff work with researchers to prepare proposals, routes them to appropriate sign-offs, and formally submits proposals to external funding sponsors on behalf of the institute for researchers.

As discussed in Chapter 2 (Figure 2.3), a proposal development has multiple steps, such as selecting subject, literature review, drafting, and revisions. Figure 7.11 shows the details on the reviews, revisions, compliance, and approvals needed before submitting to external sponsors.

For reviews and approvals, the RA office routes a final proposal revision through an internal routing system of the organization. These reviews and approvals ensure the compliance with institutional and government regulations. After approvals at the department, unit, and institution levels, the RA staff will provide the authorized institutional signature and officially submit the research proposal.

7.3.1.2 Support to Proposal Development

There are several support sources for proposal development. One important source is research proposal samples that were successfully funded. Studying such samples, we can gain insight of the proposal development by peers or the professionals in other institutes. Such proposal samples are normally not accessible by public; researchers need RA's assistance with this.

RA support also includes training and workshops. The training topics include sharing successful proposal development cases, available funding sources, data management, budgeting considerations, requirement updates of sponsors, and so on.

For external funding applications, the RA office sometimes has internal support available to partially cover preparation costs, including a compensation to a proposer, hiring expert consultants, assembling and hosting brainstorming sessions, travel expenses for meetings, etc. This small amount, usually several thousand dollars, of an internal funding can be effective to encourage and empower researchers for new proposal development and improved proposal quality.

With an innovative idea and supported from RA, a researcher's first task is to find appropriate funding opportunities. There are various funding portals and resources available. There are two types of funding overall: internal funding and external funding.

7.3.1.3 Internal Funding

Internal funding is a type of internal financial support to research. A main function of internal funding is to support the areas of research and scholarship where external funds are scarce or highly competitive. Internal funding can support a pilot project or a feasibility study for new ideas to garner preliminary results before seeking external funding. Internal funding normally has multiple levels, say college and university, for different types of researchers, such as undergraduate, graduate, new faculty, or for special subjects, e.g. material engineering.

Internal funding can also serve as an internal competition space for external funding proposals. Sponsors, including federal agencies and charity foundations, sometimes limit the number of applications per institution. In the cases with multiple research proposals, an internal review committee at the institutional level screens and determines which proposal as the most competitive one that can move to submit for the institute. In general, there is a timeframe for announcement, internal submission, review, and decision of the limited submission opportunities.

The RA office manages the internal funding in online databases. Researchers can log in the databases and

- View program-specific funding guidelines
- Download application materials
- Submit research proposals
- Receive notifications on application status changes

7.3.1.4 External Funding Search

External funding sponsors include federal agencies, nonprofit organizations, and industrial companies. Table 7.4 lists some funding databases. Most databases require institutional subscription. We as researchers can get help from our RA staff for access to these databases.

Major research universities have external funding information in their websites. For example, University of California Berkeley has a comprehensive funding list with a brief introduction on the federal agencies and nonfederal agencies: https://spo.berkeley.edu/Fund/newfaculty.html. In addition to searching the databases, researchers can receive regular updates from the RA staff about the latest funding information and updates.

Table 7.4 Some funding information databases.

Name	Website	Brief information
US Grants	https://www.grants.gov/	The main U.S. Federal Government resource for finding and applying for about 26 different federal grant-making agencies.
US FedBizOpps	https://www.fbo.gov/	A single government point of entry for federal government procurement opportunities over $25 000.
COS Pivot	pivot.cos.com	A searchable funding opportunities and expertise database. It is an aggregation of funding information that is verified for accuracy, updated for currency, and formatted for quick, targeted search.
Grant Forward	https://www.grantforward.com	A searchable information across all fields of research, including the sciences, engineering, arts, and humanities.
Foundation Center	foundationcenter.org/	A database platform employed the world over by universities and nonprofits to search for foundation grants.
Foundation Directory Online	https://fconline.foundationcenter.org/	Profiles over 105 000 U.S.-based foundations, programs, areas of funding, types of support, geographic emphasis, trustees and officers, application process, deadlines, high, low and average grants, and lists of recent grants made.
InfoEd SPIN	infoedglobal.com/solutions/spin-global-suite/	A database to search for global funding opportunities from a variety of sources, including, government, foundations, professional associations, and crowdfunding resources.
Grant Resource Center	www.aascu.org/GRC/	A not-for-profit service of the American Association of State Colleges and Universities provides personalized and comprehensive federal and private funding information.

7.3.1.5 Funding Sources

In addition to US Grants in Table 7.4, several government agencies provide research funding. Here are some additional ones:

- Department of Energy (https://www.energy.gov/energy-economy/funding-financing)

Table 7.5 Some large foundations for scientific research.

Foundation	Website	Areas	Support
Bill and Melinda Gates Foundation	https://www.gatesfoundation.org	Global development, global health, US education, global policy and advocacy	$4.72 B (2017)
The Wellcome Trust	https://wellcome.ac.uk/	Science, culture and society, innovations, strategy	£723 M (2018)
Howard Hughes Medical Institute	www.hhmi.org	Basic biomedical research, science education	$647 M (2018)
Gordon and Betty Moore Foundation	https://www.moore.org/	Environmental conservation, science, patient care, especially in the San Francisco Bay Area	$284.7 M (2017)
Simons Foundation	https://www.simonsfoundation.org	Mathematics and physical sciences, life sciences, autism research, outreach and education	$272.9 M (2017)
Alfred P. Sloan Foundation	https://sloan.org/	Research and education related to science, technology, engineering, mathematics, and economics	$98 M (2017)

- National Aeronautics and Space Administration (https://www.nasa.gov/spacebio/funding-opportunities)
- Office of Naval Research (https://www.onr.navy.mil/Contracts-Grants.aspx)

Discussed in Chapter 1, private and philanthropy foundations play an important role in scientific research. Table 7.5 lists several large foundations and their information from their annual reports.

Industrial companies have strong financial supports available to research organizations. They provide various opportunities for outside researchers, university faculty, and graduate students, for example, Google (https://research.google/outreach/). Some funding grant possibilities may depend on the partnership between the companies and research institutes, and the network between senior personnel of companies and researchers.

7.3.2 Proposal Development

7.3.2.1 Assistance from RA

We have discussed proposal development from the perspective of researchers in Chapter 2. In this section, we have a bit more discussion from the standpoint of RA staff involvement to enhance proposal development. During a grant proposal drafting, researchers contact RA staff to get their help and resources for improved proposal quality and development efficiency. Except for technical details, the RA staff can help all aspects of proposal development.

Proposals for external funding opportunities have more work for specific requirements and restrictions. Professional RA staff can provide invaluable assistance. With the experience with other research projects, the RA pre-award staff can help researchers in the following ways, such as:

- Review and identify requirements from a funding sponsor
- Suggest a plan for proposal development
- Support application items, such as proposal writing, budgeting, support needed
- Provide training and updates of funding sponsors
- Coordinate cross department and interdisciplinary for large proposals

In addition to RA functions, there are many external grant training, workshops, and resources available. Such consulting services are via an RA office without a fee to researchers. It is worth mentioning the instructors of external grant consulting firms are often experienced field practitioners and administrators, knowledgeable about the latest changes and details of grants.

7.3.2.2 Budgeting Considerations

A research proposal should include a detailed breakdown of the financial support requested from a funding sponsor; and internal fellowship programs will need a brief budget plan.

Research costs have two basic categories: direct costs and indirect costs (IDC). Direct costs are those that are directly attributable to a specific research project. IDC includes the costs of administrative and clerical services in an institute. IDC is in a percentage of total direct costs. The IDC percentage rate is based on an agreement between the federal government and the institute. Therefore, IDC rates are institute dependent normally published. For a given institute, its IDC rate varies depending on the locations and types of research. For example, an IDC rate is 55% for on-campus research and 26% for an off-campus project. An institute may extend the IDC rates to nonfederal sponsored projects. For internal projects, there is no IDC.

For budget estimation in proposal development, an RA office may have a calculation workbook or template, which can help allocate various cost items when considering applicable policies, such as IDC, student tuition, and research staff compensation.

For nongovernment sponsored projects, such as industry contracts and grants from private foundations, contracts, or agreements may be in the form of:

- *Fixed Price Contract.* A project contract has a clear scope of work. This type of research project may need the back up from researcher's unit, for just in case a cost overrun.

- *Time Contract.* Under this type of agreement, a sponsor will pay by an hourly rate for specified types of work and reimburse other associated costs, such as materials and facilities. A similar type is cost reimbursement contract.

7.3.2.3 Proposal Checklists

As discussed, we closely work with RA staff during proposal development. Researchers may refer to the following two checklists during the development with RA staff. Table 7.6 is for the overall readiness of a proposal. While Table 7.7 is to check key items satisfactory in a proposal.

The items in the checklists and corresponding answers for each item depend on sponsor specific requirements. For example, the NASA updates their *Guidebook For Proposers* every year recently (NASA, National Aeronautics and Space Administration 2018). Make sure check the latest guidelines and requirements of a sponsor and revise the checklists before using them.

Table 7.6 A proposal development checklist – overall readiness.

No.	Proposal submission readiness	Yes/No
1	Is the Project Summary completed?	
2	Is the project description completed?	
3	Are the budget section with details and justification completed?	
4	Are PI and key personal eligible and bio sketches provided?	
5	Is the facilities and resources section completed?	
6	Is research compliance (IRB, COI, etc.) completed?	
7	Are the recommendation/support letters received, if needed?	
8	Is the required assurance and certification completed?	

Table 7.7 A proposal development checklist – key items.

No.	Key item satisfactory	Yes/No
1	Has needed emphases (on environmental, career, minority, etc.)?	
2	Meets the format requirements (pages, fonts, graphics, etc.)?	
3	Specifies the start and end dates?	
4	Meets sponsor's specific requirements?	
5	Meets the budget restrictions (max, items, salary, IDC, etc.)?	
6	Are internal reviews and approvals complete?	
7	Has the Unit commitment and cost sharing, as applicable?	

7.3.3 Human Subjects (IRB)

7.3.3.1 Human Subjects Related
Mentioned in Chapters 1 and 2, it is possible that a research project is related to human subjects. Research with human subjects in the United States is governed by the US federal laws. According to the US Code of Federal Regulations 45 CFR 46.102, a human subject is (CFR, Electronic Code of Federal Regulations 2018a):

> *"a living individual about whom an investigator (whether professional or student) conducting research:*
>
> i. *Obtains information or biospecimens through intervention or interaction with the individual, and uses, studies, or analyzes the information or biospecimens; or*
> ii. *Obtains, uses, studies, analyzes, or generates identifiable private information or identifiable biospecimens."*

If a research project falls within its jurisdiction, we need an approval of IRB before data collection. Most funding sponsors require an IRB approval or human subject certification before a proposal submission. An IRB is often housed within a research institution, such as universities and hospitals. They may have additional institutional policies as well. The requirements for an IRB membership are stated in the US Code of Federal Regulations 45 CFR 46.107 (CFR, Electronic Code of Federal Regulations 2018b).

If an organization has no IRB, the organization may arrange an outside IRB for a review, which should be documented in writing. Such cases may be entitled "non-local IRB review" and "cooperative research."

Most engineering studies, focusing on technology or machines, are not human subjects related. However, some engineering research involves human interactions with the factors in the technology. To have a preliminary clue whether research is related to human subjects and needs an IRB review, we can examine our proposal by ourselves. A key is whether our research will obtain information about living individuals. In addition, research activities may either involve interactions or collect identifiable information. Using a survey in research is to get the information as related to human beings. Therefore, an IRB review and approval are required.

7.3.3.2 IRB Reviews
Before an IRB review, the PI needs to develop a complete proposal with the consent by the research participants with (electronic) signatures. In general, the PI and main team members are required to complete a basic training about how to protect the rights and welfare of human subjects recruited to participate in research

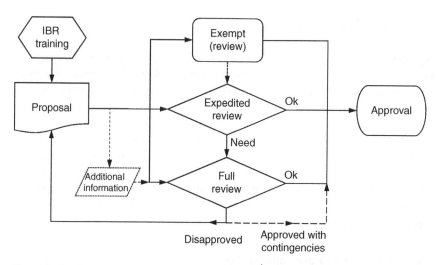

Figure 7.12 An overall IRB review process.

activities. Figure 7.12 shows an overall process of an IRB review for research proposal applications.

An IRB approval ensures that a research proposal is ethical. An IRB makes one of three reviews and approvals:

- *Exempt (from Review).* Meaning the research project involves very minimal or no risk to human subjects, meeting the criteria for a federal exemption in 45 CFR 46.104 (GPO, Government Publishing Office 2018). The IRB simply needs to be informed of such a study.
- *Expedited Review.* This means a minimal risk to human subjects, so that the project has no need for a full IRB review. The IRB chairperson, a designee, or a subset of the IRB does the review and has the authority to decide.
- *Full Review.* This is about a project deemed to be fully reviewed and voted by the board. The review can either approve the project, require modifications and/or addition information, or reject the project. An "approved with contingencies" evaluation is not really an approval, but means it is approvable if the specific changes requested are submitted. A full review is not required for most engineering and technical studies.

Applying for the Exempt decision, the PI should send a research proposal to IRB. Some institutes require all project PIs submit their projects to the IRB, not allowing "self-exempt." Some institutes allow "self-exempt" if the responses to key questions meet the criteria and the RA office follows up with a validation audit later on. Granting an "Exempt" status may take a few days, while a full review may take several weeks to decide.

7.4 Post-Award Management

A proposal approval by a sponsor is a starting point for post-award management. A research proposal approved by the funding sponsor means the proposal is worthy of being supported with their money. With the approval, the PI and the research team should kick off the research work soon.

A proposal approval by a sponsor is a starting point for post-award management. On the RA side, the post-award staff needs to work with researchers on managing a funded research project, ensuring compliance, and considering future additional funding. The main aspects of research post-award management include award acceptance, operating functions, e.g. hiring and managing team members, purchasing equipment, appropriate expenditures, and deliveries, such as regular reports. The PI and RA jointly work on most of them.

7.4.1 Project Acceptance and Set Up

7.4.1.1 Award Acceptance
In principle, a research institute has the authority for the negotiation and acceptance of contractually binding obligations in a research grant or agreement. Accepting an award on behalf of an institute, its RA office needs to review the award. The acceptance of a research award is relatively simple if there are no changes from the original proposal and no additional restrictions from the sponsor.

A research organization may be concerned about insurance requirements, reporting requirements, audit requirements, etc., as such items may or may not be stated in the grant application materials. Therefore, the RA with the PI may review and consider other items for some cases, such as,

- Amount of funding change
- Change of an IDC rate
- Cost-sharing changes
- Additional IP and publication restrictions

7.4.1.2 Project Set Up
Once a research proposal is approved, the research project is about ready to start. The accounts of the research project need to set up in the RA and accounting offices. We may establish sub-projects if a project involves the collaboration between two or more investigators or units.

Before the fund of an approved project is transferred to the institute, the researchers may need to start working on the project and spending funds. In such cases, we may require an advance account with departmental or unit approval as all advance project expenditures are guaranteed by the department.

If there are major changes at the beginning, such as a PI change, re-budgeting, equipment changes, or change of research scope, the RA offices and PI must follow the specific procedure to document and get appropriate approvals from the sponsor.

7.4.1.3 Project Reports

Project reports, for example a final project report for closeout, are the required documents of a research project. Depending on the requirements, size, and nature of a project, multiple reports may be required. For instance, if research is a multiyear project, an annual report is conditional to continued funding for the next year. Clearly, all types of reports have predefined deadlines. For government and some foundation-funded projects, a special version of summary reports, for example the project outcomes reports by the NSF, may be needed for public release. Here is an example of project outcomes report (some context replaced by … for a simplicity purpose) (Das 2016):

> *"This project was motivated by the goal of understanding how information gets aggregated and disseminated in novel venues that have been enabled by the internet, like Wikipedia, social media, and prediction markets. We have worked on multiple different aspects of this problem, including theory development, algorithm design, and empirical analysis of real-world datasets.*
>
> *In terms of intellectual merit, many outcomes of this project contribute to the literature. To highlight a few:*
>
> *… …*
>
> *In terms of broader impact, this project has enabled the training of several Ph.D. students who have worked on collective intelligence and on problems at the intersection of computer science and economics. It has sparked research directions that have led to several undergraduate and master's projects as well, and those students are all now either in the high-tech workforce or in Ph.D. programs. Many of the students at both Ph.D. and undergraduate levels have been women. The project has also enabled the development of several new classes at both the undergraduate and graduate levels. Software tools from the project are available via open-source licenses, and some are being used by other groups already, for example for quantifying controversy in Wikipedia. Last but not least, many of the algorithmic and computational ideas from this work are having impacts and influencing directions in disciplines outside of computer science."*

7.4.1.4 Project Changes

During project execution, some major changes may happen as well, such as key personnel, budget, and no-cost extension. All these changes need to follow the project agreement and sponsor's guidelines. In general, the PI should prepare a request letter with supporting materials, review with the RA staff, and go through a certain process to the sponsor.

7.4.2 Application of Invention Patents

7.4.2.1 Considerations for Patent

If the research outcomes of a research project not only contribute to knowledge but also have potential industrial applications and commercial values, it is a good idea to consider applying for patents before publishing the results. We may consider a patent protection as a way to recover research and development costs.

Applying for a patent means disclosure of research results for the patent right. Different from scholarly papers and presentations, an approved patent grants its owners the exclusive rights to make, use, sell, and import the research results for a limited period. The period is 20 years in the United States and European countries for the functional-feature patents in most cases.

The five requirements for patentability are as follows: subject matter, usefulness, novelty, non-obviousness, and enablement. We should conduct a thorough review for novelty to avoid infringement on another patent, which would have legal ramifications. In addition, a review on the benefits to justify a patent application is worth the filing fee.

If a patent is work related, researcher's intellectual achievement to patentable findings is from the paid work time and with the equipment and resources provided by the organization. Therefore, organization that researchers work for will pay patent filing costs and own patents if granted in the United States.

As inventors, the PI and team members may receive bonuses, reasonable compensation, and/or additional projects. The compensation can be case and institution dependent. For example, the University of Pittsburgh policy states that income from licensing or sale of patent rights after all costs is 30% to the inventors (UP 2005). Another example is the University of Denver. Its patent policy specifies the inventors can have 100%, 40%, and 33.3% if the net income is first $25 000, of the amount over 50 000, and of the amount over $50 000 from patent royalties, respectively (UD n.d.). In addition, there may be a noncompete clause researchers need to sign, preventing researchers from collecting possible patents, quitting the organization, and then applying for them with another organization.

7.4.2.2 Types of Patent

Two main types of patents are related to engineering research and development. The third type is plant patent.

1. *Utility Patents.* This type of patent is for the useful inventions of new and useful machines, processes, systems, products, or technology. A utility patent protects

the functional aspects of an invention for 20 years. Here are some important characteristics of utility patents.

    a. If an inventor is not 100% ready to file a full patent application, he/she may apply a provisional utility patent, which gives one year to file a formal patent application.

    b. The protection can also cover methods and software.

    c. It is possible to protect different variations of a product with a single utility patent.

2. *Design Patents.* This type of patent is for a specific ornamental design of a product or item. The protection for a design patent is mainly on appearance for 15 years. It is less expensive to apply a design patent than a utility patent.

The applications and grants of utility patents are much more than design patents. Table 7.8 shows the latest statistics by United States Patent and Trademark Office (USPTO) (USPTO, United States Patent and Trademark Office 2019). Figure 7.13 shows the overall trends of US patent applications and grants in recent 20 years (data source: USPTO, United States Patent and Trademark Office 2019).

Sometimes, an invention would need patent protection for both functionality and appearance. In other words, the inventors may need to apply for both a utility patent and a design patent.

7.4.2.3 Patent Application Process

The procedure of filing a patent application is different from academic publication. The practice of patent applications varies according to national laws and international agreements. We often file for patents through patent attorneys, but may prepare our own patent applications and file them by ourselves. Patents are formatted into sections, including background, summary, a detailed description, diagrams, and claims, among others.

For US patent applications, researchers should follow the six steps (USPTO, United States Patent and Trademark Office n.d.-a) for an application (see Figure 7.14), which is recommended by the USPTO (https://www.uspto.gov/).

1. *Select a Type of IP to Apply.* In addition to patents, there are other types, such as trademark, copyright, marketing plan, and trade secrets.

Table 7.8 Statistics of US patent applications and grants.

2018	Utility patent, US origin	Utility patent, foreign origin	Design patent	Plant patent	Total
Applications	285 095	312 046	45 083	1079	643 303
Grants	144 413	163 346	30 497	1208	339 992

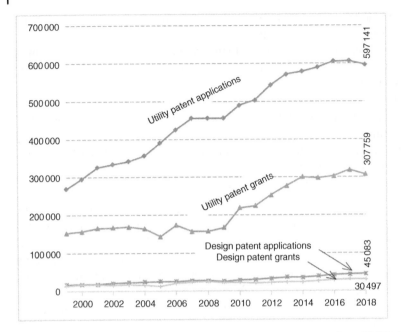

Figure 7.13 Trends of patent applications and grants. Source: Data from USPTO, United States Patent and Trademark Office (2019).

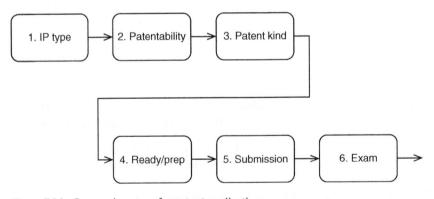

Figure 7.14 Preparation steps for patent applications.

2. *Study and Determine Patentability.* A starting point is to visit USPTO (https:// www.uspto.gov/help/patent-help#patents) to have an initial evaluation. The US patents must have identifiable benefit and be capable of use; while European patent law requires industrial applicability.

3. *If Patentable, Decide Which Kind of Patent.* Engineering and technical research is mostly for utility patents.
4. *Get Ready to Apply.* Working with or without a patent attorney, researchers need to understand the cost, timing, options, etc.
5. Prepare the materials and submit the application.
6. *Exam by the USPTO.* The examiner reviews the application to determine if the patent application meets the requirements of 35 U.S. Code §111(a) – Application (USPTO, United States Patent and Trademark Office n.d.-b).

Researchers or patent attorneys will receive a notice of allowance after the examiner determines that an application meets the requirements. A granted patent is issued after paying issuance fees. In addition, there is a maintenance fee beyond 4, 8, and 12 years for utility patents.

If more protection is needed for multiple countries, researchers should consider applying international patents (often called Patent Cooperation Treaty (PCT)). The World Intellectual Property Organization (https://www.wipo.int), an agency of the United Nations, administrates the PCT. However, a PCT application is different. For example, a PCT patent only gives us access to patent protection in about 150 countries. For remaining countries, we would have to apply PCT directly to the country or countries. In addition, PCT applications are only for utility patents.

Scheduling with patent applications is critical. A late application may allow other inventors to obtain a similar or same patent earlier, which may place us at risk of infringing other patent and costly litigation.

7.4.3 Project Closeout

7.4.3.1 Basic Process
Project closeout is the final phase of a research project's life cycle. A closeout is also an administrative process with certain steps to ensure a research project meets all the requirements based on the contract to terminate officially the award relationship. The RA staff and PI work together for the closeout process.

The requirements for a project closeout are established by contractual provisions and/or sponsor regulations. The procedure, steps, and items are institute and sponsor dependent. The timely closeout is generally the PI and institute's joint responsibility. Preparation for a closeout normally begins in three months before the end of a project. For a large research grant, the closeout process may start with three reviews at 120 days, 90 days, and 30 days prior to the end date to prepare on-time submission of final reports. The final decision on a research project closeout is the funding sponsor.

There may be various items to review and check for a project closeout. The two main documents are required for the PIs via their institute:

1. Final technical report on all deliverables of the project in required format, including title page, table of contents, abstract, etc.
2. Fiscal (financial) closeout report on the funds authorized and actually expended.

In addition, an equipment and inventory report are required for some projects. The sponsors may require other reports, such as invention and property, as well.

Defined in the project agreement, the due dates of final reports, final invoice, and other documents may be in 60–90 days after the end data of a project. The sponsor reviews the documents and determines that all administrative requirements specified in the grant have been met. A research project is administratively terminated when all obligations to the sponsor have been fulfilled. For some projects, site visits can be arranged to ensure project completion.

7.4.3.2 Final Reports
The detailed requirements are specified in project agreements. A final technical report, in general, addresses the following items:

- Predefined objectives and deliverables
- Significant results and solutions
- Research impacts
- A few examples to support
- List of publications

Fiscal (financial) report should include:

- Purchase expenditure
- Personnel (PI and labor) expenses
- Cost-sharing if applicable
- Sub-award if applicable
- Indirect cost, etc.

Research project is financially closed based on final costs reported. Sometimes, a final audit is performed prior to the closeout.

Record retention is another task. Final narrative reports may need to be retained permanently. Other records, supporting documents, and various records pertinent to a research project should be retained for a certain years according to the sponsor.

7.4.3.3 Other Administrative Tasks

Research Equipment

After a research project completes, the equipment purchased and used may rest with the institute if no specific predefined arrangement in the research contract. As a property of the institute, the equipment that completes its duty in a specific project may be transferred between departments within the institute.

No-cost Time Extensions

As the name suggests, additional time may be granted without additional fund from the agency/sponsor. The time extension may be approved either by the sponsor or within institute. With the time extension, researchers can have more time to finalize a project with the remaining fund.

Summary

Basics of Project Management

1. The life cycle of a research project has four phases: (i) Proposal Development, (ii) Pre-award Review and Submission, (iii) Post-award and Project Management, and (iv) Closeout, Reporting, and Future Work.
2. Research project management mainly addresses six aspects of a research project: (i) timeline, (ii) cost, (iii) team, (iv) external/international relations, (v) regulation compliance, and (vi) innovation.
3. It is common that realistic research efforts are more uneven and heavier than the planned.
4. The three performance indicators of a research project are its outcomes, timeline, and costs.
5. Research outcomes may be in format of knowledge, technology, artifacts, improvement, or their combinations.
6. For project issues, two key reviews address the impacts of the issues on project timing and potential risk in future.
7. Project task adjustments often involve additional cost and recourses.

Research Administration

8. The functions of research administration include policy development and enhancement, help on funding information, processing grant applications, project coordination, regulation compliance, and technology transfer, and so on.

9. Close work relationship between research PI, RA, and unit is a foundation of a research project success.
10. Three types of research misconduct are fabrication, falsification, and plagiarism. They can be serious but avoidable.
11. Conflicts of interest are mainly on financial side but may include non-finance interests.
12. Export controls is a US law on imposing restrictions on commodity to foreign persons. Fundamental research may be excluded from the export control regulations.

Pre-Award Management

13. The functions of research pre-award management include external funding search, proposal development support, review and compliance, routing and approvals.
14. Budget planning is an important part of a proposal and needs RA review and assistance.
15. Research with obtaining, using, analyzing, or generating identifiable private information is human subjects related.
16. Human subject related research should have an IRB's approval beforehand.

Post-Award Management

17. The functions of research post-award management include award acceptance, project set up, project reports, and change management.
18. There are two main types of invention patents: utility patent and design patent. The former is much more common for engineering and technical research.
19. Patent application process has six steps of preparation. International patent applications are different from domestic applications.
20. Project closeout is to manage the final reports and other required tasks by a funding agency.

Exercises

Review Questions

1 Briefly review the main tasks in one of the four phases of research life cycle.
2 How to understand that a new idea or direction may be provoked at the end of a research cycle?
3 Briefly describe the main aspects of research management for a PI.

4 Explain the three key elements of research project performance with an example.

5 In the three key elements of research project performance, which one may be challenging?

6 Review the impact analysis for a major issue in project management.

7 Review the risk assessment for a major issue in project management.

8 Discuss the trade-off between project schedule and cost of a task adjustment.

9 Briefly discuss the main functions of research administration in an institute.

10 Review the tasks of pre-award and post-award management.

11 Explain a type of research and report misconduct.

12 Discuss the issue with "text-cycling" and how to avoid it.

13 Understand and review the definition and basic requirements of COI.

14 Search and list a few internal or external funding opportunities and their eligibility requirements.

15 Discuss the basic requirements if research is involved human subjects.

16 Explain the possible decisions of an IRB review.

17 In an interview, the interviewer states she is doing a preliminary study pending its IRB approval. What is your suggestion for her?

18 Review the characteristics of two major types of patents.

19 Explain the overall application process of a patent.

20 Talk about the overall process of research project closeout.

21 List the tasks of closeout for your research project or master's thesis research.

Mini-Project Topics

1 Review a research project and use the two-step diagnostics suggested by Czarniawska (in Section 7.1.1) to consider a new research direction.

2 Analyze the actual efforts vs. planned efforts of a project and discuss the reasons.

3 Visit one of the funding information databases and summarize its functions and contents.

4 Study the work relationship between a PI and RA office on their responsible and supporting functions.

5 Search a case of research misconduct published and review the main points (hint: search a government agency OIG website for its report).

6 Study a recent paper in your discipline and find a sentence in the introduction section of the paper. Put the sentence into quotation marks and paste it into Google Scholar or an academic database of your library to search. What do you have in the search result? Any thought about the results?

7 Review the applicability of export controls on an engineering research project.

8 Discuss the possible different views or answers from a PI and RA office on some items of the checklists suggested in the chapter with examples.

9 Visit your institutional IRB website and complete the basic human subjects training if available.

10 Find a recently complete research project and discuss the patentability of the research results.

11 For an ongoing research project, develop a closeout plan with main tasks and overall timing.

References

Braverman, B. and Wong, B. (2018). Congress enacts the export controls act of 2018, extending controls to emerging and foundational technologies. https://www.dwt .com/insights/2018/09/congress-enacts-the-export-controls-act-of-2018-ex (accessed November 2019).

CFR, Electronic Code of Federal Regulations (2018a). Title 45: public welfare, part 46—protection of human subjects. Government Publishing Office. https://www .ecfr.gov/cgi-bin/retrieveECFR?gp=&SID=83cd09e1c0f5c6937cd9d7513160fc3f& pitd=20180719&n=pt45.1.46&r=PART&ty=HTML#se45.1.46_1102 (accessed May 2019).

CFR, Electronic Code of Federal Regulations (2018b). Title 45: public welfare, part 46—protection of human subjects. Government Publishing Office. https://www .ecfr.gov/cgi-bin/retrieveECFR?gp=&SID=83cd09e1c0f5c6937cd9d7513160fc3f& pitd=20180719&n=pt45.1.46&r=PART&ty=HTML#se45.1.46_1107 (accessed November 2019).

CFR, Electronic Code of Federal Regulations (2019). Title 22: foreign relations. §120.11 Public domain. https://www.ecfr.gov/cgi-bin/text-idx? SID=a8cf9a715c25c462244b2c7a8a153cdf&node=22:1.0.1.13.57.0.31.11&rgn= (accessed 7 July 2019).

Czarniawska, B. (2015). Chapter 41: moving on? In: *The SAGE Handbook of Research Management* (eds. R. Dingwall and M.B. McDonnell). SAGE Publications Ltd. ISBN-13: 978-1446203187.

Das, S. (2016). The dynamics of collective intelligence, project outcomes report. https://www.research.gov/research-portal/appmanager/base/desktop?_ nfpb=true&_windowLabel=T31400570011264188753337&wsrp-urlType=blockingAction&wsrp-url=&wsrp-requiresRewrite=&wsrp-navigationalState=eJyLL07OL0i1Tc-JT0rMUYNQtgBZ6Af8&wsrp-interactionState=wlpT31400570011264188753337_action%3DviewRsrDetail

%26wlpT31400570011264188753337_fedAwrdId%3D1414452&wsrp-mode=wsrp %3Aview&wsrp-windowState= (accessed 12 May 2019).

GPO, Government Publishing Office (2018). Part 46—protection of human subjects, electronic code of federal regulations. https://www.ecfr.gov/cgi-bin/retrieveECFR? gp=&SID=83cd09e1c0f5c6937cd9d7513160fc3f&pitd=20180719&n=pt45.1.46& r=PART&ty=HTML#se45.1.46_1104 (accessed 19 May 2019).

Gralka, S., Wohlrabe, K., and Bornmann, L. (2019). How to measure research efficiency in higher education? Research grants vs. publication output. *Journal of Higher Education Policy and Management* 41 (3): 322–341. https://doi.org/10.1080/ 1360080X.2019.1588492.

Gunsalus, C.K., Marcus, A.R., and Oransky, I. (2018). Institutional research misconduct reports need more credibility. *JAMA* 319 (13): 1315–1316. https://doi .org/10.1001/jama.2018.0358.

HHS (2018). U.S. Department of Health and Human Services. Case Summary: Narayanan, Bhagavathi. https://ori.hhs.gov/case-summary-narayanan-bhagavathi (accessed 12 May 2019).

HHS (2019a). U.S. Department of Health and Human Services. Case Summary: Cruikshank, William W. https://ori.hhs.gov/content/case-summary-cruikshank-william-w (accessed 12 May 2019).

HHS (2019b). U.S. Department of Health and Human Services. Case Summary: Fox, Edward J. https://ori.hhs.gov/content/case-summary-fox-edward-j (accessed 12 May 2019).

HHS (n.d.). U.S. Department of Health and Human Services. The Office of Research Integrity. Definition of Research Misconduct. https://ori.hhs.gov/definition-research-misconduct (accessed 12 May 2019).

Jeracki, K. (2018). Research administration: it's complicated. Colorado State University, 4 October 2018. https://source.colostate.edu/research-administration-its-complicated/ (accessed 15 October 2019).

Krugler, P.D. (2008). Performance measurement tool box and reporting system for research programs and projects. The Transportation Research Board, The National Academies of Sciences, Engineering, and Medicine. https://www.nap.edu/read/ 23093/chapter/5#14 (accessed June 2019).

MIT, Massachusetts Institute of Technology (2017). Financial conflicts of interest in research – policy. https://coi.mit.edu/policy (accessed August 2019).

Moskovitz, C. (2019). Text recycling in scientific writing. *Science and Engineering Ethics* 25 (3): 813–851. https://doi.org/10.1007/s11948-017-0008-y.

NASA, National Aeronautics and Space Administration (2018). NRA or cooperative agreement notice proposers' guidebook – final. https://www.hq.nasa.gov/office/ procurement/nraguidebook/ (accessed July 2019).

NSF (2005). 510 Conflict of interest policies. Chapter V – grantee standards. Grant Policy Manual NSF 05-131. https://www.nsf.gov/pubs/manuals/gpm05_131/gpm5 .jsp#510 (accessed May 2019).

NSF (2019). Semiannual report to congress, (OIG-SAR-60), office of inspector general (OIG). https://www.nsf.gov/oig/_pdf/NSF_OIG_SAR_60.pdf (accessed 6 June 2019).

OMB, Office of Management and Budget (2018). 2 CFR PART 200, Appendix XI compliance supplement. https://www.whitehouse.gov/wp-content/uploads/2018/ 05/2018-Compliance-Supplement.pdf (accessed 19 May 2019).

OU (n.d.). Technology transfer office. Ohio University. https://www.ohio.edu/ research/tto/ (accessed May 2019).

Resnik, D.B. and Stewart, C.D. (2012). Misconduct versus honest error and scientific disagreement. *Accountability in Research* 19 (1): 56–63. https://doi.org/10.1080/ 08989621.2012.650948.

Smith+Brown, P.C. (2019). Performance audit of incurred costs–University of Delaware, OIG 19-1-011. NSF Office of Inspector General. https://www.nsf.gov/ oig/_pdf/19-1-011_University_of_Delaware.pdf (accessed 17 May 2019).

Stanford University (n.d.). 13.1 Gift vs. sponsored projects and distinctions from other forms of funding. https://doresearch.stanford.edu/policies/research-policy-handbook/definitions-and-types-agreements/gift-vs-sponsored-projects-and-distinctions-other-forms-funding#anchor-1101 (accessed 31 May 2019).

UC, University of California (2018). Policy on export control. Issuance date: 21 June 2018, UC-AA-18-0391_ExportControlPolicy_Accessible. https://policy.ucop.edu/ doc/2000676/ExportControl (accessed 7 July 2019).

UCSC (n.d.). Welcome to the office of research. https://officeofresearch.ucsc.edu/ about/index.html (accessed 22 May 2019).

UD (n.d.). University of Denver patent policy. https://www.du.edu/research-scholarship/media/documents/patent-policy.pdf (accessed July 2019).

UM (n.d.). UMOR Vision, mission, values. https://research.umich.edu/research-u-m/office-research/vision-mission-values (accessed 22 May 2019).

UP (2005). University Of Pittsburgh policy 11-02-01, patent rights and technology transfer. https://www.cfo.pitt.edu/policies/policy/11/11-02-01.html (accessed July 2019).

USC (n.d.). Office of research. https://research.usc.edu/about/vp/ (accessed 22 May 2019).

USPTO, United States Patent and Trademark Office (2019). U.S. Patent statistics chart calendar years 1963–2018 (updated 4/2019). https://www.uspto.gov/web/offices/ ac/ido/oeip/taf/us_stat.htm (accessed February 2020).

USPTO, United States Patent and Trademark Office (n.d.-a). Patent process overview. https://www.uspto.gov/patents-getting-started/patent-process-overview (accessed March 2019).

USPTO, United States Patent and Trademark Office (n.d.-b). Manual of patent examining procedure. https://mpep.uspto.gov/RDMS/MPEP/current#/current/d0e18.html (accessed March 2019).

WSU (n.d.). Technology commercialization. Wayne State University. https://research.wayne.edu/techtransfer (accessed May 2019).

8

Research Report and Presentation

8.1 Introduction to Academic Writing

8.1.1 Academic Writing

8.1.1.1 Academic Writing Overall

On the research proposal development, discussed in Chapter 2 write-ups are an integral part and outcome of innovative thinking and effort. Specifically, we utilize write-ups to summarize research achievements and report results.

For the most part, do not consider a research project complete until the research outcomes are reported and the report itself is accepted by the funding sponsor or government agency. For example, industry-sponsored research projects often require that we submit the reports and any artifacts to our sponsors. For research projects supported by government and non-profit organizations, we are normally required to publish papers in scholarly journals or professional conferences, in addition to regular and final reports. For large projects spanning multiple years, we need to submit annual reports as required for continued funding in subsequent years.

A research report is more so expository writing rather than creative writing, which largely relies on a writer's inspiration and imagination. Academic and scholarly writing have standard formats, expectations, and stylistic guidelines to follow. We should decide on two things before writing: (i) the type of report and (ii) writing plan.

Research writing and oral presentations are essential skills for all researchers. It is recommended that novice researchers keep practicing writing and presentations to improve. As beginning researchers, you may face challenges on properly summarizing and presenting your work in reports and oral presentations, due to the complexity of technical contents. This chapter introduces the basics of academic

Engineering Research: Design, Methods, and Publication, First Edition. Herman Tang.
© 2021 John Wiley & Sons, Inc. Published 2021 by John Wiley & Sons, Inc.

writing for research projects. The next chapter will have additional discussions on the writing for publications.

8.1.1.2 Requirements of Academic Writing

The purpose of academic writing is to translate scientific and technical information to a targeted audience interested in knowing work outcomes. Therefore, the purpose and key are to convey the research information.

When writing a research report and paper, we should keep in mind of a few general requirements of academic writing. They include:

- *Objectiveness.* To report technical contents, our objective statement is based on data obtained from technical work and free from any kind of ambiguities. The process, derivation, interpretation, discussion, and conclusions should all be supported by the domain principles and study results.
- *Directness.* We should avoid any type of implicitly and directly state as much as possible on all assumptions and considerations.
- *Descriptiveness.* We need to explain the important aspects of research, such as experiments, process steps, and parameters, detailed enough to be repeatable to get the same results.
- *Clarity.* We must write from the perspective (background and needs) of the readers. We keep them in mind and convey complex technical information in an appropriate manner.
- *Format.* Professional associations, such as Institute of Electrical and Electronics Engineers (IEEE), American Society of Mechanical Engineers (ASME), and American Institute of Aeronautics and Astronautics (AIAA), have discipline preferred formats and styles. We should use them. In addition, we use the terminology that is commonly used in the particular fields.

8.1.1.3 Elements and Their Significance

A research report or paper is composed of several sections. Figure 8.1 shows a typical framework of the sections in a report or paper. However, not every report has to have all these sections. In terms of their significance, we may categorize these sections into three levels: key, core, and supporting.

This model in Figure 8.1 is purpose built with the reader in mind. When reviewing, readers normally read the title, abstract, and conclusions first to know an overall picture of a report. In other words, these three sections give the critical impression to readers. If they were interested, they would likely continue to read the sections of problem statement, methods, results, and discussion, which are the main efforts of research work and writing-up. We will discuss all of them more in the following subsections of this chapter.

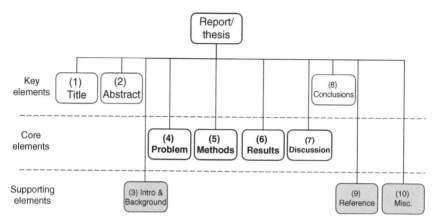

Figure 8.1 Composition of research reports and papers.

8.1.2 Common Types

8.1.2.1 Thesis and Dissertation

A master's student may need to complete a thesis and a doctoral student must do a dissertation based on their research. A thesis or dissertation is a detailed research report to show student's research achievements and qualification to obtain an advanced degree.

Figure 8.2 shows a typical structure for theses and dissertations. A master's thesis normally has one research topic to cover, while a doctoral dissertation has up to four chapters on specific research topics.

A master's thesis often looks like a long research report, while a doctoral dissertation may be more analogous to a book – a few related topics (chapters) bound by an introduction, literature review, and conclusions. A chapter of a thesis and dissertation may be further developed into a scholarly paper for publication.

Many graduate schools have specific requirements and format guidelines for a thesis and dissertation. For example, Rackham Graduate School at the University of Michigan published the Dissertation Handbook (Rackham 2017) that contains detailed format requirements, such as on the cover page, author page, dedication, acknowledge, preface, and table of contents. Similarly, the Massachusetts Institute of Technology has online specification guidelines for thesis and dissertations (MIT, Massachusetts Institute of Technology 2016). The students must follow the guidelines and meet the requirements to complete their final manuscripts.

Additionally, doctoral students need to defend orally their research process and outcomes. On the other hand, the necessity of master's students to present their research work is often determined by the academic department.

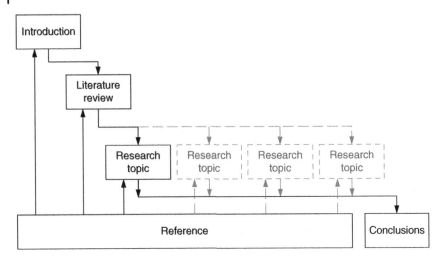

Figure 8.2 A typical structure of thesis and dissertation.

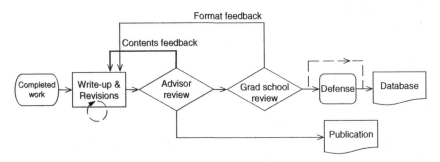

Figure 8.3 A process flow of thesis and dissertation preparation.

Figure 8.3 shows an overall preparation process of a thesis and dissertation. The reviewers of a thesis or dissertation are advisory committee members, consisting of faculty members and industry experts. They review, evaluate, and judge the details of the student's research, including research methods, findings, deliverables, the conclusions, and possibly the thinking behind the student's research.

For doctoral students, publication of two to three journal papers is required for the successful completion of a dissertation. The number and quality of publications are very important for their career as a scholar in an academia or research institute. Master's students, if conducting research projects, are often expected to publish one paper.

8.1.2.2 Project Report

A research project report is a summary for internal review and reporting purposes. The reports often give a statement, process, and the results of research, as either a final or periodic report, to upper management or a sponsor. For student research, students need to submit project reports after finishing their research projects.

As an internal document, research reports may have various requirements regarding length, structure, contents, format, etc. For external reporting, there are normally prescribed requirements. For example, the National Science Foundation (NSF) has formal and mandatory requirements for final performance progress reports with detailed instruction for each category of a report for their research projects (NSF n.d.):

1. Cover Page Data Elements
2. Accomplishments
3. Products
4. Participants and Other Collaborating Organizations
5. Impact
6. Changes/Problems
7. Budgetary Information
8. Demographic Information for Significant Contributors

Research reports may be published in an organization's intranet with limited access. The abstracts of many sponsored project reports are often open to the public, but few full-length internal project reports are available to the public.

Research reports may be distributed to external parties after appropriate approvals. In many cases, we further develop reports into scholarly papers for external professional conferences and journal publication. In this respect, a report serves as the basis or draft of a scholarly paper. More discussion on paper publication is in the next chapter.

8.1.2.3 Case Study Report

As discussed earlier, case studies are a common form of R&D projects, particularly in industrial applications. A case study is an in-depth investigation on a specific problem; interesting and relevant to other professionals in other circumstances.

Relative to theses and dissertations, a case study report is short. The structure and format of a case study report can be somewhat different from a full project report and less standardized. For example, a case study may consist of:

1. Executive summary
2. Introduction (background and problem description)
3. Methodology (procedure and approach) and data
4. Results, evaluation, and discussion, e.g. on causational relationship and alternatives

5. Concluding remarks (lessons learned, assessment, and implementation/recommendation)

Due to using proprietary data and for specific purposes, like internal project reports, most industry case studies are not published.

8.1.3 Reports and Papers

8.1.3.1 Types of Research Articles

There are several forms of research articles with varying intents and formats, addressed in different chapters listed below. This chapter primarily discusses reports and presentations. Non-research-related technical write-ups, such as training materials and textbooks, are not in the scope of this book.

- Research proposals (Chapter 2)
- Literature reviews (Chapter 3)
- Theses and dissertations (Chapter 8)
- Project reports (Chapter 8)
- Presentation files (Chapter 8)
- Scholarly papers (Chapter 9)

These types of research write-ups are different in terms of technical details and writing formality. Figure 8.4 shows a relative comparison of academic writing-ups in formality and detail. Unquestionably, they vary much and have many exclusions based in the purpose, audience, and requirements. For example, scholarly papers have a significant variation on the technical details. From this relative comparison, we can know how to develop our write-up based on what we are writing.

In addition, some technical reports, like internal documents, do not undergo a peer-review process. That is one reason that a technical report needs to be further developed into a scholarly paper, if appropriate. Scholarly paper writing will be discussed further in-depth in the next chapter.

8.1.3.2 Technical Reports Vs. Scholarly Papers

We will briefly talk about the differences between technical reports and scholarly papers. Both documents explain the process and report the outcomes of research exploration activities. Therefore, the fundamental contents of both technical reports and scholarly papers are very similar and share many common facets.

However, there are also significantly different features between the two types of documents. Commonly, reports and papers serve different purposes. Table 8.1 is a reference of different characteristics between technical reports and scholarly papers. If a paper is to be developed from an existing internal report, the main tasks are to translate a report into a paper in structure, focus, and formatting.

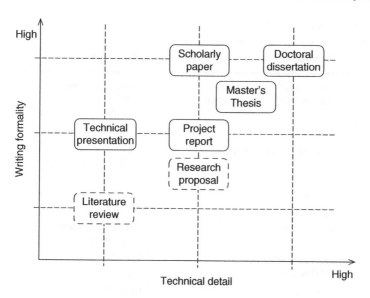

Figure 8.4 Formality and detail of technical reports.

Due to the large diversity of technical reports and scholarly papers, their characteristics overlap. For example, scholarly papers may have future potential commercial value. Many corporations, particularly large companies, conduct basic research but do not publish all their research outcomes.

In this book, we often use papers as examples rather than reports since the latter are not often published and have a large variety. Bear this consideration in mind when we learn academic writing-up with the examples of published papers.

8.2 Elements of Report and Thesis

Discussed in a previous subsection, the breadth of research reports includes the key elements, core elements, and supporting elements. Let us start our discussion from the key elements.

8.2.1 Key Elements

(1) Title
Almost all readers view the title of a report or paper first. If the title is clear and relevant, they will go to the abstract, etc. A paper title with an abstract can make a difference between your article being read or not. Thus, a paper title should be very

Table 8.1 Technical reports vs. scholarly papers.

Aspect	Technical report	Scholarly paper
Purpose	Internal reference, may related to financial benefits	Contribution to body of knowledge, not have immediate financial benefits
Accessibility	Restricted and limited	General public
Review	Internal reviews by superiors	External peer-reviews, normally in blind, with a specific rubric
Quality	Various	Consistently high
Proprietary	Normally yes for data and conclusions	No sensitive/confidential information, approval of publication maybe needed
Length	Normally less than 3000 words	4000–7000 words
Format	Various	Consistent and fixed
Summary	One-page executive summary	Abstract, 100–200 words
Author	Team and department	Individual author(s)
Copyright	Organization	Individual authors or organization

clear. In addition, a title should be sufficiently specific where readers can understand the research subject and focus without reading the paper contents.

The length of the title of a report and paper may not be critical. The length is usually 10–15 words and rarely more than 20 words. Three samples of paper titles are cited below.

> *"Torsion sensing based on patterned piezoelectric beams"* (Cha and You 2018)
> *"Autonomous vehicles: challenges, opportunities, and future implications for transportation policies"* (Bagloee et al. 2016)
> *"Development of a weighted probabilistic risk assessment method for offshore engineering systems using fuzzy rule-based Bayesian reasoning approach"* (Ung 2018)

These titles have 7, 10, and 18 words, respectively. Upon reading the titles, one might say the first one could be a bit more specific as the torsion sensing using patterned piezoelectric beams is a large subject that can be multiple projects with various focuses. How about the second one? It seems a clear statement of what the

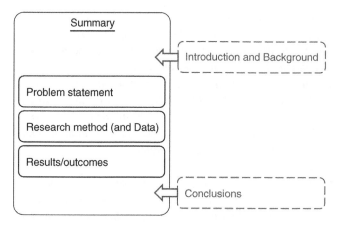

Figure 8.5 Key information in an executive summary.

paper is about. The third title seems very specific on the research objective, application area, and approach used. Is it perhaps necessary to shorten the title? For example, are the words "Development of" necessary? Many researchers consider general words, such as "study of" and "investigation of," overly used and add little value to reader's understanding. Hence, such words may not be necessary.

For some types of reports, such as master's theses, doctoral dissertations, and internal reports, a title page is required, and templates are normally available.

(2) Summary and Abstract

An executive summary provides brief information for a quick assessment. For internal reports, an executive summary or abstract is required, which is summarized from the sections of *(4) Problem Statement, (5) Research Methods, and (6) Results* without details and explanation (refer to Figure 8.1). In other words, an executive summary has at least three building blocks, as shown in Figure 8.5. When necessary, a brief information from *(3) Introduction/Background and (8) Conclusions* may be added in. For scholarly papers and theses, the summary is called abstract, which will be more discussed in the next chapter.

Due to the limited space, a summary is pure text, without figure, symbol, equation, or reference. Abbreviations used should be limited to the well-established ones, such as CAD and AI, by the professionals in disciplines.

Below is an example of executive summary (Ludwig 2012), which has five paragraphs and total about 600 words in one page (some details are removed and replaced with "..." for simplicity). Without reading the entire report, readers can get a clear picture about the study's objective, methods, and findings.

> *"Large scale multi-user research infrastructure is a critical component of the federal science and engineering research enterprise. ...*

The National Science Foundation (NSF) supports large research facilities that are created in response to community need and that span a broad range of disciplines including physics, astronomy, materials research, geosciences, ecology, engineering, nanotechnology, and polar research. ...
Multiple policy questions surround federal investments in large research facilities: who is benefiting from these investments? What is the best way to maximize scientific productivity across the research enterprise? How should investments in big science be balanced with support for individual or small group research? Ultimately, decisions on these issues ...
This study provides the first known analysis of facility utilization at NSF. ... Results show that there is a broad spectrum of users who interact with each facility in different ways and that NSF is likely serving many more users than previously thought. New users discover facilities through different mechanisms; ...
This study suggests that ... Analyses in this study show that ... This work establishes a foundation for ..."

(8) Conclusions

Similar to the executive summary or abstract at the beginning of the paper, this section is a summary to the overall paper. However, the main difference is that the conclusion section focuses more on research achievements. It also connects the broader context of the research work to the research outcomes, without introducing any new information or insights. Readers often go to the conclusion section after reading the abstract of a report or paper for more information on the research results. Therefore, a conclusion section should be significantly different from an executive summary or abstract.

A conclusion section highlights the main findings and concise interpretations based on results and discussion. It would not be a good idea to repeat or add more discussion in a conclusion section. The key information in a conclusion section consists of (see Figure 8.6):

1. Restatement of the research problem and concise summary of work (one paragraph)
2. Summary of results and findings to the problem (three to four bullet points or sentences)

In addition, it is beneficial that the conclusion section contains the information of (one or two paragraphs):

3. Contributions to the body of knowledge and/or professional community
4. Research limitations under a specific circumstance or weaknesses
5. Viewpoints on future work to do

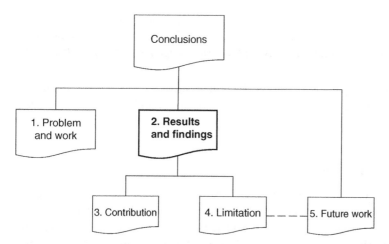

Figure 8.6 Composition of conclusion section.

For research limitations, we know that results are valid in specific situations and under certain assumptions. Therefore, mentioning the limitations of the study and future work can add a noticeable value for other researchers and impress peer reviewers. Thus, it is recommended to include one or two sentences addressing the limitations in the conclusion section.

Here is an example of conclusion section (Scheffler et al. 2018) (some details are removed and replaced with "..."). There are three parts in this conclusion. The first serves as a work summary. The next part shows the outcomes and findings of the study. The third part talks about the limitation of the research.

> *"7. Conclusion*
> *This paper presents the use of the discrete element method (DEM) to predict the bruising of fresh produce. A visco-elastic contact model was used and the material...*
> *The model accurately predicted the impact forces during drop tests to within 11%. Using multi-sphere particles, the bruise volume of 50 apples in a rotating drum was accurately predicted with an error of 28%–31%, ...*
> *Spherical particles were also used, and although less accurate than the multi-sphere particles, they could be used to accurately predict the damage of apples in the rotating drum. However, ..."*

Furthermore, future work suggestions, if mentioned, may show researcher's in-depth thinking and advance vision. For example, a paper states (Choudhury et al. 2018) in the conclusion section,

"However it has not been taken care of in the present work and therefore, an attempt will be made to see if the correction factor proposed for estimation of G_{fr} for short crack growth is applicable for components like pipes, elbows etc."

For regular reports and papers, including student's theses, the conclusion section should be about a half page long. For doctoral dissertations, they often have one or two pages.

8.2.2 Core Elements

Core elements are the main body of a research report or paper. Typically, each core element or section consists of a few subsections and each subsection may have two to four paragraphs. To have a balanced look, the lengths of subsections and paragraphs are preferably about the same. You may consider dividing a long subsection into two subsections.

(4) Problem Statement

In this section, we define a research problem to solve as well as sub-problems if applicable. The arguments and motivation to solve the problem can come from innovative ideas, requirements from fieldwork, continuation of a prior work of the research group, and/or from a literature review. The basic requirement of this section is the clarity of the addressed problem.

Problem statement can be a single sentence, a paragraph, or even an entire section. From a reviewer's standpoint, the problem statement is preferably concise, in one or two sentences. Here are three examples that a problem statement is presented concisely.

"The problem can be stated straightforward: Control the bounding box of an SCQC by controlling the distillation frequency." (Paler 2018)
"The mechanical problem statement is as follows. Parallelepiped-shaped beams made of incompressible materials in the unstrained (initial) state are subjected to initial strain (tension or compression) and are passed to the intermediate state." (Levin et al. 2015)
"Edge cracking is a common problem in sheet metal forming, mainly for Advanced High-Strength Steels (AHSS). The noticeable increase of AHSS in the automotive sector, due to the Corporate Average Fuel Economy (CAFE) regulations, makes this fact an interesting problem." (Behrens et al. 2018)

Many authors, however, use a paragraph or even a section for their problem statements. In such cases, authors not only state the problem but also provide the background, justification, and explanation of the problem. Here is an example, which well serves the purpose of Problem Statement section with research proposal (Garcia et al. 2018). The first sentence of the Problem Statement section is about the problem:

> *"This work proposes a decision-making model, motivated by emerging CPS technologies, that solves the problem of identifying the most profitable and most environmentally-friendly production plans of a product while ensuring manufacturers achieve optimal costs and equipment uptimes."*

Immediately following the problem statement, the authors claimed the needs of working on this problem (some details are removed and replaced with "..." for a simplicity purpose):

> *"In this problem, the product designer wishes to ...."*
> *"In this problem, product designers submit details of their part...."*

Then, the authors mentioned the factors associated the problem:

> *"Raw material costs and processing costs are considered."*
> *"The problem is defined with a gate-to-gate system boundary and requires..."*

Following, the authors proposed some details of the problem solving:

> *"To solve this problem, the following parameters/data are required: ..."*
> *"Noteworthy assumptions include: ..."*
> *"Major decisions include: ..."*
> *"Following precedent set in the manufacturing supply chain literature, a bilevel programming (BP) model is developed to represent the Stackelberg game structure ...."*

There is another style to address a problem statement. Within the section of problem statement, authors explained why the research is important and worthy of research (Osman and Sahraoui 2018) (some details are removed and replaced with "..." for a simplicity purpose).

> *"1.1. Problem Statement*
> *In order avoid this failure or try to minimize rate a lot researcher make studies to define critical success factors (CSF)/critical failure factors (CFF) of ERP implementation with deep analysis and categorization.*
> *One of the first articles of ERP failure defines failure reasons as lack of education, business process reengineering (BPR), project management and...*

Other studies categorize CSF to operational, organizational and cultural factors. …

But unsatisfactory user requirement is mostly due CSF and with increasing rate of ERP implementation requirement satisfaction still problem. …

We come to the point that the essence is how to gather requirements for ERP; this should multiple ways of gathering. …

All previous cases related to requirements issues and most of ERP literatures say little about requirements engineering (RE)…"

In summary, a problem statement should be clear and may be in different styles, such as a single sentence, a paragraph, or even a section. The problem statement can be either a standalone sentence or accompanied with its background, justification, and explanation in a section.

(5) Research Methods

In this core section, we discuss a method for specific considerations, assumptions, parameter settings, procedures, etc. If a method is not commonly used, then we should describe it in appropriate amount of detail so that readers can fully understand the study.

As discussed, the data used in a research project can determine the type of analytical method to use. Therefore, the discussion of the data, the source, experiment setup, measurements, and so on, should be included to support and justify a specific method to use. In some types of research, such as establishing a new model, the method itself may be the focus of study.

Concerning later publication, we should also pay more attention to the method section. In a peer-review process, referees may be more interested in this section to learn the application of method and decide whether the results are valid. In addition, the novelty of a method can be a strong selling point of a paper. If a paper proposes a novel method, we should emphasize it in the title, abstract, introduction, and conclusion sections.

In most cases, we divide this method section into three to five subsections – each dedicating to a specific item, such as data measurement, computational technique, analysis, etc. There are large variations in data analysis and method utilization, depending on the nature, objectives, and focus of a research project. Therefore, researchers often label the method section differently. Figure 8.7 shows two examples (Oks et al. 2018; Ribeiro et al. 2018) of method sections. In the two examples, one proposes a new method; the other explains the experiment used.

As a main part of research and reporting, this section contains figures, tables, equations, and/or examples. Any ethics related items, such as human subjects with an Institutional Review Board (IRB) approval, are normally reported in the method section.

(6) Results and (7) Discussion

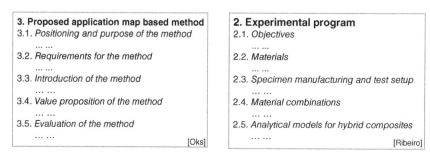

Figure 8.7 Examples of research method section.

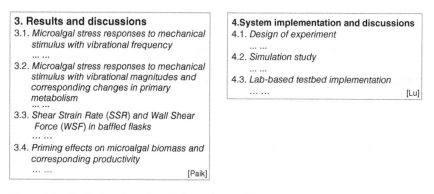

Figure 8.8 Examples of result and discussion section.

Technically, these two sections can be presented separately. In a results section, we describe results without much interpretation, with a dedicated discussion section following. For a relatively large research project, the discussion section may start with a summary paragraph of an overview of the completed work.

However, many authors prefer combining the two sections into one section, labelling it as "results and discussion," based on the main achievement points. In such a combined section, we show research achievements following the analysis and results. Then, we offer our understanding and interpretation of the intrinsic meanings of those analysis results. To generalize our conclusions, we need to provide validation and justification in discussion.

Every research is unique and so are the section organizations and section titles of results and discussion in reports and papers. Figure 8.8 shows two examples of the subsection titles of the results and discussion section (Paik et al. 2018; Lu et al. 2018).

We should keep one important item in mind that we report the main and useful findings rather than everything from all research efforts. The excessive presentation of data and analysis results should be avoided. The main findings

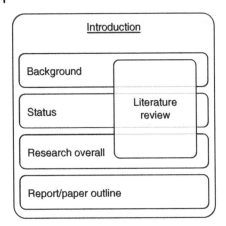

Figure 8.9 Main information in introduction section.

should be organized in a logical way, i.e. by a process sequence or by outcome significance.

Furthermore, if some results are not 100% certain or complete, we should not make a conclusive statement. Otherwise, it may open the door for challenging questions or even a negative review from reviewers. We may report preliminary results and have remarks for future work rather than a grandiose statement, which may beckon an assertion. Similarly, there may be some unexpected findings or discrepant results from research work. If that is the case, we may need specific discussions to address them, which may lead to further work.

8.2.3 Supporting Elements

(3) Introduction

The introduction of a research report and paper is the overall context and background of the research for readers. An introduction should convey four pieces of information: background, status, research overall, and report outline (see Figure 8.9). The literature review is often embedded in the introduction sections. Sometimes, literature review can be a standalone section.

In the introduction section, we normally start opening phrases with the background of the research project and then talk about the status of the research topic. With the solid support of both background and status from a literature review, we introduce our research as a problem to address: justification of the study. Most researchers also briefly present their paper outline at the end of an introduction section.

Here is an example of an introduction section (Dokhanchi et al. 2018) (some details are removed and replaced with "…" for a simplicity purpose). The authors clearly stated the background, status, and their research work in the three paragraphs, respectively.

"The wide availability of high-performance GPU-based hardware has led to an explosion in the applications of Machine Learning (ML) techniques to real-time image recognition problems, especially using deep learning... There is, thus, an urgent need for techniques to formally reason about the correctness and performance of such driving applications...
The key challenge in formal verification is that.... In this work in progress, we focus on deep learning algorithms.... In order to evaluate the performance of the perception algorithms over time, we need to provide quality requirements... Most importantly, going beyond ad-hoc validation and testing, we need a formal framework...
In this paper, we consider temporal logic based quality requirements for... Then, a quality monitor considers... We consider evaluating timed object data..."

(9) Reference

We have discussed the role and significance of references in Chapter 2 for research proposals. Similarly, the reference section is an integral part of a technical report and academic paper, too. Using references, we not only honor other researchers' credit but also have a support to our research. Unfortunately, the reference section is omitted in many internal reports. Appropriate citation is important to avoid unintended plagiarism.

Generally, only the references with in-text citations should be listed. Authors may cite their own previously published works, but the self-cited items should be limited as a very small portion of the total references. The reference section does not normally include personal communications, un-refereed conference proceedings, or commercial (.com) website addresses, unless necessary.

We need to cross-reference all citations and references in the required format for formal reports. The following are two examples in the IEEE style and American Psychological Association (APA) style, respectively. In the context of a paper, they should be cited as Ando et al. (2018) and Hu (2013), respectively. Publishers may have their own house styles.

- K. Ando, et al., "BRein Memory: A Single-Chip Binary/Ternary Reconfigurable in-Memory Deep Neural Network Accelerator Achieving 1.4 TOPS at 0.6 W," in IEEE Journal of Solid-State Circuits, vol. 53, no. 4, p. 983–994, April 2018. doi: 10.1109/JSSC.2017.2778702
- Hu S.J., (2013). Evolving Paradigms of Manufacturing: From Mass Production to Mass Customization and Personalization. Procedia CIRP, 7, 3–8. doi: 10.1016/j.procir.2013.05.002

As seen, these formats or styles are not much different. The formats of many journals fall somewhat between the two styles. The formats for referencing books, web pages, e-books, public documents, and dissertations or theses are slightly different from those for reports and papers.

If we use someone else's copyrighted work, particularly figures, we need to obtain a written permission to reuse that material, unless it is considered "fair use." Most usages of a small quantity for non-commercial research are deemed a fair use. For more details, readers may refer to the fair use guidelines (https://fairuse.stanford.edu/overview/fair-use/) developed by Stanford University Libraries (Stanford University Libraries n.d.) and the fair use checklist (https://copyright.columbia.edu/content/dam/copyright/Precedent%20Docs/fairusechecklist.pdf) developed by Columbia University Libraries (Columbia University Libraries n.d.).

Recent electronic documents come with a digital object identifier (DOI). The DOI scheme is administered by the International DOI Foundation (www.doi.org) based on ISO 26324. As a permanent identifier, it should be used for all references when possible.

(10) Miscellaneous

There may be additional small subsections in research reports and papers, such as,

Acknowledgements. In this optional subsection, we thank our funding sponsor and recognize other people who contributed to the reported work. The acknowledgement may be just one sentence or a short paragraph at the end of a report and paper. For a master's thesis or doctoral dissertation, the acknowledgement goes to the advisor, committee members, and researcher's family, placed in a preface.

Appendices. An appendix may be included in a report or paper if necessary. Some researchers use appendices to provide supporting details and keep the main text concise. Most reports and papers do not have an appendix. For students' theses and dissertations, an appendix is considered chapter equivalent regarding the format.

Lists of Figures, etc. For a long report, such as a master's thesis or doctoral dissertation, a list of figures, a list of tables, and a list of acronyms are often necessary. Such lists offer the reader's convenience to aid in comprehending the file.

Copyright Page. In theses and dissertations, a copyright page is needed in a typical format of © 2020 Your Name. For reports and papers, normally there is no copyright page or section. The copyright of a published paper may either stay with authors or transfer to publishers.

Footnotes. Authors may use them when necessary for reports and papers. Some researchers do not prefer having footnotes as they make a reading process interrupted.

8.3 Development of Research Report

8.3.1 Process of Write-ups

Write-ups themselves are a cumulative and iterative process; a small project with a breakdown of tasks and corresponding target dates. General recommendations for effective writing below are a summary from the experiences of many researchers and authors. Adopting these rules as best practice may significantly improve our effectiveness and efficiency of academic writing process.

8.3.1.1 Writing Sequence

Researchers have different preferences of where to start writing. Interestingly, a writing process should not be in sectional sequence. According to the practice of many experienced researchers, we start the core elements (i.e. Problem, Methods, Results, and Discussion) first. After drafting core elements, we can work on the three key elements (that is, Title, Abstract, and Conclusion), often the conclusion section first. We then work on remaining supporting elements. In general, we should revise all sections several times. Figure 8.10 shows a recommended report preparation sequence. Readers may try it and see if it improves your writing effectiveness.

The first effort to start writing can be the *(5) Methods* section, which is mostly derived from the initial research proposal. If there is no significant change from

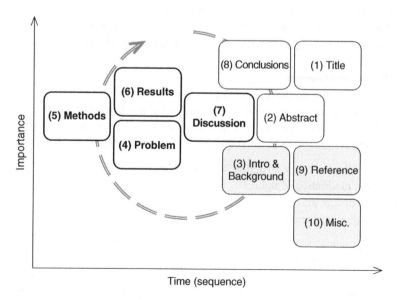

Figure 8.10   A recommended writing process.

the proposal, the methods section may look similar and with added details and updates from the research work.

Then, we can summarize the outcomes of our work and write the *(6) Results* section. With the methods and results, we are in a better place to update the *(4) Problem Statement* section, which is similar to that in the original proposal. Then, we develop the *(7) Discussion* section based on the results and our thoughts and interpretation. As discussed earlier, many researchers like to combine Results and Discussion into one section.

With a draft of the core sections above, we can go into the key elements: the sections *(8) Conclusion, (2) Abstract,* and *(1) Title.* In other words, we take care of the report title and abstract in the end to reflect the revised research outcomes and discussion.

8.3.1.2 Timing and Efforts

When the research is near completion, we may start writing with accumulated research materials. In addition, we need reasonable time for possible additional work. As a process, writing should be integrated into the research process instead of a standalone task at the end. It is better to keep thinking about the final report in the course of research. We may prepare the structure (sections and subsections) of the final report and accumulate the materials for the report as early as possible.

During research and before a formal write-up, we must identify the main points to present and then organize the ideas with supporting information. The outline may follow the same elements discussed above but with the defined titles of sections and subsections. The number and headings of sections may vary with specific requirements according to the thesis guideline of a graduate school.

8.3.1.3 Update and Revision

The final report should go through multiple iterations of updates and revisions, including by authors themselves and based on the feedback from advisor and colleagues. Some experienced researchers suggest at least five iterations of revision.

We may consider four tasks in the sequence of revising below:

1. On the overall report flow, structure, and headings
2. On the data, methods, and figures
3. On the conclusion, abstract, and title
4. On the wording and grammar

Whenever possible, we should seek feedback from others and revise our manuscript accordingly. We may request others to address the four tasks mentioned above and seek for their input. In addition, during drafting and revision, we may find ourselves in need of additional work and/or redo some tasks to support the statements in our manuscript.

Bear in mind that we should compose and review our manuscript from a reviewer and audience's viewpoint. A report or paper should be easy to follow with logical statements and with clear technical values. It is even better to make reviewers enjoy reading our report.

Another recommendation by many researchers is to plan a few days apart between each revision. A short pause makes us be able to read our own manuscript critically with less residual thoughts on the original writing. In addition, due to some changes on the research data and figures, it is important to keep all revisions, because a removed portion could be needed again later.

8.3.2 Writing Format

8.3.2.1 Common Writing Styles

There is no universal style/format for research report writing. Commonly used styles include APA (APA, American Psychological Association n.d.; Purdue Online Writing Lab n.d.) and IEEE (IEEE, Institute of Electrical and Electronics Engineers n.d.). For most reports, master's theses, and doctoral dissertations, typical recommendations on format include: the font size for body text from 11- to 12-point, minimum margins 25 cm on all sides, line spacing of the body text either 1.5 lines or double, and so on.

Many scholarly journals have their own format requirements, which are similar with one of the common formats. There are also variations of the format requirements in different disciplines and journals. Thus, it is wise to know the appropriate style when starting to write a manuscript to avoid later make-up efforts for formatting.

The format requirements also include language wording. For example, spelling can be either American English or British English but should not be mixed. The definitions of terms and units may be different, such as date format and unit of length.

8.3.2.2 Sections and Headings

The headings of sections and subsections should be brief and specific. In most cases, we use two to four words for headings. In addition to being concise, the length of all headings should also be balanced, spanning a report and paper. For example, it seems unbalanced if most headings are three words and one heading is six-word long. During the final revision process, headings should be reviewed again and revised as necessary.

In most cases, each section may have a summary by itself. It is often an issue where the transition from one section to the next is abrupt and the overall flow of a paper is not smooth. The transition between sections and between subsections can be a focus in report revision.

8.3.2.3 Figures and Tables

Figures and tables are very effective to present ideas, process, and results. A majority of technical publications have figures, such as images, diagrams, and flowcharts, to present the experimental setups, analysis procedure, research outcomes, etc. Figures visualize statements and processes and help readers understand. If the numbers are relatively simple, using tables or even a list is a good alternative. In most cases, figures are easier to follow than tables.

When preparing figures and tables, we need to keep in mind of a few points to make figure and table presentation effective:

- The captions of figures and tables should be self-explanatory, so readers do not have to read the context to understand the figures and tables.
- Every figure must be referenced in the main body of the text. In other words, an explanation is required for each figure and table.
- Figures and graphics may be generated using the figure functions of software, such as Microsoft Excel and PowerPoint.
- Color figures should have distinguished legends in terms of shape and size, such as *, +, ×, •, ¤, and Δ, which is important as the publication may be printed and copied in black and white.
- If a couple of figures are to be compared, they should be designed using the same scale.

8.3.2.4 Equations and Special Symbols

Equations are an integral part in many technical reports. It is a common practice using the Microsoft Equation function, embedded in MS Word and Power-Point, for equation creation and formatting. Google Docs has similar functions. In a Microsoft Word setting, the font settings in the equation function and in the text are separately determined.

Special symbols, such as $\geq$, Δ, and vector symbols, as well as special abbreviations, are often necessary for technical contents. All symbols used in a manuscript must be defined when it first time appears in the context. Similarly, special symbols may be inserted using the Symbol function of Microsoft Word and Google Doc. If many symbols are used in the text, a list of symbols (nomenclature) may be needed.

8.3.3 Other Considerations

Here are also some other practices, which are widely accepted in technical reports, papers, and presentations. Many universities and research institutes provide tutoring materials and workshops for technical and research writing. Here are a few examples. Readers may visit their websites for the details.

- Writing Research Papers (www.dgp.toronto.edu/~hertzman/advice/writing-technical-papers.pdf) (Hertzmann n.d.)
- How to Write a Great Research Paper (https://www.microsoft.com/en-us/research/academic-program/write-great-research-paper/) (Jones n.d.)
- Basic Steps To Write An Outstanding Research Paper (https://collegepuzzle .stanford.edu/basic-steps-to-write-an-outstanding-research-paper/) (Wilson 2017)
- Keys to Designing Effective Writing and Research Assignments (www.jsums .edu/academicaffairs/files/2012/08/Keys-to-Designing-Effective-Writing-and-Research-Assignments.pdf?x19771) (Weimer n.d.)

8.3.3.1 Statements and Limitations

In the Results, Discussion, and Conclusion sections, we present the principles, relationships, and interpretations based on our results. Again, due to limited resources and particular conditions, researcher's findings and interpretations are likely only valid for the given situations and under certain assumptions. We may infer to extended conclusive remarks or predict for other situations with rationalization. However, it would be inappropriate and even irresponsible to draw a conclusion that is beyond the data and without proper justification and validation.

As discussed, it is applaudable and beneficial that we acknowledge the limitations and/or weaknesses of our research in a report. Based on the objectives, assumptions, and resources of a research project, we may discuss a validated scope of research. We can also consider possible implications of the results for future studies as a visionary recommendation rather than a conclusion. We can discuss the uncertain meanings of research results and possible influences by known and unknown factors on the results as well. By reading such discussions, readers can better know how far our research effort and results may extend.

In most cases of technical writing, we should avoid subjective statements and overestimated contribution as they may result in reader's dislike.

8.3.3.2 Conciseness and Wording

Academic writing and presentation should be clear and simple, do not overwhelm readers with needless complex words. We carefully select jargon based on audience. In most cases, simple words are preferred. Table 8.2 lists a few examples.

To be concise, two words are not preferred when one word can work for the same purpose. In some cases, spoken English uses more words than written English. Table 8.3 lists several examples, not a complete list, for the hints to readers.

Contractions, such as "it's" and "weren't," are not used preferably in technical and professional writing. Another suggestion is not to use "to be" words, such as

Table 8.2 Examples of considering simple words.

Complex word	Simple word
Apprise	Inform
Aforementioned	Mentioned
Endeavor	Try
Heretofore	Previous
Individualized	Individual
Modicum	A small quantity
Utilize	Use

Table 8.3 Examples of considering concise phrases.

Redundant word	Concise word
Already existing	Existing
At the present time	At present
Completely eliminate	Eliminate
Different entities	Entities
Due to the fact that	Because or since
Had completed previously	Had completed
Introduce a new	Introduce
Make a decision	Decide
Mix together	Mix
Period of time	Period

"was," "were," and "has been" too many times. Instead, try to use active verbs as they may speak powerfully.

We would also suggest using neutral (nondirectional) language in most places of a technical report. Here are a few examples:

- Concerning the data reliability in conclusions, we may use phrases like "it seems that…", "it is likely that…", "one leads to believe that…"
- Talking about a fact, we may use the phrases "Evidence indicates…" rather than "Everyone agrees…"
- For a remaining disagreement, we can say, "Some would argue/contend the point…"

Table 8.4 Examples of word selection for research statements.

Preferred	Questionable	Example: usage in sentence
Few	None	_____ of subjects has been investigated.
Rarely	Never	The characteristics are _____ addressed.
Often	Always	The data _____ support the principle.
Most	All	_____ research on this topic agrees...

Strong words, such as "prove," may not be appropriate in the conclusions for most applied research and R&D. With limited resources and efforts, it can be challengeable if we claim something is "proven." There have been discussions on in what conditions "prove" or "proven" may be used in research conclusions (Cooper 2016), but no widely accepted criteria. However, "prove" may be good for some basic research and theoretical analysis.

In addition, Table 8.4 comments a few words in research writing in most cases of avoid fully definitive statements. Without an exhaustive, comprehensive investigation, the statements with "never" or "always" may not be best to use.

8.3.3.3 Other Tips on Writing Style

Using appropriate verb tense style can be subtle in academic writing: most authors use the past tense for the research work done and results in a manuscript. For the existing situation, known facts, and established knowledge, we use the present tense. When possible, we should avoid shifting tenses within a paragraph.

Active voice is often preferred as it is straightforward and stronger than passive voice. In addition, active voice is less wordy, clearer, and more forceful. Here is an example for reader's review. In the introduction of a paper, authors wrote (some details are removed and replaced with "..." for a simplicity purpose) (Dokhanchi et al. 2018),

> "In this work in progress, we focus on deep learning algorithms.... In order to evaluate the performance of the perception algorithms over time, we need to provide quality requirements... Most importantly, going beyond ad-hoc validation and testing, we need a formal framework that facilitates....
> In this paper, we consider temporal logic based quality requirements for scoring or grading the results of perception algorithms. ... We consider evaluating timed object data..."

Some researchers prefer writing in an impersonal style. For example, authors use the passive voice to show the facts in technical papers and sound objective. In general, the excessive use of personal nouns, such as I and me, may lead readers to feel the study result was subjective.

Here are some additional tips:

- Using altered expressions when appropriate for the same meaning, such as "This analysis provides…," "The paper analyzes…," and "This paper presents an analysis of…" is suggested.
- Transition words, such as "accordingly," "by comparison," "equivalently," and "on the contrary," can make a paper smooth for reading comprehension.
- To make reading easier, subjects should sometimes be repeatedly used in sentences instead of using their pronouns. It may be difficult to identify the antecedent when using "it" or "this" where there are several subjects in a paragraph.

8.4 Research Presentation

8.4.1 Presentation at Conference

8.4.1.1 Attending Conferences

Attending technical conferences is a common professional exercise for both experienced and novice researchers. First, we have opportunities to learn the latest updates and to meet new colleagues by attending conferences. Moreover, many people consider conferences a good professional networking platform and an opportunity to find a new job. In addition to learning from others, we also can present our research findings at a conference. Compared to written reports and papers, an oral presentation is an interactive learning opportunity with other professionals from the world. Many conferences also have a poster session.

Attending conferences can be expensive, which will be discussed later. Therefore, we need to select which conferences to attend to effectively learn new research findings and present our own. The tradition and core of our discipline is the first consideration. Newly developed conferences on emerging subjects can be good choices as well.

8.4.1.2 Presentation and Keynote

The aim of a research presentation is somewhat different from general technical presentations. The former is to report detailed research work and findings, while the latter is for lecturing and insight sharing. Figure 8.11 shows typical presentation examples: (a) research work and (b) a keynote speech on a similar subject at a professional conference.

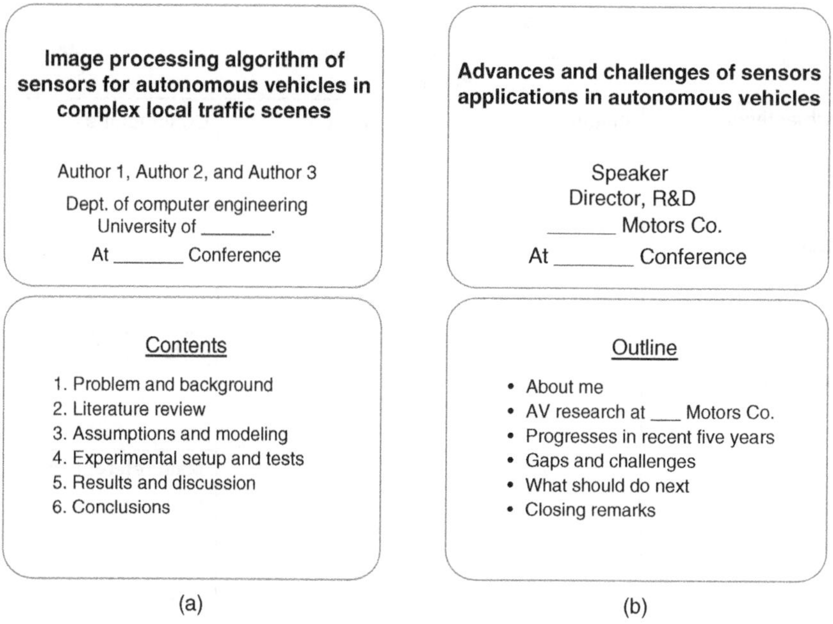

Image processing algorithm of sensors for autonomous vehicles in complex local traffic scenes

Author 1, Author 2, and Author 3

Dept. of computer engineering
University of _____.

At _____ Conference

Advances and challenges of sensors applications in autonomous vehicles

Speaker
Director, R&D
_____ Motors Co.

At _____ Conference

Contents

1. Problem and background
2. Literature review
3. Assumptions and modeling
4. Experimental setup and tests
5. Results and discussion
6. Conclusions

Outline

- About me
- AV research at ___ Motors Co.
- Progresses in recent five years
- Gaps and challenges
- What should do next
- Closing remarks

(a) (b)

Figure 8.11 Research presentation vs. keynote speech. (a) Research presentation. (b) Keynote speech.

8.4.1.3 Poster Presentation

There are often poster sessions at professional conferences. Posters display research and support individual discussion with interested parties. Advantages of using posters to present research projects include overall impression, individual discussion, and time flexibility to visitors. In a poster session, we stand by our posters available for visitors' questions and discussion. We may also ask for the contact information of visitors for later contact and possible research collaboration.

A major difference between a poster and oral presentation is the design. For example, the poster's shape and size are generally limited to one large page or tri-fold display design in size of A0 (841 mm × 1189 mm) and need to put all information into an available space. Conference organizers often provide a template for the layout of posters, referring to Figure 8.12 as an example. The font sizes and format are recommended in a poster template.

We have a few recommendations for a poster design:

- More graphics and less text
- Readable size of fonts from one meter away

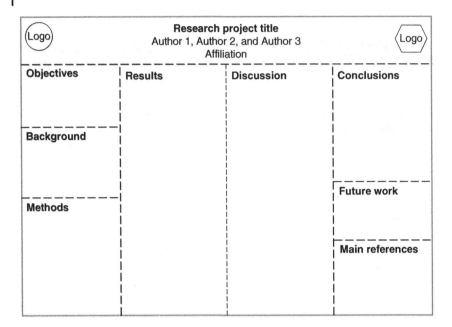

Figure 8.12   An example of poster layout.

- Using colors if possible
- Leaving appropriate empty space overall

8.4.1.4   Conference Costs

As briefly mentioned earlier, attending a professional conference can be costly. For example, a registration fee is from several hundred dollars to two thousand dollars depending on location, sponsors, etc. The total attending cost, including registration fee, lodging, and travel, etc. Table 8.5 may be used to budget the total cost.

The total attending cost of a nonprofit professional conference in North America may be $2000–$2500. We should check and comply with our institutional policies on transportation, airfare, lodging, etc. For student researchers, there may be student discounts from the conference or scholarships from their university and a government agency for well-established conferences.

8.4.2   Presentation Design

We should design a presentation to prompt others to ask questions and give us valuable feedback, which can help advance and strengthen our own research. Bearing in mind these considerations, we have the following recommendations for a presentation design:

Table 8.5 Considerations on conference costs.

Item	Additional information	Estimation ($)
Registration	Discount (about 15–20%) for early registration	
	Significant discount for students	
Air and ground transportation	Including shuttle, car rental, gas, toll fees, taxi, etc. if applicable	
Lodging	Special rates for conference	
	Lower rates if offsite, but with additional walk time	
	Parking may not free, especially downtown areas	
Meals	Most meals included	
	Organization's policy	
Educational sessions	Beneficial to new researchers and graduate students	
	Often before conference, maybe parallel	
Tours	For additional learning on nonconference sites	
	With minor charges for transportation	

8.4.2.1 Number of Slides

The number of slides of a presentation (using Microsoft PowerPoint or Google Slides) file is mainly determined by presentation time. As a rule of thumb, one slide takes 1.5–2 minutes on average. In other words, a 20-minute talk should consist of around 12–16 slides.

Similar to a research report and paper, a presentation file has several sections for research objectives, methodology, and achievements. No slide should have two sections. Table 8.6 lists a typical design of a presentation file and the number of slides for each section for a 20-minute presentation.

The number of the slides for each section varies due to content, style, and format of a presentation. For example, some slides may have just a few words and take less time. As a result, more slides may be needed to fill the time. If a presentation is a doctoral dissertation defense, the presentation file generally has more slides as the process may take about one hour. Students can design the structure of a long

Table 8.6 A recommended presentation file planning.

Section	Number of slides
Cover (title, name, and affiliation)	1
Outline	1
Introduction and lit review	1–3
Data, experiment, or model	3–4
Analysis and discussion	3–4
Conclusion (or summary)	1
Reference	1
Thank you and Q&A	1

presentation into several main sections, most likely following the corresponding chapters of a dissertation. The conclusion section should integrate all the findings from the doctoral research into two to three slides.

8.4.2.2 Slide Layout

A presentation audience not only listens the presenter but also reads the slide show on a projector screen. Therefore, the layout design of presentation slides should be friendly to the audience. Accordingly, the font size of context should be large enough, say at least 20-point. If there is no room for a large font size, it may tell us that there are too many details on one slide. The content should be either simplified or extended to two slides. The graphics should be large as well, including the smaller letters in a graphic. Figure 8.13 shows an example of presentation layout design.

Presentation slides should be as self-explanatory as possible, which is beneficial not only for an effective verbal presentation but also for file distribution later. During and after a conference, some attendees may contact a presenter and say, "Sorry, I missed your presentation but am very interested in your study. Can you just share with me your presentation file?" A well-designed, self-explanatory presentation file can lead to more interest and a further discussion.

During a presentation, a presenter should not read the contents of a slide. Rather, a presenter should use slides as a reminder of what to say and supporting facts and evidence. Therefore, do not put too many words in a slide, aiming for about 50 words or fewer.

In addition, most technical professionals use simple backgrounds, rather than the fancy color schemes, to avoid becoming distractive or difficult to read. It is also

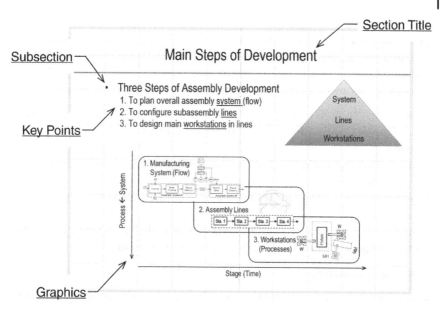

Figure 8.13 An example of presentation slide design.

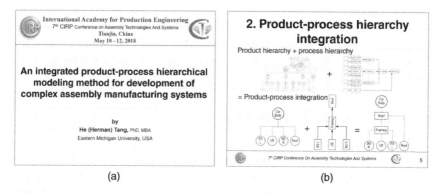

(a) (b)

Figure 8.14 An example of presentation in a template. (a) Cover slide. (b) Content slide.

a good idea that each slide has a plenty of "white space" rather than being full of contents. Often, the organizers of a professional conference suggest a template for the preparation files. Figure 8.14a,b show an example of a presentation.

8.4.2.3 Graphics and Fonts

In general, technical professionals prefer using graphics to convey information, such as diagrams, layout maps, and pictures. Some slides may have only graphics without text.

To show data and information, we consider the commonly used graphics:

- Pie Charts: To show percentages and comparisons if multiple pie charts used.
- Bar Graphs: Good to illustrate relative amounts, may be more obvious than line charts.
- Line Charts: Appropriate to demonstrate continuous changes and patterns; and to illustrate the interactions if multiple lines in a chart.
- Scatterplots: For a large amount of data to show individual points rather than lines or bars.
- Statistical charts, such as control charts, box-and-whisker plot, etc. Depending on the target audience, a certain level of explanation is necessary.

We follow some rules:

- Graphics and pictures should be in a high resolution with an appropriate contrast.
- Certain text letters or phases may be colored for emphasis.
- Colors and context in the graphics should be kept simple.

In general, cool colors, such as blue and green, work well for backgrounds. On the other hand, a warm color, like orange and red, is good for the foreground, e.g. text. If a presentation is in a dark room, then dark background works fine. However, if a presentation is in a lit-on room, a white or light-color background may have a better effect.

8.4.3 Considerations for Presentation

8.4.3.1 Practice for Overall Flow

Presenting a research project is like telling a story: we need a logical flow to convey key points clearly and concisely. It is important to keep in mind who our audience are and what they are most interested. In most cases, they are professionals in the same fields. Therefore, presenters can know the professionals' background, expectations, and possible questions and develop a presentation accordingly, focusing on main achievements and preparing for possible questions.

In a storytelling style, we may consider the flow below to present our research story (see Figure 8.15).

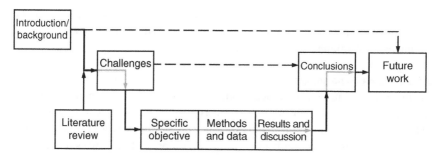

Figure 8.15 Section connection and flow of research presentation.

Before reporting our specific research, we start with a big picture – interesting and worthwhile research efforts. Then, we introduce what challenging barriers were to the big picture based on literature review (and our insight). When closing, we state what we have contributed to the challenges and what future directions are in the big picture.

It is a good idea to review the presentation file in a slide sorter view when revising the presentation. In the sorter view, we can see how the presentation logical flow proceeds. Based on the review, we can easily re-arrange the slides and check the balance of contents and the level of details of each slide. Sometimes, we might need to remove some pieces of information or add visual data to increase visual clarity for improved communication.

Usually, we should practice at least two times before a formal presentation. With practice, we can know how well we manage presentation time and handle the transition between sections. A second-time practice should check improvement since the first time and ensure good time management. During preparation, the research team and presenter(s) can brainstorm possible questions and prepare the ways to answer them.

8.4.3.2 Professional Presenting

During a presentation, we should have good eye contact with the audience. In addition, we keep our eyes uniformly around the entire audience, rather than staying on any particular person. As a formal presentation, common word fillers in daily spoken English, such as "well," "you know," and "I mean," should be avoided as well.

For dissertation defense, a doctoral student may pause the presentation at the end of each section/chapter for comments and questions from his/her dissertation committee, and then proceed. By doing so, the student may better deliver the information and make materials easier to follow.

We want to be open to share our work. Attendees may take pictures or record videos during our presentation without our consent. This can be a good sign that attendees show interest in our research.

8.4.3.3 Q&A Management

During a presentation, experienced audiences may ask challenging questions. If we do not immediately know the answer to a question, we may say something like, "Thank you for the question. I actually don't know the answer at this moment, but I will look into it and get back to you." After a presentation, we should make connections with interested audience members and follow up with them after the conference.

When discussing with an audience, we should avoid coming across as argumentative or overly defensive. We can state we had different considerations and understanding, which is considered impolite. However, it is inappropriate that a presenter argues with an audience member regardless of who is correct. We may arrange a healthy discussion and detailed explanation after the presentation section or after the conference.

8.4.3.4 Student's Projects

In addition to oral presentation and posters, student's projects can be presented in the form of artifacts, which is a common practice for the R&D projects of undergraduate students. The artifacts show student's inventions. Besides the

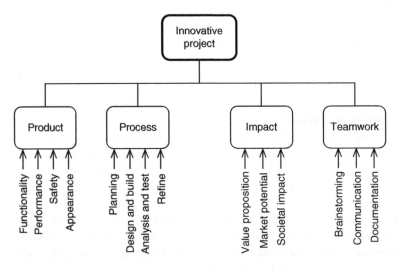

Figure 8.16 Evaluation factors of student invention projects.

preparation work similar to posters, students keep the project judging criteria and rate weights, which vary significantly with projects, when preparing a project presentation. Figure 8.16 lists the common factors of project evaluation as a reference.

Summary

Introduction to Academic Writing

1. The main requirements for academic writing include objectiveness, direct, descriptiveness, clarity, and certain formatting.
2. The structures of master's thesis and doctoral dissertation are in a similar format but different in length. University graduate schools normally have detailed format requirements.
3. Research articles have different types, categorized and viewed in terms of technical details and writing formality.
4. Technical reports and scholarly papers have similar contents but different features and focuses.

Elements of Report and Thesis

5. All the elements of research articles can be categorized into key, core, and supporting parts.
6. The key elements are title, summary/abstract, and conclusions. The core elements are problem statement, method, results, and discussion. The supporting parts of a research article include introduction, reference, and optional acknowledgement, appendix, etc.
7. The title of a research article is normally 10–15 word long.
8. The summary or abstract of a research article addresses problem statement, method and data, and outcomes.
9. The problem statement of a research article can be one sentence, paragraph, or section; preferably in a straight and concise manner.
10. The method section of a research article should be specific, with appropriate detailed information.
11. The result and discussion sections of a research article may be either combined or separated.
12. The conclusion section of a research article briefly reports research results and findings, including contributions, limitations, and future work.

Development of Research Report

13. The writing sequence of a research article differs from the order of article sections. Some core elements (method, results, and problem statement) are developed first. While, the key elements (title, summary/abstract, and conclusions) are composted later.
14. Research articles should be in a professional fashion and language, different from non-technical writing and oral style, following certain formats.
15. There are often requirements and recommendations for the format, sectional structure, sectional headings, figures, etc. of research writing.
16. Wording, voice, and style are also important, following professional practices.

Research Presentation

17. Presentations at professional conferences are beneficial for exchanging ideas and networking.
18. Attending a conference incurs costs of registration, travel, lodging, etc.
19. The number of presentation slides depends on the given presentation time, about 1.5 minutes a slide on average.
20. There are some recommendations for presentation design and oral preparation.

Exercises

Review Questions

1 Talk about the different requirements between academic writing and non-technical writing.
2 Discuss the essential and optional elements of a research report.
3 Compare two types of technical articles (e.g. project report, thesis, review, paper, etc.) for their formality or structure.
4 Summarize the main requirements of a master's thesis of your graduate school in terms of format.
5 Find a master's thesis and doctoral dissertation in your major and discuss their similarity or differences on structure, format, and length.
6 Review a case study for its common structure and main elements.
7 Review the title of a paper in your discipline. Can you rephrase it to be better?

8   The title of a research article is suggested 10–15 word long. Find a shorter or longer title and justify its length.

9   Find a research project report and review its (paragraph or information) structure of the executive summary.

10 Find a paper and review its conclusion section. Should any information be added in the section?

11 Should the study limitations mentioned be in a research report? Why?

12 If you are to write a problem statement, do you prefer it in one sentence, one paragraph, or one subsection? Why?

13 One states that in general it is best to make statements as definitive as possible. Do you agree with the statement? Why?

14 Do you prefer result section and discussion section of a research article separate or combined? Why?

15 For research reporting, what writing (sectional) sequence do you prefer? Why?

16 Show a couple of example of a word or phrase in daily communication that is not suitable in a technical report.

17 To write a technical report, how many times of revisions you would need. Why?

18 One suggests that a presentation file should include as many figures as possible. Do you agree? Why?

19 Find a presentation poster in a professional conference and review its structure design.

20 For a 20-minute presentation, how many slides can you handle based on your experience?

Mini-Project Topics

1   Find three master's theses in your area, review their structures (sections and subsections), and comment their logic flow.

2   Interview an experienced researcher and ask his/her sectional writing sequence when preparing a research article.

3   Review the writing sequence suggested in the chapter, provide your insight and justification to agree or disagree with the sequence.

4   Find three papers in the same subject, review their titles and abstracts, and comment their effectiveness of conveying research information.

5   Find three papers in the same subject you are familiar with and compare their method sections.

6 Find a paper that has not mentioned possible future work, read the paper, and suggest a future research direction for the paper.

7 Review the composition of the conclusion section of three papers and provide your comment on their effectiveness of conveying research achievements.

8 Review several technical papers, analyze their (active or passive) voices used and provide your comments.

9 Study the research writing guide of your (or another) university and summarize the main suggestions on writing style.

10 Interview an experienced researcher for s/he presentation experience and preference.

11 Review a presentation file and provide your suggestions to improve its structure design.

References

Ando, K., Ueyoshi, K., Orimo, K. et al. (2018). BRein memory: a single-chip binary/ternary reconfigurable in-memory deep neural network accelerator achieving 1.4 TOPS at 0.6 W. *IEEE Journal of Solid-State Circuits* 53 (4): 983–994. https://doi.org/10.1109/JSSC.2017.2778702.

APA, American Psychological Association (n.d.). APA style CENTRAL. www.apastyle.org/ (accessed August 2018).

Bagloee, S.A., Tavana, M., Asadi, M. et al. (2016). Autonomous vehicles: challenges, opportunities, and future implications for transportation policies. *Journal of Modern Transportation* 24 (4): 284–303. https://doi.org/10.1007/s40534-016-0117-3.

Behrens, B., Diaz-Infante, D., Altan, T. et al. (2018). Improving hole expansion ratio by parameter adjustment in abrasive water jet operations for DP800. *SAE International Journal of Materials and Manufacturing* 11 (3): 241–251. https://doi.org/10.4271/05-11-03-0023.

Cha, Y. and You, H. (2018). Torsion sensing based on patterned piezoelectric beams. *Smart Materials and Structures* 27 (3): 035010. https://doi.org/10.1088/1361-665X/aaa931.

Choudhury, S., Acharyya, S.K., and Dhar, S. (2018). Simulation of ductile behaviour of 20MnMoNi55 steel using unsaturated G_{fr} values for short crack growth estimated by introduction of correction factor to critical fracture energy, G_{fr}. *International Journal of Pressure Vessels and Piping* 168: 174–182. https://doi.org/10.1016/j.ijpvp.2018.09.013.

Columbia University Libraries (n.d.). Fair use checklist. https://copyright.columbia
.edu/content/dam/copyright/Precedent%20Docs/fairusechecklist.pdf (accessed
May 2019).

Cooper, H.M. (2016). *Research Synthesis and Meta-Analysis: A Step-by-Step Approach*,
5e, ISBN-13: 978-1483331157. Thousand Oaks, CA: SAGE Publications.

Dokhanchi, A., Amor, H.B., Deshmukh, J.V., and Fainekos, G. (2018). Evaluating
perception systems for autonomous vehicles using quality temporal logic. The 18th
International Conference on Runtime Verification, Limassol, Cyprus. Lecture
Notes in Computer Science. Vol. 11237. https://doi.org/10.1007/978-3-030-03769-
7_23.

Garcia, D., Mozaffar, M., Ren, H. et al. (2018). Sustainable manufacturing with
cyber-physical discrete manufacturing networks: overview and modeling
framework. *Journal of Manufacturing Science and Engineering* 141 (2): 17. https://
doi.org/10.1115/1.4041833.

Hertzmann, A. (n.d.). Writing research papers. University of Toronto. www.dgp
.toronto.edu/~hertzman/advice/writing-technical-papers.pdf (accessed August
2019).

Hu, S.J. (2013). Evolving paradigms of manufacturing: from mass production to mass
customization and personalization. *Procedia CIRP* 7: 3–8. https://doi.org/10.1016/j
.procir.2013.05.002.

IEEE, Institute of Electrical and Electronics Engineers (n.d.). IEEE article templates.
https://ieeeauthorcenter.ieee.org/create-your-ieee-article/use-authoring-tools-
and-ieee-article-templates/ieee-article-templates/ (accessed August 2018).

Jones, S.P.A. (n.d.). How to write a great research paper. Microsoft Research Lab,
University of Cambridge, United Kingdom. https://www.microsoft.com/en-us/
research/academic-program/write-great-research-paper/ (accessed August 2019).

Levin, V., Zubov, L., and Zingerman, K. (2015). An exact solution for the problem of
flexure of a composite beam with preliminarily strained layers under large strains.
International Journal of Solids and Structures 67–68: 244–249. https://doi.org/10
.1016/j.ijsolstr.2015.04.024.

Lu, S., Wu, C., Zhong, R., and Wang, L. (2018). A passive RFID tag-based locating and
navigating approach for automated guided vehicle. *Computers & Industrial
Engineering* 125: 628–636. https://doi.org/10.1016/j.cie.2017.12.026.

Ludwig, K. (2012). Characterizing the utilization of large scientific research facilities.
National Science Foundation. https://www.nsf.gov/attachments/134059/public/
CharacterizingUtilizationofLargeFacilitiesReport_LudwigK2012.pdf (accessed
April 2019).

MIT, Massachusetts Institute of Technology (2016). Specifications for Thesis Preparation, Prepared by the MIT Libraries as prescribed by the Committee on Graduate Programs and the Committee on Undergraduate Programs. https://libraries.mit.edu/archives/thesis-specs/index.html#title (accessed July 2019).

NSF (n.d.). Final format research performance progress report. https://www.nsf.gov/bfa/dias/policy/rppr/format_ombostp.pdf (accessed June 2019).

Oks S.J., Fritzsche, A. and Möslein, K. (2018). Engineering industrial cyber-physical systems: an application map based method. 51st CIRP Conference on Manufacturing Systems, Procedia CIRP. Vol. 72, 456–461. https://doi.org/10.1016/j.procir.2018.03.126

Osman, N. and Sahraoui, A. (2018). A software requirement engineering framework to enhance critical success factors for ERP implementation. *International Journal of Computer Applications* 180 (10): 32–37.

Paik, S.M., Jin, E., Sim, S.J., and Jeon, N.L. (2018). Vibration-induced stress priming during seed culture increases microalgal biomass in high shear field-cultivation. *Bioresource Technology* 254: 340–346. https://doi.org/10.1016/j.biortech.2018.01.108.

Paler, A. (2018). Controlling distilleries in fault-tolerant quantum circuits: problem statement and analysis towards a solution. Proceedings of the 14th IEEE/ACM International Symposium on Nanoscale Architectures (July 2018), 147–152. https://doi.org/10.1145/3232195.3232224

Purdue Online Writing Lab (n.d.). APA formatting and style guide. https://owl.purdue.edu/owl/research_and_citation/apa_style/apa_formatting_and_style_guide/general_format.html (accessed August 2018).

Rackham, H. (2017). The dissertation handbook. University of Michigan Rackham Graduate School. www.rackham.umich.edu/downloads/oard-dissertation-handbook.pdf (accessed May 2019).

Ribeiro, F., Sena-Cruz, J., Branco, B., and Júlio, E. (2018). Hybrid effect and pseudo-ductile behaviour of unidirectional interlayer hybrid FRP composites for civil engineering applications. *Construction and Building Materials* 171: 871–890. https://doi.org/10.1016/j.conbuildmat.2018.03.144.

Scheffler, O.C., Coetzee, C.J., and Opara, U.L. (2018). A discrete element model (DEM) for predicting apple damage during handling. *Biosystems Engineering* 172: 29–48. https://doi.org/10.1016/j.biosystemseng.2018.05.015.

Stanford University Libraries (n.d.). Fair use. https://fairuse.stanford.edu/overview/fair-use/ (accessed May 2019).

Ung, S. (2018). Development of a weighted probabilistic risk assessment method for offshore engineering systems using fuzzy rule-based Bayesian reasoning approach. *Ocean Engineering* 147: 268–276.

Weimer, M. (ed). (n.d.). Keys to designing effective writing and research assignments. Jackson State University. www.jsums.edu/academicaffairs/files/2012/08/Keys-to-Designing-Effective-Writing-and-Research-Assignments.pdf?x19771 (accessed August 2019).

Wilson, A. (2017). Basic steps to write an outstanding research paper. Stanford University. https://collegepuzzle.stanford.edu/basic-steps-to-write-an-outstanding-research-paper/ (accessed October 2019).

9

Scholarly Paper and Publication

9.1 Considerations For Publication

9.1.1 To Publish, or Not To Publish

9.1.1.1 Possible Outlets

As discussed in Chapter 8, reporting results is necessary for all research projects. Good results from a research project may have different types of outlets. If the outcomes of a research project are promising, for example, they can lead to additional tasks and/or projects. Table 9.1 lists these possibilities and corresponding efforts.

An outlet of research largely depends on the nature and agreement of a research project. There may be regulations and policies of a research organization or department, which tell us where to go next. In addition, some of these outlets may be combined either concurrently or sequentially. Obviously, "cold storage" or "do nothing" is neither in the best interest of researchers nor in that of their organization.

Publication is encouraged and required for basic research conducted in universities and research institutes, particularly for government-sponsored and foundation-supported projects. Publications can build a better research credential to gain a competitive edge for future research opportunities. In applied research and engineering R&D, we implement the outcomes and may publish some of them, which will be further discussed in a following subsection.

Publishing in high-quality journals is a tough task for all and can be even more challenging for new researchers. Therefore, firmly understanding the objectives, process, requirements, and methods of publishing helps all researchers.

9.1.1.2 Objectives of Publications

The objective of research, particularly basic research and most applied research, is to further a body of knowledge and advance new technology. Publication means that research results are open and permanently accessible to the public and can be

Engineering Research: Design, Methods, and Publication, First Edition. Herman Tang.
© 2021 John Wiley & Sons, Inc. Published 2021 by John Wiley & Sons, Inc.

Table 9.1 Destinations of research outcomes.

	Outlet	Effort
Research outcomes to	Publication	Revise and submit
	Patent	Prep and file application
	Implementation or artifact	Develop process, plan, and act
	New base for next project	Review and propose
	In a "cold storage"	Do nothing (or forget)

used and cited by anyone in the world. Therefore, the main purposes of publication are to contribute to an ongoing body of knowledge and share newest information with other professionals in the discipline.

Without access to completed and reported research, it is likely that other researchers would repeat the same or similar research efforts. In other words, publishing is vital to the effective advancement of science and technology. Many public funding agencies require researchers to disseminate their findings to broader audiences and research communities. Publications are also important for research work in the national laboratories and the R&D departments of large companies.

Publishing research results can also get work recognized by the professional community. A quality publication is an important indicator of the professional achievements of a researcher. For example, scholarly publication is one of the basic requirements for job offerings, tenure status, and professional advancement in academia and research world. Publishing research results can help build research credits in the professional field, which may increase opportunities and funding offers for future research. Graduate students may need to publish theses for their graduation.

9.1.1.3 Overall Publication Status

Academic and research publishing has been in place for several hundred years. Table 9.2 lists the science and engineering articles published in 2014 (Data Source: White et al. 2017). Table 9.3 lists the research papers of all research subjects published in the scholarly journals that were selected by SCImago in 2018 (Data Source: SCImago 2019).

Many researchers consider the H index a relative indicator of measuring the quality of scholarly publication based on the citation in other publications (Hirsch 2005). The H index is widely used but its accuracy has been debated by other researchers (Costas and Franssen 2018; Oravec 2019). There are several other indexes, and a study shows there exists weak correlation between various indices

Table 9.2 Science and engineering articles in 2014.

Region or country	Quantity	Percentage (%)
World	2 290 294	100.0
United States	431 623	18.8
China	395 588	17.3
Germany	107 747	4.7
India	106 574	4.7
Japan	103 793	4.5
United Kingdom	101 536	4.4
France	74 269	3.2
Italy	70 453	3.1
South Korea	63 748	2.8
Canada	60 916	2.7
Spain	56 604	2.5
Brazil	53 152	2.3
Australia	52 269	2.3
Russia	43 487	1.9
Iran	36 539	1.6

(Raheel et al. 2018). Table 9.3 shows both the information of quantity and H index of engineering research papers, which were published in quality scholarly journals selected by SCImago.

9.1.1.4 Publication of Industrial R&D

For engineering R&D, outcomes may be important to keep the company competitive in the market. In such cases, research activities and results may be viewed as a type of trade secret. Thus, the publication of engineering R&D or industry-sponsored research carries some risks. In addition, revealing proprietary data may harm a company's competitiveness in the marketplace. Therefore, companies may not be willing to publish their R&D results. The reason for non-publication is obvious for military applied research.

Companies have different publication policies. Some companies encourage employees to submit papers to professional conferences and even scholarly journals. One benefit can be that the company influences technology in the right direction. For example, an industrial company requires approval from department managers for an abstract submission to a professional conference. If the abstract is

Table 9.3 Engineering papers in scholarly journals selected by SCImago in 2018.

Country	Quantity	H index
China	188 249	475
United States	95 958	915
India	51 663	284
Germany	29 478	465
United Kingdom	28 291	494
Japan	27 824	417
Russian Federation	23 768	193
South Korea	22 569	346
Italy	20 369	357
France	19 345	392
Canada	16 530	403
Iran	15 237	175
Australia	14 242	340
Spain	13 516	316
Malaysia	11 821	160

accepted, then the research work and paper draft will be reviewed by supervising engineers and then by the head of R&D before it can be submitted for conference publication.

Unless there are patents to protect their intellectual property (IP), many companies do not share their research information with the public. In other words, publications can be delayed by filing patent applications. Even if a company can share research information, the original proprietary data can be modified, such as using a ratio or different scales, to protect the original information.

If industrial R&D achievements are not allowed for publication, they still have reference value for others within the organization. Learning from research work, organization employees can think differently and generate new ideas to keep the organization competitive. However, the influence of the research can be limited as there is no public access.

For joint research projects between industry and academia, or industry-sponsored projects conducted at research institutes, the ownership of IP and future publication should be discussed and agreed upon in their research contracts. In most cases, the IP is owned by a research institute. However, publication may be restricted and require a preview and approval. Submission approval

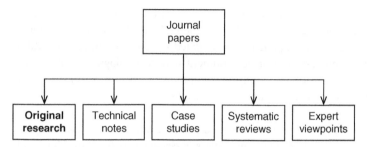

Figure 9.1 Types of scholarly journal publications.

and its process depend on the nature of the research, institutional policy, and the project agreement with sponsor. If an agreement includes delayed publication, the research results should be held for the agreed period.

9.1.2 Types of Publication

9.1.2.1 Types of Journal Papers

Scholarly journal papers are considered as the backbone of academic publishing because, if peer reviewed or refereed, they can be seen as a contribution to the field. There are several types of journal papers (see Figure 9.1).

Original Research Papers. These are the majority of papers published in scholarly journals. An original paper reports new study and results, which have an archival value to a professional community. The key feature of original papers is their originality and innovation of work. Discussed in Chapter 3 literature review, original research papers are innovative in terms of outcome, methodology, and/or approach in detail.

Technical Notes. Also called technical brief articles or letters, technical notes are based on preliminary results for quick publication, without including too much data or details. Thus, technical notes are short with approximately 2000 words. Some journals are more included to lean toward such technical notes. For example, Manufacturing Letters (ISSN: 2213-8463) and Nano Letters (ISSN: 1530-6984), are highly cited journals. Here are a few examples of short papers:

> *"Detachable Bronchoscope 2ith a Disposable Insertion Tube"* (2 pages) (Carlson et al. 2015)
> *"Current Pass Optimized Symmetric Pass Gate Adiabatic Logic for Cryptographic Circuits"* (2.5 pages) (Koyasu and Takahashi 2018)
> *"A New Method of Bottleneck Analysis for Manufacturing Systems"* (3.5 pages) (Tang 2019)
> *"Knockdown Factor for the Buckling of Spherical Shells Containing Large-Amplitude Geometric Defects"* (4 pages) (Jiménez et al. 2017)

Case Studies. This type of article is usually based on applied research and/or R&D and reports specific instances of phenomena. Instead of focusing on a new knowledge or theory, such application-orientated papers have an archival value on a new design, process, and/or development with technological implications. Research work in form of case study is a common practice in engineering and technical fields. Here are a few published examples:

> *"Application of the preventive maintenance scheduling to increase the equipment reliability: Case study - bag filters in cement factory"* (Ebrahimi et al. 2018)
>
> *"Power Quality Impact of Charging Station on MV Distribution Networks: A Case Study in PEA Electrical Power System"* (Pothinun and Prem 2018)
>
> *"Development of a series hybrid electric aircraft pushback vehicle"* (Cash et al. 2019)
>
> *"Correlation between acute phase head injury to cyclists and consciousness disturbance: a case study in reconstruction of single-cyclist accident"* (Oikawa et al. 2019)

Systematic Literature Reviews. A technology review article is often in a specific, emerging subject and based on over 100 recently published papers. The articles summarize others' research results that have relevance to a professional community, not necessarily providing new information and knowledge. A review article may provide critical and constructive analyses of existing literature and make some recommendations for future research, discussed in Chapter 3 as well.

Another type of review article is called *Expert Viewpoints*, where well-known domain experts offer such reviews with the invitation by a journal. Materials Science and Engineering: R: Reports (ISSN: 0927-796X) (R Reports n.d.), a top scholarly journal, states, "publishes invited review papers covering the full spectrum of materials science and engineering. The reviews, both experimental and theoretical, provide general background information as well as a critical assessment on topics in a state of flux."

9.1.2.2 Other Types of Publication

Conference papers are a publication at a professional association meeting, symposium, or similar gathering. Professionals present their latest, albeit preliminary results at conferences. There may be hundreds of conferences in a field. Attending conferences is also an excellent professional networking opportunity, as discussed in the previous chapter.

Some conferences go through a peer review or refereed process, while others do not. Thus, conference papers may or may not give the same reference value as journal papers, particularly if no formal proceedings are published after a conference.

However, a main advantage of conference papers is that they report new research and studies quicker than journal papers.

Furthermore, other types of publications can enhance the distribution of research findings and knowledge.

- *Authored Book*. This provides a broad treatment of a subject and is considered a significant professional accomplishment. Major publishers do rigorous reviews for book proposals on sample manuscript chapters before a publishing agreement. Clearly, composing a book manuscript can take a lot of time. For an academic book project with a good proposal, a two-year writing plan is a reasonable expectation working as a part-time writer.
- *Edited Book*. A collection of chapters or separate papers on a focused subject from different authors and compiled by an editor or editors. The editors are often significant figures in the field. One effective way to initiate a book project is to have a writer's workshop with the goal of a thematic edited volume in mind.
- *Book Chapter*. In many cases, it can be a summary of recent research projects. The authors of book chapters are often invited to an edited volume.

In addition, there are some other types of publications, such as book reviews, commentaries, articles in trade magazines, and newspaper articles. These types of non-refereed articles are not considered research publications, and their quality varies significantly. Thus, researchers normally do not cite such publications.

9.1.3 Paper Quality

9.1.3.1 Basic Requirements for Publication

A publication, in the format of a paper, report, or presentation, should give readers a clear picture of the problem, work, outcomes, and the merits of research. In other words, a publication should provide technical value to the professional community. Accordingly, the organization of ideas, processes, and results in a paper is important for effective communication. The data should substantiate the interpretations and conclusions precisely and completely. There should be a clear link between the data, the research problem, and the results. Scientific interpretation on data analysis and results should be solid and convincing.

There are several matrices to evaluate paper quality. Here is a brief list:

- Innovation (originality or contribution, etc.)
- Significance of findings (benefits, predicted impacts)
- Quality of scientific work (completion, assumptions, issue/error, etc.)
- Reference value (to the professional field, potential applications, future work, etc.)
- Acknowledgement (citation, relevant, etc.)
- Presentation (organization, clarity, language, format, etc.)

Clearly, these factors are not equally important. For example, academic publication emphasizes the originality and significance of research work. For originality, a submitted manuscript should be created by the researchers and have not already been published. It should also not be under consideration for publication elsewhere. If a manuscript is not original, it may not be acceptable to publish regardless of other aspects.

The reference value of a paper is very important as well. The value may be judged on its innovation, significance, completeness, etc. The reference value of a paper may be measured by the number of citations after publication.

Many academic professionals do not consider a specific application or case study appropriate for publication because such a study may be exclusive to a particular situation. However, the academic value of a case study may not always be specially limited to the case. The research procedure, method, and findings can have a good reference value to other professionals in the field or even to those in other areas. It is often worth extrapolating a successful case study into other situations. Therefore, the findings from case studies may be good for publication, which will be discussed later.

9.1.3.2 Preparation for Reviews

Understanding the quality criteria of journal publication can help us prepare our manuscripts, even though the quality criteria of journals vary. Figures 9.2–9.4

Journals Publications	
Paper Review Confirmation	
Paper No. MANU-17-■■■ (Research Paper) cross check 18% Match	Date Assigned: 22 Jun 17
Title: ■■■■■■■■■■■■■■■■■■■■■■	Date Due: 13 Jul 17
Paper Profile	Date Reviewed:
Originality	Good
Significance	Acceptable
Scientific relevance	Acceptable
Completeness	Good
Acknowledgment of the Work of others by References	Good
Organization	Good
Clarity of Writing	Acceptable
Clarity of Tables, Graphs, and Illustrations	Good
In your opinion, is the technical treatment plausible and free of technical errors?	Yes
Have you checked the equations?	Yes
Are you aware of prior publication or presentation of this work?	No
Is the work free of commercialism?	Yes
Is the title brief and descriptive?	Yes
Does the abstract clearly indicate objective, scope, and results?	Yes
Recommendation	
This paper is Acceptable (Suggested changes (changes not mandatory)), for publication as Full research paper. The quality of the paper is Good.	
Comments to Author:	

Figure 9.2 Paper review criteria and result – Example 1.

* 1. Does the paper make a new and significant contribution to the Production Research literature?	No ▼
* 2. Does the paper provide evidence of real or potential application for Production Systems?	No ▼
* 3. Is adequate credit given to other contributors in the field and are references sufficiently complete?	Yes ▼
* 4. Does the paper appropriately compare the performance of proposed methodologies with those found in the published literature?	No ▼
* 5. Does the paper state what the author(s) propose to do in the future?	Yes ▼
* 6. Are the character and contents of the paper clear from the title and abstract?	Yes ▼
* 7. Is the paper clearly, concisely, accurately, and logically written?	No ▼
* 8. Could it benefit from condensing or expansion? If yes, please explain why in the comments to author section.	No ▼
* 9. Is the subject matter of relevance to Production Research and appropriate for IJPR?	Yes ▼
* 10. Are all references relevant? If not please indicate in your review not relevant references.	Yes ▼

Recommendation

◯ Accept
◯ Minor Revision
◉ Major Revision
◯ Reject but allow Resubmission
◯ Reject

Confidential Comments to the EIC

Figure 9.3 Paper review criteria and result – Example 2.

show three examples of scholarly paper evaluations. These criteria are often self-explanatory. The knowledge of paper reviews may serve as a reference for paper preparation. Based on the paper evaluation criteria of a journal, readers may do a self-evaluation before submission.

9.2 Publication Process

9.2.1 Overall Publication Process

9.2.1.1 Main Steps

For journal publication, we need to find an appropriate journal for consideration, prune the research report or manuscript draft to the content, structure, format based on the journal requirements, and submit the manuscript to the journal.

Reviewer Confidential Comments to Editor:
For each statement, please place an x in the space provided next to relevant answer:

RELEVANCE

Is this paper relevant to actual manufacturing systems or manufacturing processes problems? If the relation to current problems is weak, does it at least have future potential?

Highly Relevant _x_   Possibly Relevant ___   Has Future Potential ___   No Clear Relevance ___

CONTRIBUTION

To Theory (specify area) _____
High ___ Average _x_ Low ___ Not Sure ___

To Practice (specify application)
High _x_ Average ___ Low ___ Not Sure ___

To Synthesis (tutorial, review)
High _x_ Average ___ Low ___ Not Sure ___

Other (specify) _____
High ___ Average ___ Low ___ Not Sure ___

ORIGINALITY
High _x_ Average ___ Low ___ Not Sure ___

QUALITY OF WRITING AND WRITTEN PRESENTATION
High ___ Average _x_ Low ___ Not Sure ___

Comments on Writing/Grammar_____

DETAILED EVALUATION

EVALUATION OF CONTENT
Title
Good ___ Adequate _x_ Poor___ Not Sure ___

Abstract
Good ___ Adequate _x_ Poor___ Not Sure ___

Introduction and Motivation
Good _x_ Adequate ___ Poor___ Not Sure ___

Review of Related Work
Good _x_ Adequate ___ Poor___ Not Sure ___

Technical Soundness of Body of Paper
Good ___ Adequate _x_ Poor___ Not Sure ___

Have you checked the equations? _____

Conclusions
Good ___ Adequate _x_ Poor___ Not Sure ___

EVALUATION OF PRESENTATION

Quality of Writing/Language
Good ___ Adequate _x_ Poor___ Not Sure ___

Figures (incl. captions and legends)
Good _x_ Adequate ___ Poor___ Not Sure ___

Tables (incl. captions and legends)
Good ___ Adequate _x_ Poor___ Not Sure ___

References
Good _x_ Adequate ___ Poor___ Not Sure ___

Length
Good ___ Adequate _x_ Poor___ Not Sure ___
Indicate Suggestions:
Shorten Sections _____
Expand Sections _____

Figure 9.4   Paper review criteria and result – Example 3.

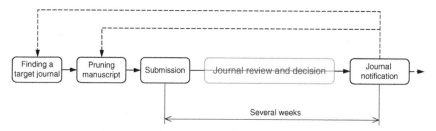

Figure 9.5 Author's work for paper publication.

Figure 9.5 illustrates the workflow. After submission, we wait for the response from journal editors.

After receiving a paper manuscript, a journal editor team takes several steps to evaluate the manuscript. Figure 9.6 shows a typical process of (a) refereed conference publication and (b) refereed journal. The main steps include:

1. Authors to submit a manuscript (to an electronic edit management system)
2. Editor-in-Chief (EIC) to scan suitability and overall quality, if fits, assign Associate Editor (AE)
3. AE to review the manuscript and select reviewers
4. Reviewers to review and submit review results
5. AE to summarize review results and make a recommend to EIC
6. EIC to make the final decision and notify authors

Normally, EIC's decision is final. No higher governing body exists to investigate or change EIC's decisions. If a manuscript is rejected, authors may communicate with the EIC. However, it is rare that EIC would reconsider their decision.

9.2.1.2 Copyright Paperwork

Journals normally request an author's warranty and/or transfer of copyright agreement. Authors ensure that the manuscript submitted for publication is original and has not been submitted or published elsewhere. Authors should not use substantial verbatim text from another copyrighted work without the written permission of the copyright holder.

Researchers should keep in mind copyright requirements when preparing a manuscript. Once a manuscript is submitted, a copyright form should be ready to sign. Often, authors retain the rights to any IP developed with research, but not the verbatim text of the manuscript.

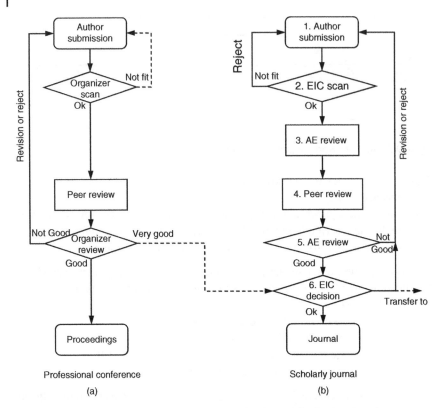

Figure 9.6 Overall review process of paper publication. (a) Professional conference. (b) Scholarly journal.

A copyright form can be signed either by the first author on behalf of all co-authors or an authorized representative on behalf of an employer. In a copyright form, signing authors may need to identify their status whether they are (USA, UK, Canadian, or Australian) government employees or contractors, stipulated by given copyright laws.

The detailed items of a copyright form vary among journals and publishers. For example, some journals keep the right to edit the work for the original edition and for any revision, provided that the meaning of the content is not materially altered.

9.2.2 Peer Review Process

We know who assesses our paper quality. For student research reports, the advisors normally decide based on academic requirements; for journal publication, the quality evaluation is based on peer review.

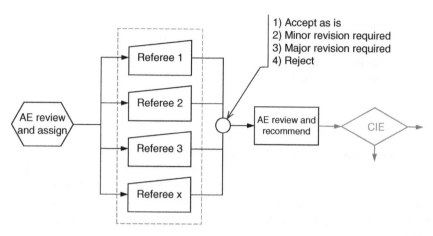

1) Accept as is
2) Minor revision required
3) Major revision required
4) Reject

Figure 9.7 A peer review process managed by AE.

9.2.2.1 Peer Review Overview

Almost all scholarly journals require peer reviews, also known as refereed journals. The most important and time-consuming function in a journal publication process is the peer review. To authors, it is a blind process. However, understanding the overall process can help us prepare manuscripts.

A peer review is a process to obtain referee's viewpoints on manuscripts. Peer reviewers help editors determine the merit and quality of a paper in terms of originality, validity, significance, and writing. With reviewer's feedback, authors can revise and improve a manuscript.

Due to limited time and resources, reviewers normally focus on study design, methods, and how reported data supporting author's assertions. Reviewers are not able to determine whether data are accurate. Therefore, routine peer review is not effective in detecting possible research misconduct (Vastag 2006).

It is recommended that new researchers become a reviewer after they have published a couple of papers. Being a reviewer is not only a voluntary contribution to a professional community but also a learning opportunity from reviewing papers.

One additional note is that non-refereed journals and their papers are not considered as scholar publications, so they are not good to be used for graduate student graduation, promotions, and job applications in academia and research institutes.

9.2.2.2 Review Process and Ratings

For journal publication, an AE, working on a volunteer basis, acts as a manager to assign reviewers, summarize review results, and make publication recommendations to the journal EIC. AE's work in a peer review process is shown in Figure 9.7.

Reviewing a paper manuscript, referees or reviewers recommend how a paper suitable for the journal and rate a paper in one of four levels to AE:

1. Accept as is
2. Minor revision required
3. Major revision required
4. Reject

The "accept as is" is rare for high-quality journals but may be common for some conferences. With both "minor revision requested" and "major revision requested," a manuscript is to return to the authors for revision consideration. For minor revisions, AE reviews the manuscript revision and the responses from the authors and makes publishing recommendation to EIC. A second round of peer review, preferably by the original reviewers, is often required for the situations of a major revision.

AE mainly relies on the recommendations from peer reviewers. Usually, an AE makes a publication recommendation to the EIC for decision if all reviewers reach a consensus, even though they may not be in a perfect agreement. It is uncommon that an AE's publication recommendation is against the results from peer reviewers. In addition to the four ratings, AE may also suggest that a manuscript transfer to a more suitable journal.

Furthermore, some journals ask AEs to evaluate the reviewers for their peer review efforts and performance. Low-rated reviewers may not likely be invited to review manuscripts again. Among the possible recommendations, rejection is common for high-quality journals. The publish acceptance rates, which combines the numbers of papers "accept as is" and after revisions, of high-quality journals can be lower than 20%.

9.2.2.3 Characteristics of Peer Review

The referees or reviewers are normally not part of the editorial staff of a journal. Reviewers are volunteers without compensation and giving up their own time to serve a research community. Almost all of them are authors too.

There are three types of peer review processes:

1. *Double-blind Review*: The identities of both reviewers and authors remain anonymous throughout a review process.
2. *Single-blind Review*: The author does not know reviewers; but the reviewers know authors' names and affiliations.
3. *Open Review*: The identities of the reviewers and the authors are not concealed.

Thanks to the anonymity of authors and reviewers, a double-blind review is in the best interest of both authors and journals as openness can be expected in the review. However, reviewers might still identify the authors and/or their affiliations based on the contents of a manuscript. For single-blind reviews, reviewers see who the authors and affiliations are, which may cause bias from reviewer's opinion based on the authors and their organization. The ethics of single-blind peer review is a topic of discussion. On the other hand, some argue that author's identity is helpful information for reviewers to do their job. The third type, open review, is very rare in scholarly publication field as it is prone to various bias influences, such as nepotism, loyalty, retaliation, prestige, gender, origin, etc. More discussion can be found in academic publisher websites, such as https://authorservices.wiley.com/Reviewers/journal-reviewers/what-is-peer-review/types-of-peer-review.html#singleblind.

As human beings, reviewers have their various backgrounds and own personality. Sometimes, they may not be very knowledgeable to the specific topic even they are in the same field. Occasionally, their comments can be very direct and may sound harsh. Authors should view reviewers as co-workers rather than a perfect authority.

9.2.2.4 Peer Review for Conference

The publication process of a professional conference is similar to but often simpler than the process for a scholarly journal. The top-rated papers based on peer reviews of a conference may be transferred to a journal for consideration (Figure 9.6a). Sometimes, this path is called a "fast track." It is often the case when a conference belongs to the same association or organization of a scholarly journal. Most reputable conferences, such as those organized by ASME, IEEE, International Academy for Production Engineering (CIRP), and Society of Manufacturing Engineers (SME), publish their proceedings of full-length manuscripts.

On average, the peer review for professional conferences is less strict than that for a scholarly journal. Some professional conferences do not have a serious peer review process or may not even require a peer review. The articles published in such conferences are un-refereed conference papers. Due to that reason, many researchers and institutes do not view conference papers at the same level of quality as journal papers. In addition, some conferences may not be considered research-oriented.

9.2.3 Review Comment and Response

9.2.3.1 Comments and Recommendations of Reviewers

Reviewers are required to comment on the manuscript in addition to providing ratings based on the journal's criteria. Figure 9.8 shows two examples.

Most times, reviewers provide detailed comments and suggestions to authors as well as to editors. The comments and suggestions can be on any part of a

---

Recommendation

This paper is Acceptable (Suggested changes (changes not mandatory)) , for publication as Full research paper. The quality of the paper is Average.

Recommendation

This paper is Not Acceptable (Revision required; resubmit as Tech. Brief) . The quality of the paper is Inferior.

Figure 9.8 Examples of reviewer's recommendations.

manuscript, in any angle of viewpoints, and with various tones. Here are a few of examples:

- This research work is well done experimentally. However, there are a few typos in the manuscript. For example, Table 4 is mistakenly captioned as Table 3.
- I cannot recognize the necessary details in Figure 3. Probably the author can add or extend the figure to show the detail, which may be helpful in understanding the equipment actions.
- In equation (2), how did the authors decide the values of the parameters k and n? Are the test results sensitive to these values?
- The numerous spelling and grammar mistakes in the manuscript made it difficult to read.
- In the literature review, the authors cited several papers relating to one of the subjects. However, the reason for doing so and the relation to the work is not clearly established.
- Authors should describe how the outputs are compared with an appropriate interpretation.
- Did you match the theoretical way of optimization with some experiments to validate the results? Please round up your work.

Obviously, authors must carefully analyze feedback and think about how to respond the reviewers' comments and questions.

9.2.3.2 Response to Peer Review

After receiving reviewer's feedback from the journal, authors must carefully read over the referee's comments. Then, authors decide whether and how to incorporate the referee's suggestions into the manuscript revision. Authors may also decline to revise and withdraw their manuscript.

If we agree to revise our manuscript, as professionals, we may like to:

- Wait and think a couple of days before revising and responding
- Make a list of the reviewer comments and draft point-to-point responses

- Revise the manuscript based on the agreed comments
- Stick to the facts and avoid blaming reviewers when rebutting

Even if we agree to revise, we do not need to agree with all of the reviewers' viewpoints and suggestions. If we disagree with a referee on one point, we should politely explain and justify our position to convince the editor the validity of our viewpoint. We may say, "Thank you for your time and comments. However, we are afraid that we do not fully agree with your second comment because ..."

We must make sure all the comments received from the journal have been addressed, either incorporated into the revision or explained as a rebuttal with a point-to-point responsive correspondence or rebuttal to the reviewer's comments. Particularly for the disagreed items, citing relevant parts of the paper can be helpful to prove our position. When resubmitting the revised manuscript, we should also explain how the reviewers' comments have been considered in the revision. We may highlight the revised parts in the revision.

In most cases, an AE reviews the manuscript revision and correspondence and make a next recommendation to the EIC. For major revisions, as mentioned, a peer review process may be undertaken again on the revision.

9.2.3.3 Rejection Handling

Understandably, we may feel disappointed and distressed when our paper is rejected by a scholarly journal. A rejection can come for different reasons. Discussed above, a paper acceptance decision is mainly based on the results of peer reviews. It is fair to say that a peer review can be subjective and related to reviewer's background, in addition to their limited information about the research reported in a paper. In other words, we should not expect reviewer's viewpoints to be perfectly correct.

If we feel that our manuscript was unfairly or carelessly treated by reviewers, we may send an email to EIC to explain and politely request for a second run review with different referees. AE and/or EIC will review the situation again and respond with their decision.

If a manuscript is not fully in line with the aims and scope of the journal, the manuscript can be rejected from the editor without being sent out for peer review, which is called desk rejection. For example, an EIC may say, "Albeit in an area of great importance and interest, your paper is not well aligned with the scope and the areas of interest of the readership of the journal." More than one-third of manuscripts are rejected before peer review for high-quality journals. A desk rejection is not necessarily related to the manuscript quality but can be due to the type or focus of the research reported.

Regardless of the rejection reasons, we should remain professional, not complaining about such a decision. A common practice for a rejected manuscript is that authors revise their manuscript based on the review results and submit to another journal for consideration. As authors, we should appreciate the reviewers' time and input and editors' effort to make our research and manuscript better.

9.3 Target Scholarly Journals

9.3.1 Journal Selection

9.3.1.1 Overall Considerations

Selecting an appropriate journal is not a simple task, considering several factors. First, we need to know the publishers. Publishers of scholarly journals include for-profit companies and not-for-profit organizations. The former are normally academic publishers, while the latter are professional societies or associations. Some large universities have their own academic presses as well. Most publishers publish both books and journals.

Visiting the websites of candidate journals is a good way to know about the journals. We may consider a few factors at the beginning. For example, aim, scope, type of paper, and readership are the first filters to create a short list of candidate journals.

When selecting target journals, we may use Table 9.4 as a checklist. The factors in the checklist will be more discussed in the following subsections. Readers may scan and find three to five journal candidates and then use the checklist for evaluation and comparison. The rating for each journal in the checklist can be in three levels of scores: 2 (or $\sqrt{}$) = Yes, 1 (or ?) = Maybe, or 0 (or ✗) = No. Please bear in mind that the importance of these factors is case dependent.

To increase the likelihood of successful publication, it is critical to self-evaluate the research outcomes and compare the published papers in candidate journals. By doing this, we may estimate the likelihood of our paper to be accepted by a journal. It is also suggested that new researchers seek advice from senior professionals.

Table 9.4 A checklist of journal selection.

Factor	Description	Journal 1	Journal 2	Journal 3
Relevance	Fitting aim, scope, readership, etc.			
Quality	Good reputation in the field, impact factor, etc.			
Accept rate	Appropriate (likelihood) rate			
Cost	APC reasonable, if applicable			
	Sub-total (score) =			

9.3.1.2 Relevance

Every journal has its own aim and scope. We should know the criteria and expectations of a journal to identify its requirements and guidelines to ensure our paper a "good fit" with the journal. As suggested, we read already-published papers in the journal for a better understanding.

Major academic publishers provide guides for journal selection. Table 9.5 lists three examples.

Within a specific subject, there may many journals across different publishers. For example, there are over 100 journals related to the subject of "mechanical design." Table 9.6 lists 16 of them, sorted by the publication volume as of 2018.

Some journals focus on a type of research or methodology rather than a subject. For example, there are interdisciplinary journals dedicated to specific methodology, such as:

- *Qualitative Research Journal* (https://www.emeraldinsight.com/journal/qrj)
- *The Qualitative Report* (https://nsuworks.nova.edu/tqr/)
- Journal of Mixed Methods Research (https://journals.sagepub.com/toc/MMR/current)
- Journal of Simulation (https://orsociety.tandfonline.com/toc/tjsm20/current)
- Journal of Survey Statistics and Methodology (https://academic.oup.com/jssam)

If your research is more on a specific methodology, you may like to consider such interdisciplinary journals.

9.3.1.3 Quality Factors

As discussed, peer review is a key for publication quality. Another way to evaluate the quality of a journal is to check the editorial board of the journal. A publishing company and/or a professional association form an editorial team for a journal. The editorial team consists of domain experts in given subject areas, and managing staff. Their names and credentials should be published on the journal's website. The managing staff, including managing editor, editorial assistant, and IT support, who take care of routine work, can be employees of the publisher.

Table 9.5 Some publisher's guides for journal selection.

Publisher	Title	Website
Elsevier	Find the perfect journal for your article	https://journalfinder.elsevier.com/
Wiley	Find the journal that's right for your research	https://journalfinder.elsevier.com/
Springer Nature	Journal Suggester	https://journalsuggester.springer.com/

Table 9.6 Some scholarly journals related to mechanical designs.

No.	Journal title	Publisher	Vol. (2018)
1	Journal of Materials: Design and Applications	Institution of Mechanical Engineers (UK)	232
2	Journal of Mechanical Design	American Society of Mechanical Engineering (US)	140
3	Materials & Design	Elsevier (US)	137
4	International Journal of Mechanical Sciences	Elsevier (US)	135-
5	Journal of Strain Analysis for Engineering Design	SAGE Publications (US)	53
6	Mechanics Based Design of Structures and Machines	Taylor & Francis (UK)	46
7	Journal of Engineering Design	Taylor & Francis (UK)	29
8	Chinese Journal of Engineering Design	Zhejiang University Press (China)	25
9	International Journal of Design Sciences & Technology	Europia (UK)	23
10	The Design Journal	Taylor & Francis (UK)	21
11	Journal of Engineering, Design and Technology	Emerald Publishing Limited (UK)	16
12	International Journal of Mechanics and Materials in Design	Springer (Germany)	14
13	Journal of Advanced Mechanical Design, Systems, and Manufacturing	Japan Society of Mechanical Engineers (Japan)	12
14	Journal of Advanced Design and Manufacturing Technology	Islamic Azad University (Iran)	11
15	Machine Design	University of Novi Sad (Serbia)	10
16	International Journal of Design Engineering	Inderscience (Switzerland)	7

In addition, a journal's long history may indicate its quality and reputation. The journals listed in the above table are of good quality ones since they have high volumes of publication. However, a new journal is not necessarily a low quality, particularly for emerging subjects, such as AI and nanotechnology.

There are several widely used quality indicators for academic journals, discussed in the next subsection. We need to decide which level of journal to consider for our manuscript submission. Once fortified with successful publications, built up research credentials, and better research findings, you may try higher ranked journals later.

Some journals and conferences are viewed as "predatory" by serious researchers. A common observation is that predatory journals have little peer review, publish articles as suit it, and charge substantial fees (called article processing charges or APC). Without any quality control mechanism, the papers published in such journals can be of poor quality. There however is no clear definition of predatory publishing, but there are discussion and arguments about it. Readers may consult with experienced authors and refer to dedicated publications, such as,

"*What is a predatory journal? A scoping review*" (Cobey et al. 2018)
"*Publication ethics workshop: how to cope with predatory journals*" (Park 2019)
"*An approach to conference selection and evaluation: advice to avoid "predatory" conferences* (Lang et al. 2019)
"*Hype or Real Threat: The Extent of Predatory Journals in Student Bibliographies*" (Schira and Hurst 2019)

9.3.1.4 Publishing Cost

On a nontechnical note, we need to consider publication cost as well. Most high-ranked scholarly journals do not have page charges. However, many new journals do, particularly in the Open-Access (OA) publication model.

Discussed in Chapter 3 Literature Search and Review, OA publication is an increasingly popular option. In such a format, research outputs can be available online and free of cost to the public. OA publication may help us maximize our research impact and possibly achieve a higher citation rate for the first one or two years after publication. Table 9.7 lists four models of OA publications. There are other models, such as hybrid and black OAs.

Under the bronze and green models, some academic journals provide OA after an embargo period, which varies from six months to two years. Self-archiving papers, even with peer-reviews, in the green model are not normally be considered high quality in research community.

When a journal is designed as conventional access, the publication costs are covered by journal subscriptions. For OA publication, the publication costs, including typesetting, marketing, etc. are covered by either the authors or subscription fees. Sometimes, the costs may be partially covered by the professional society or academic institute who hosts the journal. If an OA publisher charges authors, the

Table 9.7 Variant models of OA publication.

Aspect	Gold	Bronze	Green	Diamond
Access	Free, immediately	Free after embargo period	Free, may have embargo period	Free, immediately
Cost	Author or funding	Subscription	Institute	Funding
Repository	Journal	Journal	Self-archiving	Journal

APC varies from $50 to $2000. For many cases, the cost is reimbursable from the author's organization.

Requiring payment from authors, there is an argument against OA publication regarding the possible impact on the quality of scholarly papers because of some predatory OA publishers. Many OA publishers make editorial decisions ahead of a request for payment, providing APC discounts and waivers under certain conditions.

9.3.2 Journal Quality Indicators

Citation-based impact metrics, as the average impact of all the articles in a journal, are often used as a proxy for the impact of a journal itself. Such main metrics include Journal Citation Reports (JCR) and CiteScore.

9.3.2.1 JCR Impact Factor

JCR is a resource tool published annually by Thomson Reuters. JCR evaluates and compares over 8600 scholarly journals of sciences and technology. JCR publishes the Impact factor (IF), which is an indicator of reference value of the papers published as a journal-level metric. IFs are calculated only for science and engineering journals indexed in the Science Citation Index Expanded (SCI) for the previous two years. For example, the IF score of a journal in calculated in 2018.

$$IF = \frac{\text{Cites in 2018 in items published in 2016 and 2017}}{\text{Number of items published in 2016 and 2017}} = \frac{326}{141} = 2.31$$

Other JCR metrics include a five-year IF (over the past five years), total cites (total number of citations to the reference year), and so on. In general, the higher the factor, the better the journal is.

In addition, to compare journals across subject areas, JCR provides a journal IF percentile as either Q1, Q2, Q3, or Q4. The Q1 indicates that the journal is in the top 25% of its subject category. For example, a Q1 journal in information and communications technology (ICT) is Bell Labs Technical Journal (ISSN: 1538-7305). The journal publishes "peer-reviewed articles from researchers solving the current challenges in information and communications technology (ICT)." (Bell 2018). As another example, the International Journal of Mechanics and Materials in Design (ISSN: 1569-1713), which is listed as 12 in Table 9.6, is considered as a Q1 journal in 2017 (SCImago 2018).

9.3.2.2 CiteScore

Similarly, Elsevier has a measure reflecting the yearly average number of citations in a journal, called CiteScore. It is relatively new and considers the citations in three-year publications instead of two in IF. For example, the CiteScore of a journal in 2018 is calculated:

$$\text{CiteScore} = \frac{\text{Cites in 2018 in items published in 2015 thru 2017}}{\text{Number of items published in 2015 thru 2017}}$$
$$= \frac{889}{205} = 4.34$$

Even though they are based on similar concepts, IF and CiteScore cannot be directly compared to each other due to different evaluation periods. In addition, these indicators are available only for the journals selected in their own database. CiteScores are available on more than 23 830 journal, book series, and conference proceedings in 330 disciplines for free access (McCullough 2019). For example, the highest CiteScore of engineering publication of 2018 is "Proceedings of the IEEE Computer Society Conference on Computer Vision and Pattern Recognition" (ISSN: 1063-6919). Its CiteScore was 37.26.

9.3.2.3 Other Indicators

One well-known indicator is SCImago Journal Rank (SJR) indicator, which expresses the average number of weighted citations by documents published in the journals selected in the three previous years. For example, in industry and manufacturing engineering fields, the top three scholarly journals are Journal of Operations Management (ISSN: 0272-6963), Production and Operations Management (ISSN: 1937-5956), and International Journal of Machine Tools & Manufacture (ISSN: 0890-6955). Their SJR indicators are 5.739, 3.379, and 2.700, respectively, for 2017.

Other indicators are also based on citations. For example, Eigenfactor Score shows the citations made to a journal over time, but the Eigenfactor Score gives more weight to highly ranked journals over other ones. In addition, Source Normalized Impact per Paper (SNIP) measures the contextual citation impact by weighing citations based on the total number of citations in a subject field.

It is worth noting for novice researchers that there are a few controversial IFs in the market. These factors look similar with the well-known ones but no credible information in their websites. Some new journals cited the questionable IFs.

However, some serious authors and editors have been worried and provided their observation and concerns, such as:

> *"New corruption detected: Bogus impact factors compiled by fake organizations"* (Jalalian and Mahboobi 2013)
> *"Academic Journals Plagued by Bogus Impact Factors"* (Dadkhah et al. 2017)
> *"Many so-called "predatory journals" manufacture fake impact factors (also known as "impact factor rigging") to boost their reputation."* (Cuellar et al. 2019)

9.4 Writing For Publication

9.4.1 From Report to Paper

9.4.1.1 Additional Revision

With completed research work and understanding of the importance and process of paper publication, we are about ready to start manuscript preparation. We have examined the report writing in Chapter 8. Preparing reports and papers shares many same or similar rules and practices, of which some requirements differ. For example, journal papers are normally of a certain length (4000 to 6000 words) in original research papers. Keep in mind that some differences may be subtle.

Compared with writing a report, we need to do a little extra revision work for a publication. A research report may not be the best-written piece of literature. However, a scholarly paper needs to be the best we can write because it will be read by a much larger audience and serve as a reference for a longer period. In addition, there are some specific considerations for a revision to meet journal requirements.

The structure and organization of a scholarly paper are important since a paper needs to pass peer review for publication. We should organize an entire paper, proceeding smoothly and logically from one section to the next, and making main points clear to reviewers.

9.4.1.2 Elements of Research Paper

A research paper normally has the following sections, similar to the sections of reports and thesis discussed in Chapter 8:

1. Title
2. Abstract
3. Introduction
4. Methods
5. Results
6. Discussion

7. Conclusions
8. Acknowledgements
9. References

Among these nine elements, the Introduction, **Methods, Results, and Discussion,** sometimes called IMRaD (Sollaci and Pereira 2004), are considered essential. The IMRaD structure has been adopted and used in scientific publications by many researchers for recent years.

Some professionals depict academic writing structure IMRaD as an hourglass (see Figure 9.9). The hourglass model provides hints and guides for writers about specific content and scope of each section. The hourglass model shows that the Introduction and Discussion sections have a broad perspective while the Method and Results should be specific and focused.

Discussed in the last chapter, it is worth iterating that we normally prepare and revise a manuscript in a specific sequence, say working on the method and results first and finalizing the title and abstract last, rather than following a sequential order.

9.4.1.3 Convert Thesis to Paper

Analyzing the similarities and differences, we can turn a research report or thesis into a scholarly paper for publication efficiently. Table 8.1 Technical Reports vs. Research Papers in Chapter 8 shows their different aspects. From a conversion standpoint, here are a few more minor differences to consider for novice researchers (see Table 9.8).

Knowing the differences, student researchers can convert their thesis and dissertations to publishable papers without much difficulty. However, the preparation

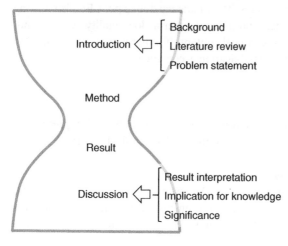

Figure 9.9 An hourglass model of technical writing.

Table 9.8 Style differences between theses and papers.

Thesis and report	Paper
In chapters	In sections
Varied length	With word limit
Format and structure of Graduate School	Format of journal
Details in method, data, etc.	Simplified descriptions of method, data, etc.
All research findings	Selected findings
Long list of references	Selected references

and conversion work can be time-consuming. For example, shortening the length of a thesis to a journal article neither is a simple copy/paste effort nor just simplifies the statements and descriptions. They may have to do careful selection of contents and reorganization to rephrase some statements, etc.

Recall other considerations and tasks, such as authorship, copyright, publishing model (e.g. OA), submission, etc. for publications are discussed in separate subsections of this chapter.

9.4.1.4 English Writing

In Chapter 8, we discussed some writing considerations, such as word selection and writing style. These tips are applicable to paper writing as well.

In addition, for journals in English, non-native-English authors may get native English speakers to proofread and edit their manuscripts. It is extra work but normally worth doing. Some referees lose their patience due to poor English writing (grammar and word usage, etc.) and judge a paper as low quality, which is not necessarily true concerning its technical content. In other words, writing quality affects reviews and possibility for publication acceptance.

9.4.2 Abstract

The abstract of a paper is a brief summary on the research work reported. An abstract can be similar to the executive summary of a technical report.

9.4.2.1 General Requirements

Abstracts are published in index databases. Almost all journals provide free access to the abstracts of published papers. Therefore, an abstract should be considered as an independent document, so that it does not rely upon any material in the main body of the manuscript.

A journal often specifies the length of an abstract. In most cases, it is one paragraph with 100–200 words. There is no standard for the structure of a paper abstract. It is a common practice that an abstract addresses three questions:

1. The objective and problem of research work (Why), in one sentence
2. The research work – method and process (How), in two sentences
3. The accomplishments and value (What), in two sentences

We begin by introducing the subject to let readers know what it is about, then introduce the work presented in the paper. If research is hypothesis based, the hypothesis should be briefly stated in the abstract.

9.4.2.2 Examples for Discussion
From the suggested three parts of an abstract, the structure and detail level has a significant deviation, given the nature and focus of research work. Here are four samples of abstracts, as shown in Figures 9.10–9.13. The lengths of the abstracts are 74, 147, 147, and 191 words, respectively. Readers can read these samples and evaluate how effectively the authors conveyed the key information in their abstract write-ups.

9.4.2.3 Structured Abstract
To be friendlier to editors and readers, some journals require abstracts in a structured format, containing a few labeled subsections to address the elements of research information in an abstract. For example, subsections of a structured abstract are:

1. Objectives
2. Method and/or Design
3. Results (Findings)
4. Conclusions/Practical Implications
5. Originality or Value
6. Research Limitations

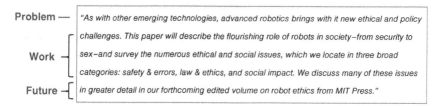

Figure 9.10 Abstract Example 1 (Lin et al. 2011).

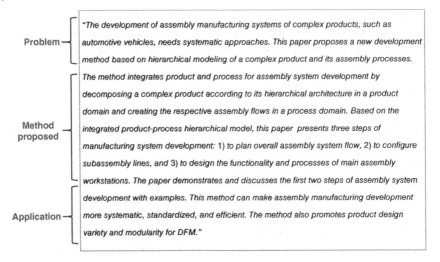

Figure 9.11 Abstract Example 2 (Tang 2018).

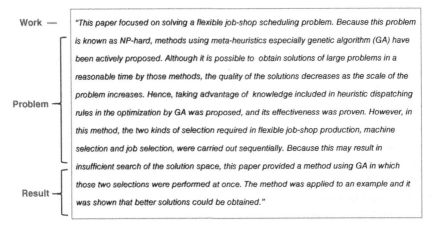

Figure 9.12 Abstract Example 3 (Morinaga et al. 2017).

Among these six subsections, the Objective, Method, Results, and Value are mostly requisites; others may be optional, depending on the journal's policy. Here is a published example (Kim et al. 2018):

> **"*Objectives*
> *To estimate the optimal bending angles in the running loop for mesial trans-
> lation of a mandibular second molar using indirect skeletal anchorage and to
> clarify the mechanics of tipping and rotating the molar.*

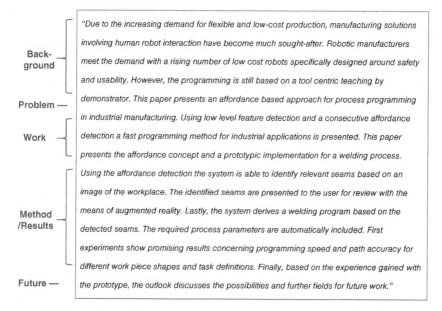

Background

"Due to the increasing demand for flexible and low-cost production, manufacturing solutions involving human robot interaction have become much sought-after. Robotic manufacturers meet the demand with a rising number of low cost robots specifically designed around safety and usability. However, the programming is still based on a tool centric teaching by

Problem

demonstrator. This paper presents an affordance based approach for process programming in industrial manufacturing. Using low level feature detection and a consecutive affordance

Work

detection a fast programming method for industrial applications is presented. This paper presents the affordance concept and a prototypic implementation for a welding process.

Method /Results

Using the affordance detection the system is able to identify relevant seams based on an image of the workplace. The identified seams are presented to the user for review with the means of augmented reality. Lastly, the system derives a welding program based on the detected seams. The required process parameters are automatically included. First experiments show promising results concerning programming speed and path accuracy for different work piece shapes and task definitions. Finally, based on the experience gained with

Future

the prototype, the outlook discusses the possibilities and further fields for future work."

Figure 9.13 Abstract Example 4 (Heimann and Krüger 2018).

Methods
A three-dimensional finite element model was developed for predicting tooth movement, and a mechanical model based on the beam theory was constructed for clarifying the force systems.

Results
When using a running loop without bends, the molar tipped mesially 14.4° and lingually 0.6°, rotated counterclockwise 4.1°, and the incisors retracted 0.02 mm and intruded 0.05 mm. These angles were about the same as those estimated by the beam theory. When the amount of tip back and toe-in angles was 11.0°, mesial translation of the molar was achieved, and incisors retracted 0.10 mm and intruded 0.30 mm.

Conclusions
Mesial translation of a mandibular second molar without any significant movement of anterior teeth was achieved during protraction by controlling the tip back and toe-in angles and enhancing anterior anchorage with the combined use of a running loop and indirect skeletal anchorage."

9.4.2.4 Highlights

Recently, some journals request a highlight for each paper, which is a list of bullet points on key points and core findings. Highlights, as a summary of research, allow readers assess the value and interests of a paper quickly. Currently, a highlight is an addition rather than replacement of conventional abstracts or keywords. In many cases, both a highlight and abstract are required.

By design, a highlight performs similarly to structured abstracts. For example, the first one or two items in the "highlight list" guide readers to the nature of a study, while the remaining two or three are a summary of main outcomes. Here is an example,

> "
> - *Combined effects of activation energy and binary chemical reaction are proposed.*
> - *Spectral quasi-linearization method (SQLM) is used for computer simulations.*
> - *Use Arrhenius activation energy in the chemical species concentration.*
> - *Validate the accuracy and convergence using residual error analysis."* (Dhlamini et al. 2019)

9.4.2.5 Keywords

Keywords are a kind of summary for an abstract, immediately following the abstract of a paper. A scholarly paper usually has three to six keywords or phrases. They should reflect the most important aspects of a paper in separated words or phrases.

Keywords as a concise indicator of a paper can be very useful for readers reviewers editors and database indexing. For example a journal AE can use the keywords of a paper to search and nominate reviewers. From keywords readers may know the main technologies used or studied in a paper. Here are four examples.

> *"Knowledge based system, Numerical design of experiments, Bayesian network"* (Blondet et al. 2019)
> *"Autonomous control; Agent systems; Factory planning; Identification"* (Kiefer et al. 2019)
> *"Industry 4.0, Industrial internet of things, Manufacturing ecosystem, Integration platform, IT requirements, Data reliability, Predictive maintenance"* (Olaf and Hanser 2018)
> *"Customization, Product/service system, PSS, Integrated solution, Design science, Platform,*
> *Literature review, Modularity"* (Hara et al. 2019)

Some keywords may be found in the title and abstract of a paper. As the title, abstract, and keywords serve different purposes, we recommend that they be designed differently to provide more information to readers thus granting the authors more power. Due to the function of keywords, we should pay close attention to choosing words. For example, when selecting keywords, we may imagine other people searching for our article in databases. The keywords should help searchers quickly locate our article. Some journals have a predefined keyword list. In that case, we have to select keywords from the list.

9.4.3 Other Sections

9.4.3.1 Introduction

As discussed in the last chapter, an introduction plays a supporting role to a report and paper. In this section, the first paragraph should get the reader's interest and giving an impression of our research background. Here is an example of the first paragraph of the introduction section (Libonati and Buehler 2017):

> *"Nature has evolved for billions of years, leading to sophisticated multifunctional materials with major structural functions, such as bone, shells, and wood. Wood has been widely used for engineering structural applications in the past, then, slowly replaced by synthetic man-made materials, providing better mechanical performance (e.g., strength, toughness). Nevertheless, many of these synthetic material shave always faced issues in satisfying both strength and toughness requirements.[1,2] This has prompted engineers to give toughness a higher priority than strength, especially in the most critical applications (e.g., gas pipelines, endoprostheses), where safety is the first requirement and failure is unacceptable."*

After the first paragraph, subsequent ones can be the current status with literature review. As discussed in Chapter 3, a literature review investigates current knowledge, revealing our research need. For research papers, their literature review section on research status should be concise – one page may be appropriate in most cases. The last paragraph of the introduction section may be designed as a self-introduction or a road map to all subsequent sections in the paper.

9.4.3.2 Discussion

The discussion section of a paper is equally important to the research results and sometimes combined with the result section, discussed in Chapter 8. It is common that we start discussion with describing results. We may need several paragraphs, each one dedicates to one detailed result or research objective. If the work is hypothesis based, the discussion should use results to provide a logical argument to support or reject the null hypothesis.

Many manuscripts are returned by a journal for revision or rejection due to weak discussions. Peer reviewers may be dissatisfied if the discussion is simply a description of results. We should try to make the in-depth discussion correspond directly to the results.

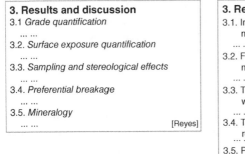

Figure 9.14 Examples of result and discussion section.

There are two examples provided in Chapter 8. Two additional examples (Reyes et al. 2018; Yoshinaga et al. 2019) are for reader's reference (Figure 9.14) here. The subsection titles of the second example seem uncommonly long, which might be due to their technical contents.

9.4.3.3 Optional Items

For scholarly papers, we normally place an acknowledgment at the end of a paper. We acknowledge the persons who helped and contributed to our research work, financial sponsors, and/or supplier as applicable. Often, the acknowledgment is a short paragraph with only one or two sentences.

Similar to technical reports, some additional subsections, such as endnotes, appendices, and supplemental information, which are not common, may be attached to the end of a paper as necessary.

A cover letter is not required for submission. However, we may send editors a short cover letter to show the significance and novelty of the research. We may convey some background information, which is not included in the paper, if necessary. Repeating the same information from the paper abstract in a cover letter is not a wise idea.

9.4.4 Publishing Ethics

9.4.4.1 Appropriate Citation

Academic integrity is a foundation of research. Academic integrity is even more important for paper publications than for internal reports. There are several aspects of academic integrity. It is extremely important that a manuscript contain no plagiarism. Appropriately crediting the work, such as statements, data, graphics, or ideas, of other people and sources is discussed in earlier chapters.

Several Internet-based tools, for example, iThenticate (www.ithenticate.com/) for publishers, government, and organizations and TurnItIn (turnitin.com/en_us/) for students are available. These tools can check a manuscript based on millions of publications and other sources and report a "similarity" rate, which is a useful reference to publisher editors and faculty. Figure 9.15a,b show two examples of similarity check reports.

Many journals require that such similarity rate be lower than 10% or 15%. If the rate of a manuscript is significantly higher than the required criteria, a journal editor should review the similarity check report. After review, the editor can either accept the manuscript to peer review process, return the manuscript for revision without peer review, or reject the manuscript based on review.

While reviewers may suggest authors modify and/or further develop their paper. It is questionable if a reviewer asks an author to cite a particular publication without academic justification. It is unethical if a reviewer or an editor demands a particular reference in a manuscript as a condition of acceptance, known as "coercive citation."

9.4.4.2 Authorship

Authorship confers work credit and has significant academic, social, and financial implications. In addition, authorship also means responsibility and accountability for published works. Essentially, authors should have substantial contributions to a paper, the associated tasks, and/or writing effort.

A research team may agree on participation and authorship earlier, before writing starts, to avoid any potential internal conflict of interest (COI). In addition, a team should agree to the appropriate order of authorship when there are multiple authors. If some team members did routine work but were less innovative to the research project (such as supportive staff), they may or may not be listed as authors. If not, they should be acknowledged at the end of a paper.

Both adding people who did not contribute significantly to an author list and missing the person who did contribute significantly in an author list are research misconduct. Inappropriately handling authorship may damage future collaboration and ruin the reputation of main authors.

For student research, the student conducts their main effort under advisement of faculty. A common practice is that the student is the first author, while the adviser serves as the corresponding author and in most cases the second author. If a faculty member provides the main idea and method, while a student conducts the detailed work, then it can be fair for the faculty to be the first author. The same

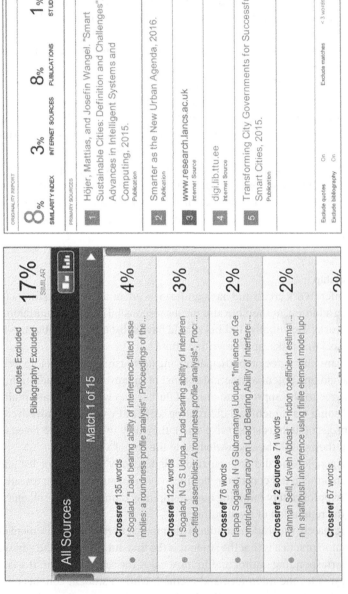

(a)

(b)

Figure 9.15 Examples of iThenticate and TurnItIn reports. (a) iThenticate report. (b) TurnItIn report.

practice applies for large sponsored projects. The PI of a research project is often the corresponding author.

The corresponding author takes primary responsibility for communication with a journal during the manuscript submission, peer review, copyright, and publication process. Typically, the corresponding author ensures all the journal's administrative requirements are met, while the first author is responsible to the details and correctness of paper contents.

Research publication authorship is very important to students as journal publication is valuable to student's degree program, career, and opportunity in the job market.

9.4.4.3 Exclusive Submission

It is unethical to submit a manuscript to two or more journals in engineering and technology fields at the same time. The submitted work should have not been published elsewhere in any form or language. Though the authors may use their own work to expand or support new work, they should be careful to avoid concerns of text-recycling.

If a paper or the same work has been presented to a conference, the authors should state it when submitting the paper to a journal if no copyright conflict. A journal may accept conference papers as is or request extended work. In addition, a few journals also require declarations if a manuscript was previously submitted to another journal and the reason for rejection.

9.4.4.4 Publishing COI and IRB

We have discussed COI and human subjects (or IRB – Institutional Review Board approval) issues in early chapters. To ensure the principles of ethical and professional conduct, we should report associated funding, potential COI, and informed consent if a research project involves human participants.

Regarding COI, we as researchers should disclose all relationships or interests that could have direct or potential influence or impart bias on the work, including nonfinancial relations. For example, a COI declaration states, "The authors of this paper have no financial or personal relationships with other people or organizations that could inappropriately influence our work" (Sanders et al. 2019). A study found, "A total of 84 presentations were available for review; 70 met the inclusion criteria, and 48 (69%) contained a COI statement." (Crawford et al. 2019).

If a study involves human participants, we must report that the study has been approved by an appropriate committee (IRB). All participants have individual rights that are not to be infringed. Authors should ensure that informed consent was obtained from all individual participants in the research work.

Summary

Considerations for Publication

1. The objectives of publications are to contribute to the ongoing body of knowledge and share latest information with others in the discipline.
2. The outcomes from engineering R&D may or may not be publishable.
3. Types of scholarly papers include original research, technical notes, case studies, systematic reviews, and expert viewpoints. Among them, the original research is the majority in journal papers.
4. The quality of papers can be viewed from different angles and measured in different ways.

Publication Process

5. There are several steps in the journal publication process. Among them, peer review is the core.
6. There are different policies and practices of copyright for published papers.
7. There are three types of peer reviews: double-blind, single-blind, and open review.
8. Authors must carefully review and respond to reviewers' comments.

Target Scholarly Journals

9. Selecting target journals involves several considerations, such as relevance, reputation, competitiveness, open access, cost, etc.
10. OA publishing normally requires a certain APC from authors.
11. There are a few journal quality indicators, e.g. IF and CiteScore.
12. Some journals are viewed "predatory" by some researchers.

Writing for Publication

13. The Hourglass model may be used for technical writing and paper preparation.
14. Additional effort is required to convert a research report to a scholarly paper.
15. The abstract of a paper states what, why, and how of the complete research in about 150 words.
16. The keywords of a paper normally have 3–6 words.
17. Appropriate citation and reference is critical to academic publishing.
18. Authorship of a paper should be handled appropriately.
19. COI and IRB are important to academic publishing in addition to research work as a whole.

Exercises

Review Questions

1 Discuss the objectives of research publication.
2 For the research results that you have, discuss the possibility of publication.
3 Explain some differences between research reports and scholarly papers.
4 Discuss the reasons of limited publications in industrial R&D.
5 Discuss the main differences between original research papers and case study articles.
6 Review the criteria of paper quality evaluation.
7 Briefly explain the overall process of journal publication.
8 Review the pros and cons of the three types of peer review (double-blind, single-blind, and open).
9 Discuss how to respond to the peer reviewer's comments and criticisms.
10 Review the factors to consider for journal selection.
11 Use the checklist of journal selection to evaluate a target journal.
12 Review open access (OA) publication with an example.
13 Find a journal and check its quality indicator.
14 Explain the writing model IMRaD with an example.
15 One states that we should refer to a structured abstract for our abstract, even though structure format is not required. Do you agree? Why?
16 Review the structure of the introduction section of a scholarly paper.
17 Discuss the keywords selected in a scholarly paper.
18 Discuss the importance of appropriate citation in publications.
19 Why not submit a paper manuscript to two journals simultaneously?
20 Find a journal paper with a COI declaration.

Mini-project Topics

1 Review the publication policy of a company and discuss the kinds of process involved in publishing R&D results outside the organization.
2 Based on the differences between technical reports (or theses) and scholarly papers, propose a few tasks to convert a report to a paper.
3 In many cases, critical comments from a peer review have more added value than commendations. Do you have an example to support or argue this statement?
4 Review the structure of a scholarly paper based on the Hourglass model.
5 Compare two quality indicators of journals, such as JCR vs. CiteScore.
6 Review and comment on the structure of the abstract of a scholarly paper.

7 Review the differences between the highlights and abstracts of a scholarly paper.

8 Check the title, abstract, and keywords of two papers and look at their differences and similarity. Provide your comments about the effectiveness of choosing keywords.

9 Find a few papers and compare the structures of result and discussion for their similarity and differences.

10 Find a paper that does not mention the limitations of the research study, provide your insight about the potential limitations of the study.

References

Bell (2018). Bell labs technical journal. https://www.bell-labs.com/our-research/technical-journal/ (accessed June 2019).

Blondet, G., Duigou, J.L., and Boudaoud, N. (2019). A knowledge-based system for numerical design of experiments processes in mechanical engineering. *Expert Systems with Applications* 122: 289–302. https://doi.org/10.1016/j.eswa.2019.01.013.

Carlson, T.P., Moen, C.W., Soli, T.D. et al. (2015). Detachable bronchoscope with a disposable insertion tube. *Journal of Medical Devices* 9 (020933): 1–2. https://doi.org/10.1115/1.4030189.

Cash, S., Zhou, Q., Olatunbosun, O. et al. (2019). Development of a series hybrid electric aircraft pushback vehicle. *Engineering* 11 (1): 33–47. https://doi.org/10.4236/eng.2019.111004.

Cobey, K.D., Lalu, M.M., Skidmore, B. et al. (2018). What is a predatory journal? A scoping review, [version 2; referees: 3 approved], F1000Research 2018, 7:1001. https://doi.org/10.12688/f1000research.15256.2, https://www.ncbi.nlm.nih.gov/pmc/articles/PMC6092896.2/pdf/f1000research-7-17518.pdf (accessed June 2019).

Costas, R. and Franssen, T. (2018). Reflections around 'the cautionary use' of the h-index: response to Teixeira da Silva and Dobránszki. *Scientometrics* 115 (2): 1125–1130. https://doi.org/10.1007/s11192-018-2683-0.

Crawford, W., Camm, C.F., Prachee, I., and Ginks, M. (2019). 115 Quality of conflicts of interest declarations in a conference setting – are audiences given a chance to adequately assess bias? *Heart* 105 (Suppl. 6): A96–A97. https://doi.org/10.1136/heartjnl-2019-BCS.112.

Cuellar, M., Trues, D., and Takeda, H. (2019). Reconsidering counting articles in ranked venues (CARV) as the appropriate evaluation criteria for the advancement of democratic discourse in the IS field. *Communications of the Association for Information Systems* 44: 188–203. https://doi.org/10.17705/1CAIS.04410.

Dadkhah, M., Borchardt, G., Lagzian, M., and Bianciardi, G. (2017). Academic journals plagued by bogus impact factors. *Publishing Research Quarterly* 33: 183–187. https://doi.org/10.1007/s12109-017-9509-4.

Dhlamini, M., Kameswaran, P.K., Sibanda, P. et al. (2019). Activation energy and binary chemical reaction effects in mixed convective nanofluid flow with convective boundary conditions. *Journal of Computational Design and Engineering* 6 (2): 149–158. https://doi.org/10.1016/j.jcde.2018.07.002.

Ebrahimi, M., Ghomi, S.M.T.F., and Karimi, B. (2018). Application of the preventive maintenance scheduling to increase the equipment reliability: case study-bag filters in cement factory. *Journal of Industrial and Management Optimization* 16 (1): 189–205. https://doi.org/10.3934/jimo.2018146.

Hara, T., Sakao, T., and Fukushima, R. (2019). Customization of product, service, and product/service system: what and how to design. *Bulletin of the JSME – Mechanical Engineering Reviews* 6 (9): 18-00184. https://doi.org/10.1299/mer.18-00184.

Heimann, O.and Krüger, J. (2018). Affordance for industrial robots in manufacturing. 7th CIRP Conference on Assembly Technologies and Systems, Procedia CIRP. Vol. 76, 133–137. https://doi.org/10.1016/j.procir.2018.01.033

Hirsch, J.E. (2005). An index to quantify an individual's scientific research output. *Proceedings of the National Academic of Sciences of the United States of America* 102 (46): 16569–16572. https://doi.org/10.1073/pnas.0507655102.

Jalalian, M. and Mahboobi, H. (2013). New corruption detected: bogus impact factors compiled by fake organizations. *Electron Physician* 5 (3): 685–686. https://doi.org/10.14661/2013.685-686. https://www.ncbi.nlm.nih.gov/pmc/articles/PMC4477750/ (accessed December 2019).

Jiménez, F.L., Marthelot, J., Lee, A. et al. (2017). Knockdown factor for the buckling of spherical shells containing large-amplitude geometric defects. *Journal of Applied Mechanism* 84 (3): 4. https://doi.org/10.1115/1.4035665.

Kiefer, L., Voit, P., Richter, C., and Reinhart, G. (2019). Attribute-based identification processes for autonomous manufacturing system – an approach for the integration in factory planning methods. *Procedia CIRP* 79: 204–209. https://doi.org/10.1016/j.procir.2019.02.047.

Kim, M.-J., Park, J.H., Kojima, Y. et al. (2018). A finite element analysis of the optimal bending angles in a running loop for mesial translation of a mandibular molar using indirect skeletal anchorage. *Orthodontics and Craniofacial Research* 21: 63–70. https://doi.org/10.1111/ocr.12216.

Koyasu, H. and Takahashi, Y. (2018). Current pass optimized symmetric pass gate adiabatic logic for cryptographic circuits. *IPSJ Transactions on System LSI Design Methodology* 12: 50–52. https://doi.org/10.2197/ipsjtsldm.12.50.

Lang, R., Mintz, M., Krentz, K.B., and Gill, M.J. (2019). An approach to conference selection and evaluation: advice to avoid "predatory" conferences. *Scientometrics* 118 (2): 687–698. https://doi.org/10.1007/s11192-018-2981-6.

Libonati, F. and Buehler, M.J. (2017). Advanced structural materials by bioinspiration. *Advanced Engineering Materials* 19 (5): 1438–1656. https://doi.org/10.1002/adem.201600787.

Lin, P., Abney, K., and Bekey, G. (2011). Robot ethics: mapping the issues for a mechanized world. *Artificial Intelligence* 175 (5–6): 942–949. https://doi.org/10.1016/j.artint.2010.11.026.

McCullough, R. (2019). CiteScore 2018 metrics now available, Elsevier. https://blog.scopus.com/posts/citescore-2018-metrics-now-available?utm_campaign=RN_AG_Sourced_300000272&utm_medium=email&utm_dgroup=RE_SCP_EG_20190528_DL_100000515&utm_acid=347203307&SIS_ID=-1&dgcid=RN_AG_Sourced_300000272&CMX_ID=&utm_in=DM529741&utm_source=AC_71 (accessed June 2019).

Morinaga, E., Sakaguchi, Y., Wakamatsu, H., and Arai, E. (2017). A method for flexible job-shop scheduling using genetic algorithm. *Journal of Advanced Manufacturing Technology* 11 (2): 79–86.

Oikawa, S., Matsui, Y., Gomei, S. et al. (2019). Correlation between acute phase head injury to cyclists and consciousness disturbance: a case study in reconstruction of single-cyclist accident. *Journal of Biomechanical Science and Engineering* 14 (1): 18-00143. https://doi.org/10.1299/jbse.18-00143.

Olaf, J.M. and Hanser, E. (2018). Manufacturing in times of digital business and industry 4.0 – The industrial internet of things not only changes the world of manufacturing. In: *Advances in Manufacturing Engineering and Materials* (eds. S. Hloch, D. Klichová, G. Krolczyk, et al.), 11–17. Cham: Springer https://doi.org/10.1007/978-3-319-99353-9_2.

Oravec, J.A. (2019). The "Dark Side" of academics? Emerging issues in the gaming and manipulation of metrics in higher education. *The Review of Higher Education* 42 (3): 859–877. https://doi.org/10.1353/rhe.2019.0022.

Park, J.H. (2019). Publication ethics workshop: how to cope with predatory journals. *Science Editing* 6 (1): 80–82. https://doi.org/10.6087/kcse.160.

Pothinun, T. and Prem, S. (2018). Power quality impact of charging station on MV distribution networks: a case study in PEA electrical power system. 53rd International Universities Power Engineering Conference (UPEC), Glasgow, UK (4–7 September 2018). https://doi.org/10.1109/UPEC.2018.8541921

R Reports (n.d.). Materials science and engineering: R: reports. https://www.journals.elsevier.com/materials-science-and-engineering-r-reports (accessed June 2019).

Raheel, M., Ayaz, S., and Afzal, M.T. (2018). Evaluation of h-index, its variants and extensions based on publication age & citation intensity in civil engineering. *Scientometrics* 114 (3): 1107–1127. https://doi.org/10.1007/s11192-017-2633-2.

Reyes, F., Lin, Q., Cilliers, J.J., and Neethling, S.J. (2018). Quantifying mineral liberation by particle grade and surface exposure using X-ray microCT. *Minerals Engineering* 125: 75–82. https://doi.org/10.1016/j.mineng.2018.05.028.

Sanders, O., Hsiao, H., Savin, D.N. et al. (2019). Aging effects of motor prediction on protective balance and startle responses to sudden drop perturbations. *Journal of Biomechanics* 91: 23–31. https://doi.org/10.1016/j.jbiomech.2019.05.005.

Schira, H.R. and Hurst, C. (2019). Hype or real threat: the extent of predatory journals in student bibliographies. *Partnership: The Canadian Journal of Library and Information Practice and Research* 14 (1), 16 pages. https://doi.org/10.21083/partnership.v14i1.4764.

SCImago (2018). SJR – SCImago Journal and Country Rank. https://www.scimagojr.com/journalsearch.php?q=26053&tip=sid&clean=0 (accessed June 2018).

SCImago (2019). SJR – SCImago Journal & Country Rank. https://www.scimagojr.com/countryrank.php?year=2018&area=2200 (accessed 2 November 2019).

Sollaci, L.B. and Pereira, M.G. (2004). The introduction, methods, results, and discussion (IMRAD) structure: a fifty-year survey. *Journal of Medical Library Association* 92 (3): 364–371.

Tang, H. (2018). An integrated product-process hierarchical modeling method for development of complex assembly manufacturing systems. *Procedia CIRP* 76: 2–6. https://doi.org/10.1016/j.procir.2018.01.023.

Tang, H. (2019). A new method of bottleneck analysis for manufacturing systems. *Manufacturing Letters* 19: 21–24. https://doi.org/10.1016/j.mfglet.2019.01.003.

Vastag, B. (2006). Cancer fraud case stuns research community, prompts reflection on peer review process. *Journal of the National Cancer Institute* 98 (6): 374–376. https://doi.org/10.1093/jnci/djj118.

White, K.E. Robbins, C., Khan, B. and Freyman, C. (2017). Science and engineering publication output trends: 2014 shows rise of developing country output while developed countries dominate highly cited publications. InfoBrief, NSF 18-300, National Science Foundation. https://www.nsf.gov/statistics/2018/nsf18300/nsf18300.pdf (accessed July 2018).

Yoshinaga, N., Uchida, S., Naito, M. et al. (2019). Induced packaging of mRNA into polyplex micelles by regulated hybridization with a small number of cholesteryl RNA oligonucleotides directed enhanced in vivo transfection. *Biomaterials* 197: 255–267. https://doi.org/10.1016/j.biomaterials.2019.01.023.

Epilogue

After students completed this research methodology class, almost all of those who were new to research agreed that the research work is interesting, challenging, and rewarding. Students said the following: "Upon completion I feel I could begin a research project on my own or at work," "this class really opened my eyes to forming research and writing a research report," "I learned a great deal about topics I never thought I would dive into," and the like. Most students founded a new enthusiastic passion in research.

After following the methodologies described in this book on how to develop proposals, conduct research tasks, and prepare for publication, everyone would be ready to take the next step towards performing research in the real world. With the newfound understanding, beginning researchers can start their own research projects and experienced researchers could improve their current research skills using the knowledge developed through this book.

Some students even asked about taking an advanced research course that would allow them dive deeper into research design. Taking an additional methodology class may be useful but we can learn much more from *doing* the actual research.

In this book, I present many aspects of research. While it is true that we can perform our research effectively and efficiently by following common methodology and implementing proven approaches, creative and critical thinking are the foundation of research success. I would encourage everyone to learn from experienced researchers and try something novel, in terms of methods, processes, disciplines, and other perspectives.

I wish you success and enjoyment on your research journey!

Engineering Research: Design, Methods, and Publication, First Edition. Herman Tang.
© 2021 John Wiley & Sons, Inc. Published 2021 by John Wiley & Sons, Inc.

Index

Engineering Research: Design, Methods, and Publication, First Edition. Herman Tang.
© 2021 John Wiley & Sons, Inc. Published 2021 by John Wiley & Sons, Inc.

Printed and bound by CPI Group (UK) Ltd, Croydon, CR0 4YY

16/04/2025

14658418-0004